S.S.R. Kumar Challa (Ed.)
Chemistry of Nanomaterials

Also of interest

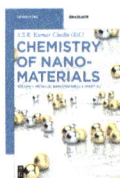

Chemistry of Nanomaterials
Volume 1: Metallic Nanomaterials (Part A)
Kumar Challa (Ed.), 2019
ISBN 978-3-11-034003-7, e-ISBN 978-3-11-034510-0

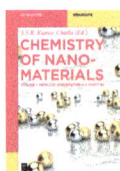

Chemistry of Nanomaterials
Volume 1: Metallic Nanomaterials (Part B)
Kumar Challa (Ed.), 2019
ISBN 978-3-11-063660-4, e-ISBN 978-3-11-063666-6

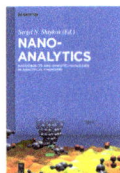

Nanoanalytics
Nanoobjects and Nanotechnologies in Analytical Chemistry
Shtykov (Ed.), 2018
ISBN 978-3-11-054006-2, e-ISBN 978-3-11-054024-6

Nanoscience and Nanotechnology
Advances and Developments in Nano-sized Materials
Van de Voorde (Ed.), 2018
ISBN 978-3-11-054720-7, e-ISBN 978-3-11-054722-1

Physical Sciences Reviews.
e-ISSN 2365-659X

Chemistry of Nanomaterials

Volume 2: Multifunctional Materials

Edited by
S.S.R. Kumar Challa

DE GRUYTER

Editor
Dr. S.S.R. Kumar Challa
Quantum Technology Group (QTG)
400 Trade Center Dr, Woburn, MA 01801
USA
kc@qtech-group.com

ISBN 978-3-11-034491-2
e-ISBN (PDF) 978-3-11-034500-1
e-ISBN (EPUB) 978-3-11-038380-5

Library of Congress Control Number: 2020931157

Bibliographic information published by the Deutsche Nationalbibliothek
The Deutsche Nationalbibliothek lists this publication in the Deutsche Nationalbibliografie;
detailed bibliographic data are available on the Internet at http://dnb.dnb.de.

Cover image: Science Photo Library / Ella Maru Studio
Typesetting: Integra Software Services Pvt. Ltd.
Printing and binding: CPI books GmbH, Leck

www.degruyter.com

Contents

List of Contributing Authors

M. Arturo López-Quintela
Facultade de Quimica, Laboratorio de
Magnetismo y Nanotecnología
University of Santiago de Compostela
Santiago de Compostela
Galicia E-15782, Spain
malopez.quintela@usc.es

David Buceta
Facultade de Quimica, Laboratorio de
Magnetismo y Nanotecnología
University of Santiago de Compostela
Santiago de Compostela
Galicia E-15782, Spain
buceta.david@gmail.com

Sumana Kundu
ECPS,
CSIR-Central Electrochemical Research
Institute
Karaikudi, India
kundusumana@yahoo.com

Vijayamohanan K. Pillai
ECPS,
CSIR-Central Electrochemical Research
Institute
Karaikudi, India
&
Indian Institute of Science Education and
Research (IISER), Chemistry, Transit campus:
Sree Rama Eng. College Tirupati, India
vijay@iisertirupati.ac.in, k.
vijayamohanan@gmail.com

Peter Schaaf
Institute of Materials Science and
Engineering and Institute of Micro- and
Nanotechnologies
TU Ilmenau
Gustav-Kirchhoff-Str. 5
Ilmenau 98693, Germany

Concha Tojo
Faculty of Chemistry
Department of Physical Chemistry
University of Vigo
Vigo, Galicia E-36310, Spain
ctojo@uvigo.es

Dong Wang
Institute of Materials Science and
Engineering and Institute of Micro- and
Nanotechnologies
TU Ilmenau
Gustav-Kirchhoff-Str. 5
Ilmenau 98693, Germany
dong.wang@tu-ilmenau.de

Yifu Yu
Department of Chemistry School of Science,
and Tianjin Key Laboratory of Molecular
Optoelectronic Sciences
Tianjin University,
Tianjin 300354, China
yyu@tju.edu.cn

Bin Zhang
Department of Chemistry School of Science
and Tianjin Key Laboratory of Molecular
Optoelectronic Sciences
Tianjin University,
Tianjin 300354, China
bzhang@tju.edu.cn

Jingfang Zhang
Department of Chemistry
School of Science
Hebei Agricultural University
Baoding 071001, China

Yunyun Zhou
National Energy Technology Laboratory
Jefferson Hills
United States of America
yyzhouzhou@gmail.com

https://doi.org/10.1515/9783110345001-203

Dong Wang and Peter Schaaf

1 Synthesis and characterization of size controlled bimetallic nanosponges

Abstract: Metallic and bimetallic nanosponges with well-defined size and form have attracted increasing attention due to their unique structural properties and their potential for many applications. In this chapter, the recently developed methods for the synthesis and preparation of metallic and bimetallic nanosponges are presented. These methods can be mainly cataloged in two groups: dealloying-based methods and reduction reaction-based methods. Different topographical reconstruction methods for the investigation of their structural properties are then reviewed briefly. The optical properties of the metallic nanosponges are clearly different from those of the solid counterparts due to the tailored disordered structure. The recent advances in the exploration of the distinct linear and non-linear optical properties of the nanosponges are summarized.

Graphical Abstract:

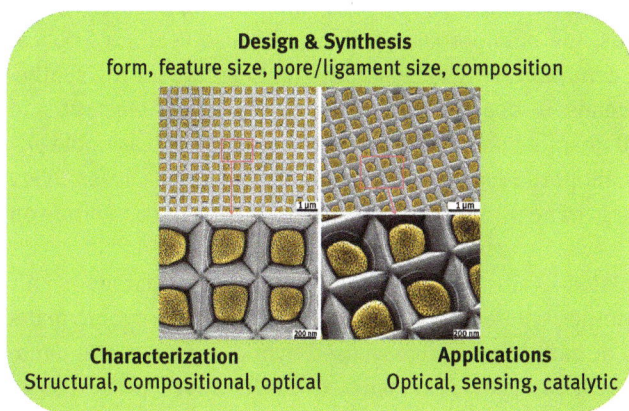

Keywords: bimetallic, nanosponges, nanoporous, nanoparticles, size-controlled

This article has previously been published in the journal *Physical Sciences Reviews*. Please cite as: Dong, W., Schaaf, P. Synthesis and characterization of size controlled bimetallic nanosponges. *Physical Sciences Reviews* [Online] **2019**, 4. DOI: 10.1515/psr-2018-0125.

https://doi.org/10.1515/9783110345001-001

1.1 Introduction

The important advantage of bimetallic alloys, which has already been taken for a long time, is tailoring the desired properties (such as hardness, ductility, strength, weight, corrosion resistance, color, electrical and thermal conductivities, catalytic properties, etc.), by simultaneously tuning their composition and their microstructures/morphologies. The pioneering work on such bimetallic nanostructures and nanoparticles (NPs) was achieved in the 1960s for tailoring the optical and plasmonic properties, and colloidal Au@Ag core/shell NPs have been synthesized using 5.9 nm of Au nuclei [1]. In addition, the marvelous catalytic properties of bimetallic nanoclusters, due to their bi-functional effects, ensemble effect and size effect (for large surface area), have further stimulated the research interest on the synthesis and properties of nanoscale bimetallic structures [2, 3].

Recently, another type of interesting metallic catalysts emerged, namely nanoporous metals, whose large surface area is very profitable for the enhancement of the catalytic performance [4–8]. The 3D bi-continuous ligament/channel structures in such nanoporous metals enable a mass transfer and a charge transfer in a more efficient way as in solid structures during the catalytic reactions. The ligament/pore size of nanoporous metals can scale in a large range from a few nanometers to several hundreds of nanometers, determined by the fabrication conditions. In addition to the catalytic applications, the nanoporous metals can be also used for applications in sensing [5, 6] and actuation [9]. The most important method for the synthesis of the nanoporous metals is dealloying. Alloys with a noble element and another less noble element are often used, and during dealloying the less noble element is removed and the nanoporous structure of the noble element evolves. A very important example for that is nanoporous gold, which can be formed easily from an Au−Ag alloy [10–13].

In addition, the percolated ligament/channel structures of nanoporous metals allow the deposition of another type of metal, inorganic material or organic material, and nanoporous gold or nanoporous metals-based hybrid materials and nanocomposites have been fabricated. They demonstrated the improved properties or even possessed new functionalities [14–26]. For instance, different oxides can be deposited into the channels of nanoporous gold for the application in lithium-oxygen batteries [17]. The performance can be clearly enhanced by using nanoporous bi-metallic PdNi catalyst instead of simple nanoporous Pd catalyst [14]. The composites of nanoporous gold and polymers demonstrated also modified electrical and mechanical properties [25].

Furthermore, combining dealloying with other nanostructuring technologies, the outer size and shape of the nanoporous gold and nanoporous metals can be well designed and controlled, and finally tailored metallic nanosponges can be obtained. Different from the dealloying-based methods, reduction reaction-based methods (including template assisted methods) are another group of methods for synthesis of metallic nansponges [27]. The important structural feature of nanosponges is that nanoporous structures with pore size of 10–20 nm are percolated over a defined 0D or 1D forms with sizes in the submicron range. The outer shapes of nanosponges can be varied with different forms like NPs [28, 29], nanodisks [30, 31], nanowires [32, 33], nanotubes [34], nanobowls [35], and many more. The nanosponges have demonstrated exciting optical and plasmonic properties [36–41], photothermal effects [31], unprecedented SERS (surface enhanced Raman spectroscopy) performance [30, 42], surface-enhanced near-infrared absorption [43], superior catalytic performance [27] and even more. Size and shape-controlled bimetallic and hybrid nanosponges can be subsequently developed and synthesized, and they show a much larger flexibility for tailoring the further enhancement of their functionalities [27, 44, 45].

The synthesis methods and the most important properties of nanoporous gold and their applications in plasmonic enhancement, analytic chemistry, catalysis and sensors have been summarized as book chapters and review papers [5, 46–52]. Nanoporous gold with structural hierarchy has been developed by applying a two-step dealloying strategy and the recent advances have been summarized and reviewed [53, 54]. Recent research activities with regard to synthesis, characterization, and properties of bimetallic nanoframes (with ordered inner structure) and nanoporous structures have been summarized, too [27, 55].

In this chapter, we specially focus on the size controlled bimetallic nanosponges, and summarize the recent advances in size and form controlled fabrication and synthesis of metallic and bimetallic nanosponges. Here, we start with the introduction of the preparation methods of nanoporous metals and nanoporous bimetallic metals, and then continue with a summary of the fabrication or synthesis methods of size-controlled metallic nanosponges and bimetallic nanosponges. The synthesis methods can be mainly cataloged in two large groups: dealloying-based methods and reduction reaction-based methods. Furthermore, it is followed with a brief review of different topographical reconstruction methods for the exploration of the structural properties. Finally, some advanced linear and non-linear optical characterization methods and application examples are presented.

1.2 Preparation methods

1.2.1 Preparation methods of nanoporous metals

The popular method for the fabrication and synthesis of nanoporous metals is dealloying. Dealloying is selective leaching of the less noble element from a binary alloy, and nanoporous structure of the noble element evolves in the meanwhile. The dealloying mechanism was well established with atomistic simulations [11, 12]. There are two important ways for dealloying, as shown in Figure 1.1: chemical dealloying and electrochemical dealloying. Acids are usually used for chemical dealloying to remove

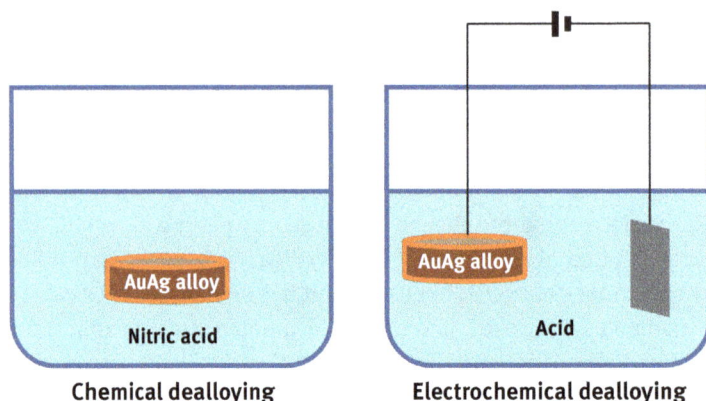

| Chemical dealloying | Electrochemical dealloying |

Figure 1.1: Schematic drawing of chemical dealloying and electrochemical dealloying for the synthesis of nanoporous gold from Au–Ag alloys.

the less noble element out of the binary alloy. Nitric acid (HNO_3) is used for synthesis of nanoporous gold by dealloying of Au–Ag alloy. Hydrochloric acid (HCl) is used for synthesis of nanoporous metals including Ag, Pd, Pt and Cu from their Al-based alloys [56, 57]. In addition, by dealloying of Al-based alloys, both acids (HCl and H_2SO_4) and base (NaOH) solutions can be applied [57]. Alternatively, dealloying can also be conducted through the electrochemical corrosion of the less noble element, when an anodic potential is applied [58]. For the chemical dealloying, the composition of leached element in the alloy must be beyond the so-called "parting limit"[59]. For the electrochemical dealloying, the applied electrical potential must be larger than the threshold potential for dealloying [58]. The important feature sizes, i.e., mainly ligament/pore size, can be controlled by changing the dealloying conditions (time, temperature, and acid concentration). Volume shrinkage and crack formation are often accompanying the dealloying [60].

Both, chemical dealloying and electrochemical dealloying, are used mostly for the synthesis of noble nanoporous metals. With dealloying of Al-based alloys, nanoporous metals of less-noble transition metals (such as Ni and Co) can be obtained, because Al is more reactive [61]. In addition, liquid metal dealloying has been applied for the synthesis of nanoporous Ti [62]. Another recently developed dealloying method for the fabrication of nanoporous non-noble metals is vapor phase dealloying, by which the vapor pressure difference between solid elements are utilized to selectively evaporate a component from an alloy [63]. Different from dealloying, a new emerging method for synthesis of less noble nanoporous metals is via solid-state conversion reactions [64]. As schematically shown in Figure 1.2, a nanocomposite of the desired metal with an ionic compound is formed, and the nanoporous metals (Cu, Fe Ni, and Co) or bimetallic nanoporous metals (Au-Cu, Cu-Co, and Fe-Co) are synthesized when the ionic compound is removed by dissolution with a common organic solvent.

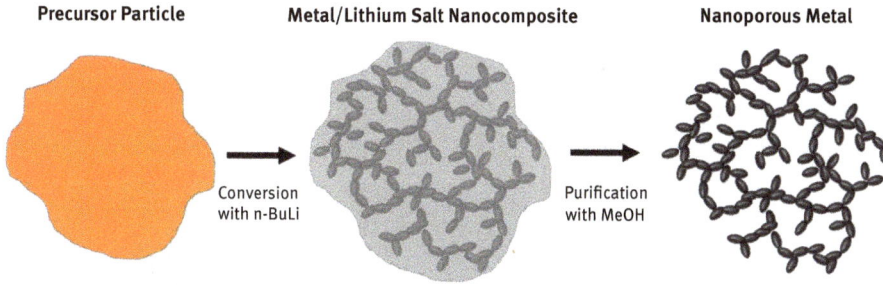

Figure 1.2: Illustration of nanoporous metal preparation *via* conversion reaction synthesis. An anhydrous transition-metal halide precursor is reacted with *n*-BuLi and converted to a metal/ lithium halide nanocomposite. The lithium halide is removed by dissolving it with methanol, leaving behind a nanoporous metal. Reproduced from Ref. [64] with permission from American Chemical Society.

1.2.2 Preparation methods of size controlled metallic nanosponges

Size and form controlled nanoporous metals (nanosponges) can be synthesized mainly by two groups of methods, namely dealloying-based methods and reduction reaction-based methods. A size and form definition process with resolution in submicron range (e.g. nanolithography) is usually combined in the dealloying-based methods. Figure 1.3 shows the synthesis route and scanning electron microscopy (SEM) images of the nanoporous gold NPs (gold nanosponges). Au–Ag alloy NPs were fabricated via solid-state dewetting of Au/Ag bilayers, and then nanoporous gold NPs were formed after dealloying [28]. The ligament/pore size can be well controlled by changing the dealloying conditions, as shown in Figure 1.4. Coarsening of pores can be also realized via post-annealing of the as-prepared nanoporous gold NPs [65]. Solid-state dewetting is a simple method for the fabrication of metallic and alloy NPs [66–72], and even ordered arrays of metallic NPs and nanosponges (Figure 1.5) can be obtained by templated solid-state dewetting [73–78]. The nanoporous gold NPs with limited particle size have often been found to be single-crystalline over the whole individual NPs [38, 45, 79].

Nanoporous gold nanodisks of different shapes (Figure 1.6) can be fabricated via combination of E beam lithography and dealloying [30, 41, 80]. Au/Ag bilayer nanodisks have been first fabricated via E beam lithography and lift-off process. E beam lithography is a powerful tool for the structure definition in the submicron range. The Au–Ag alloy nanodisks were then formed after annealing at 400°C for 30 min. After dealloying, nanoporous gold nanodisks were synthesized. Anodic Aluminium oxide nanostructures and Si nanostructures fabricated via laser interference lithography were used as template for the synthesis of nanoporous gold nanowires [32, 33]. Au–Ag alloy was at first deposited into the nanopores of the templates, and then the alloy

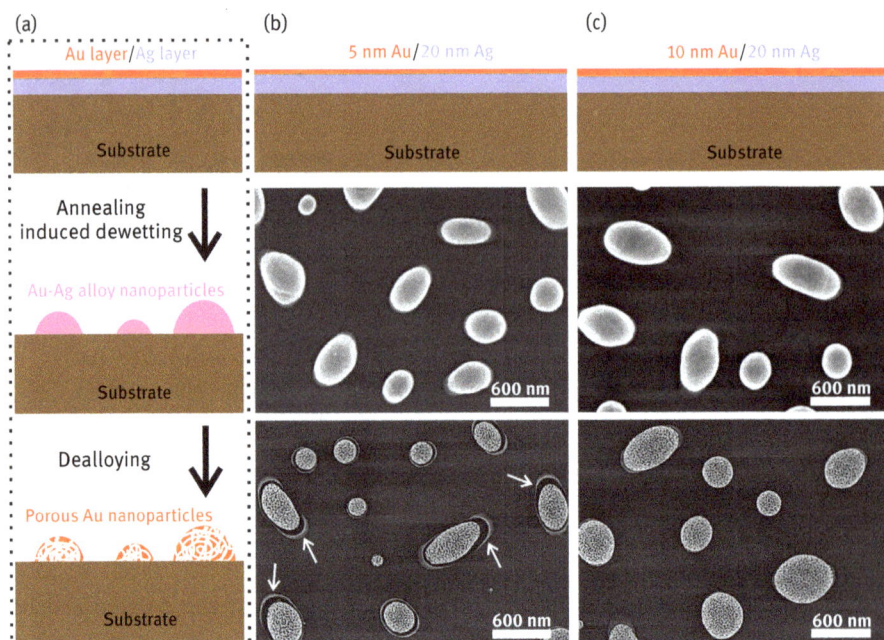

Figure 1.3: (a) Schematic drawing of the synthesis of nanoporous gold nanoparticles by combining solid-state dewetting (for synthesis of alloy nanoparticles) and dealloying (for synthesis of nanoporous gold). (b) Schematic drawing of the 5 nm Au/20 nm Ag films (upper panel) and the corresponding SEM images of the dewetted alloy nanoparticles (middle panel) and formed nanoporous gold nanoparticles (lower panel). White arrows in the lower panel SEM indicate the contour of the particles before dealloying. (c) Schematic drawing of the 10 nm Au/20 nm Ag films (upper panel) and the corresponding SEM images of the dewetted alloy nanoparticles (middle panel) and formed nanoporous gold particles (lower panel). Reproduced from Ref. [28] with permission from Royal Society of Chemistry.

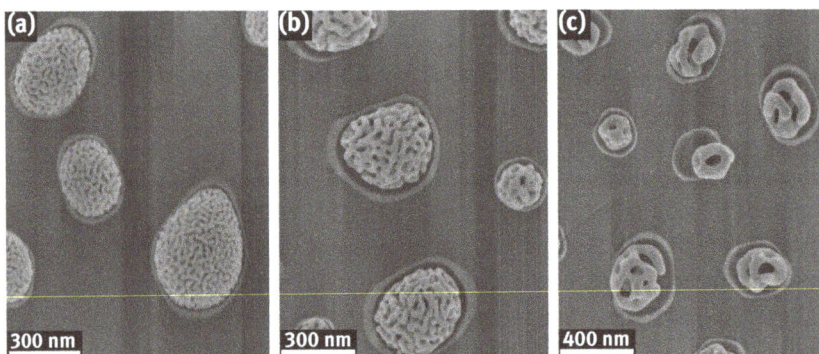

Figure 1.4: SEM images of gold nanosponges with different mean ligament/pore sizes: (a) 11.6 nm/9.0 nm, (b) 22.8 nm/14.4 nm, and (c) 50.4 nm/32.7 nm. Reproduced from Ref. [38] with permission from American Chemical Society.

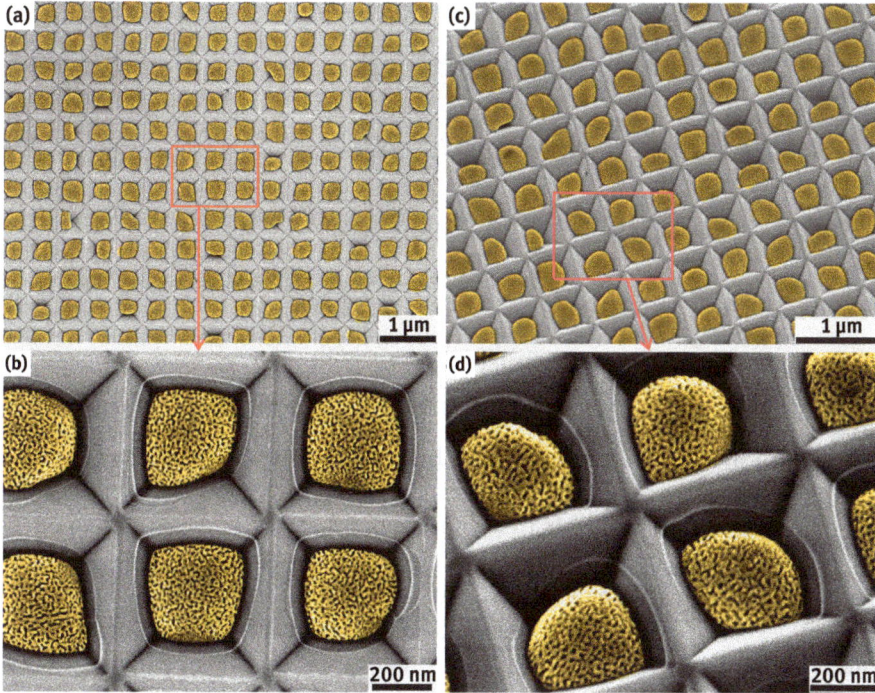

Figure 1.5: SEM images of ordered arrays of nanoporous gold nanoparticles. Reproduced from Ref. [73] with permission.

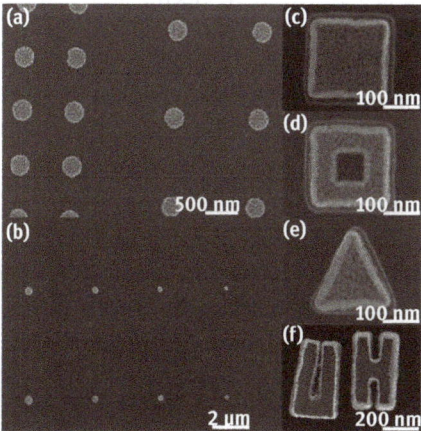

Figure 1.6: (a) Periodic arrays of 300 nm diameter nanoporous gold disks with two different interdisk distances (0.8 and 1.3 μm). (b) Periodic array of nanoporous gold disks of different diameter (200–500 nm). Nanoporous gold structure of (c) "square", (d) "hollow-square", (e) "triangular" shapes, and (f) "UH" pattern. Reproduced from Ref. [41] with permission from American Chemical Society.

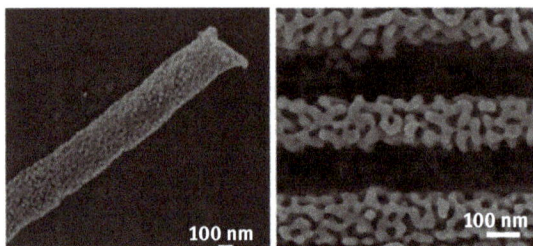

Figure 1.7: Nanoporous gold nanowires. Reproduced from Ref. [32, 33] with permission from American Chemical Society.

nanowires were removed from the template. Finally, the nanoporous gold nanowires were obtained after dealloying, as shown in Figure 1.7.

Based on dealloying methods, bimetallic nanosponges can be further realized via two different ways: (i) controlled partial dealloying, in which some part of the less noble element is remained in the nanoporous structures, and (ii) additive deposition, by which a second element can be deposited into the nanoporous structure to form the nanoporous hybrid structure. Actually, the less noble metal usually cannot be fully removed, and for example, about 7% residual Ag can be found in the formed nanoporous gold NPs [38]. Even nanoporous Au–Ag nanospheres have been deliberately synthetized for the application of SERS [45]. The most important influencing parameters for the controlled partial dealloying are: dealloying time, concentration of the less noble element in the nanoalloys, and the concentration of the acid solution for dealloying. Bimetallic PtAg nanosponges have been synthesized via partial dealloying of PtAg NPs, and show an excellent electrocatalytic performance towards oxygen reduction reaction [81].

In addition to the chemical and electrochemical methods, another method to dealloy and porosify alloy NPs is recently proposed via photoelectrochemical dealloying [82]. As schematically shown in Figure 1.8, under irradiation with visible light in water, Ag dissolves from Au–Ag alloy NPs deposited on the TiO_2 thin film, and nanopores are formed. The dissolution rate of Ag depends on the irradiation wavelength and intensity, and different pore and ligament sizes can be obtained. The related mechanism is based on plasmon-induced charge separation (PICS) at the alloy–TiO_2 interface. This method can be also referred to as controlled partial dealloying by controlling the irradiation parameters. The formed nanoporous Au-Ag NPs are shown in Figure 1.9.

The second way to synthesize the bimetallic nanosponges is via additive deposition of the second element or material into the bicontinuous porous structure of the as-prepared nanosponges. In fact, different deposition methods have already been used to add the second material into the nanoporous metals, including electroplating, electroless plating, atomic layer deposition (ALD), and chemical vapor deposition (CVD). Plasma-enhanced atomic layer deposition (PE-ALD) and chemical

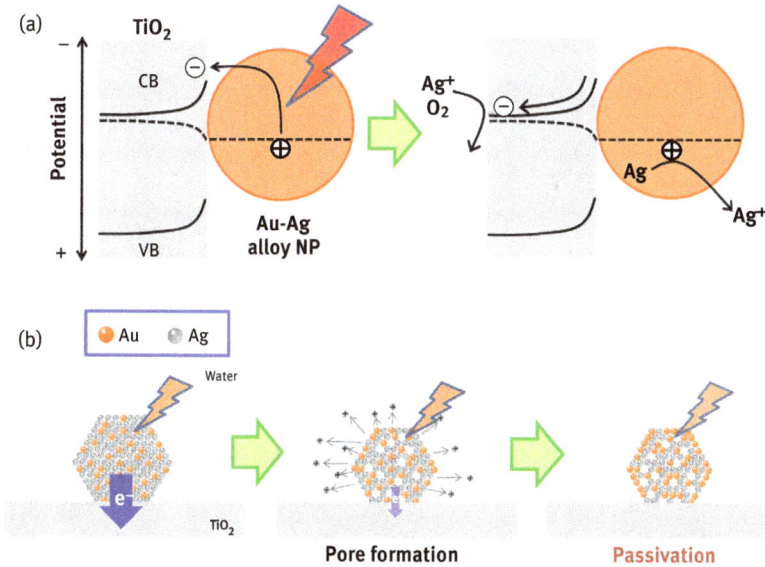

Figure 1.8: Schematic illustrations of (a) photoelectrochemical dealloying and (b) accompanying nanopore formation process of Au–Ag nanoparticles. Reproduced from Ref. [82] with permission from American Chemical Society.

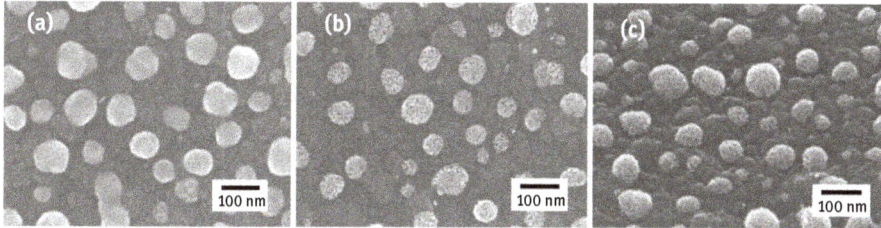

Figure 1.9: SEM images of Au–Ag alloy nanoparticles ensembles on TiO_2 electrodes (a) before and (b, c) after irradiation of broadband visible light (>480 nm, ca. 9 mW·cm^{-2}) in water at room temperature. Image (c) was obtained at a tilt angle of 45°. Reproduced from Ref. [82] with permission from American Chemical Society.

vapor deposition (PE-CVD) are preferred, as those low temperature processes avoid the coarsening of the nanoporous structure. The second material can be metals, inorganic materials and organic materials, and different nanoporous hybrid composite systems have been synthesized, such as Au–Ag [83], Cu–Au [16], Au–Pt [4, 84–86], Au–TiO_2 [87, 88], Au–Al_2O_3 [38, 88, 89], Au–MnO_2 [90, 91], Au–SnO_2 [92], Au–RuO_2 [93], Au–MoS_2 [94, 95], Au–polypyrrole [96], and Au–polyaniline [97, 98]. The bimetallic nanoporous metals fabricated via additive deposition usually have a hybrid structure, in which both the elements are mixed in mesoscale, as shown in

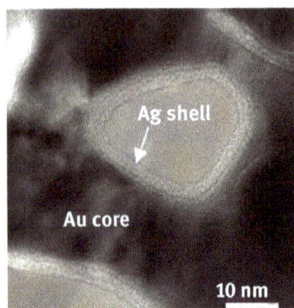

Figure 1.10: A TEM image of Ag-plated nanoporous gold. Reproduced from Ref. [83] with permission from American Chemical Society.

Figure 1.10. This is not like the bimetallic nanoporous metals fabricated via controlled partial dealloying, in which the both elements are still in alloy state and mixed at atomic scale.

Electroplating can be used for the synthesis of bimetallic nanoporous metals, but it is difficult to use it for the synthesis of bimetallic nanosponges due to the limitation of electrical contacting to the nanosponges with particle size in submicron range. Bimetallic hybrid nanoporous nanosponges can be obtained via cyclic electroless plating. Figure 1.11 shows Au–Ag hybrid nanosponges, which demonstrated enhanced SERS performance, better than pure gold nanosponges, and the amount of Ag can be well controlled by the deposition cycles [44]. Ascorbic acid (Vit C) and silver nitrate ($AgNO_3$) solutions were used for the electroless deposition of Ag into the as-prepared gold nanosponges. The sample with Gold nanosponges was first immersed in Vit C solution, and then rinsed with deionized water to remove the excess Vit C outside of the nanoporous structure. To complete one cycle deposition, the sample was then immersed in $AgNO_3$ solution and rinsed with deionized water again. The cyclic process is designed to confine the plating of Ag only in the porous structure of the gold nanosponges, and is different from traditional electroless plating by which the Vit C and $AgNO_3$ mixed solution is used [44].

Different from dealloying-based methods, size-controlled bimetallic or trimetallic nanosponges can be also realized through reduction reaction-based methods, in which metal precursors are reduced and nanosponges are formed. Templates are usually used for guiding the reduction reactions of the metal precursors within the confined structures. Three types of templates are usually used: hard (inorganic) templates, soft (organic) templates and sacrificial templates. Mesoporous silica is a very popular hard template for synthesis of a metal replica. Figure 1.12 shows the synthesized Pt nanosponges replicated from the KIT-6 and SBA-15 porous silica templates [99]. K_2PtCl_4 was used as metal precursor and ascorbic acid as reduction agent. The deposition of Pt was confined through the sequential processes: (i) immersing the porous silica powder into the K_2PtCl_4 solution, (ii) then drying the silica

as-prepared Au NSs	Au-Ag-1c hybrid NSs	Au-Ag-3c hybrid NSs	Au-Ag-6c hybrid NSs

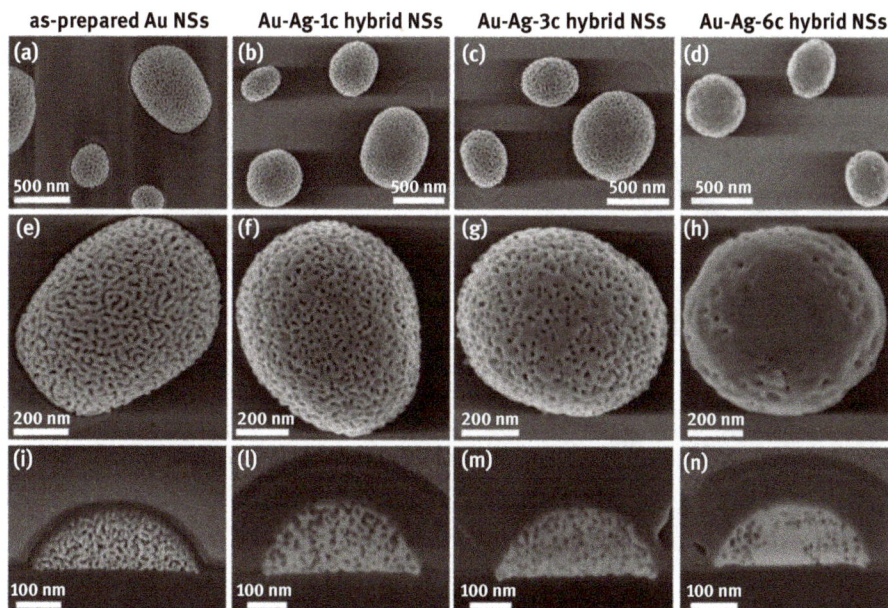

Figure 1.11: Plan view (a-h) and cross-sectional (i-n) SEM images of as-prepared Au nanosponges (NSs) (a, e and i) and the Au-Ag hybrid NSs after 1 cycle (b, f and l), 3 cycles (c, g and m) and 6 cycles (d, h and n) of Ag deposition, respectively. Pt layers (induced by electron and Ga ion beams) have been deposited on top of the nanosponges, which acted as protection layers during FIB milling (i-n). It should be noticed that not the same particles in plan view images have been milled by for recording the cross-sectional images. Reproduced from Ref. [44] with permission from American Chemical Society.

powder under vacuum condition to remove the additional metal precursor from outside of the porous structure, and (iii) immersing the treated porous silica powder in reduction agent for the confined deposition of Pt inside of the porous structure. Then Pt nanosponges with well-defined shapes and sizes can be obtained after removing the silica with HF solution. The precise control of the shape and size of the nanosponges can be achieved by tuning the morphology of the original porous silica templates and processing time.

Soft templates usually are structure-directing agents with amphiphilic molecules, with which surfactant micelles can be formed in a self-assembly way. Porous metallic nanostructures can be formed on the surfactant micelles via chemical reduction of metal precursors. If different metal precursors are added, bimetallic or trimetallic nanosponges can be also achieved. Different metallic nanosponges have already been synthesized, including Pt [100], Rh [101], PdPt [102–104], AuPdPt [102], PdPtCu [102], PtPdNi [105], PtPdRu [106, 107], which mostly demonstrated a superior electrocatalytic performance. Lyotropic liquid crystals [108–110], block copolymers (such as Brij 58® (Brij 58), Pluronic® F-127 (F-127), and Pluronic® P-123)

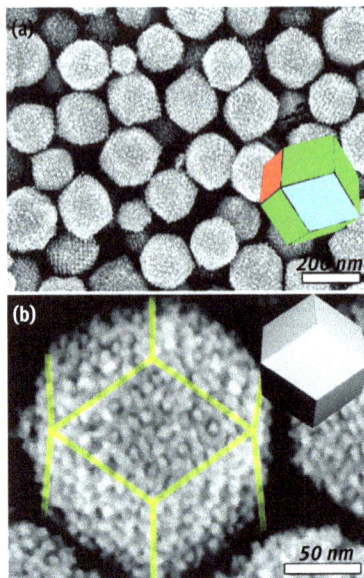

Figure 1.12: SEM images of the obtained porous Pt nanoparticles prepared with mesoporous silica KIT-6. Reproduced from Ref. [99] with permission from American Chemical Society.

[102, 103, 111], and dioctadecyldimethylammoniumchloride (DODAC) [112] are often utilized as soft templates for synthesis of metallic nanosponges. For instance, F-127, a typical block copolymer as pore-directing agent for synthesis of porous nanostructures, has been used for synthesis of porous bimetallic PdPt spheres, porous trimetallic Au@PdPt spheres, and porous trimetallic PdPtCu spheres [102], as shown in Figure 1.13. Na_2PdCl_4 solution was used as Pd precursor, H_2PtCl_6 and K_2PtCl_4 solutions as Pt precursors, and $HAuCl_4$ as Au precursor. F-127 consists of poly(ethylene oxide)-poly(propylene oxide)-poly(ethylene oxide) (PEO-PPO-PEO), where the PEO is hydrophilic and the PPO is hydrophobic. The porous structure is formed when the hydrophilic units of the amphiphilic molecules accommodate the metal atoms during the reduction of the metal precursors. The size of the nanosponges can be controlled by the processing time and concentration of the precursors. It is interesting to notice that different soft templates can have a clear morphological influence of the synthesized nanosponges. For instance, by comparing the transmission electron microscopy (TEM) images and Energy dispersive spectroscopy (EDS) mapping, the bimetallic PdPt nanosponges synthesized with two different soft templates Brij 58 and F-127, have been identified with a Pd-core/Pt-shell structure [103]. The morphology of PdPt nanosponges, prepared with Brij 58 is clearly denser than those prepared with F-127. In addition, amino acid can be also used for self-assembly synthesis, and bimetallic PtCu, and PtAu nanosponges can be obtained with this method [113, 114].

Mesoporous PdPt spheres

Mesoporous Au@PdPt spheres

Mesoporous PdPtCu spheres

Pt precursor   Pd precursor   Cu precursor

F127 micelle   Au seed

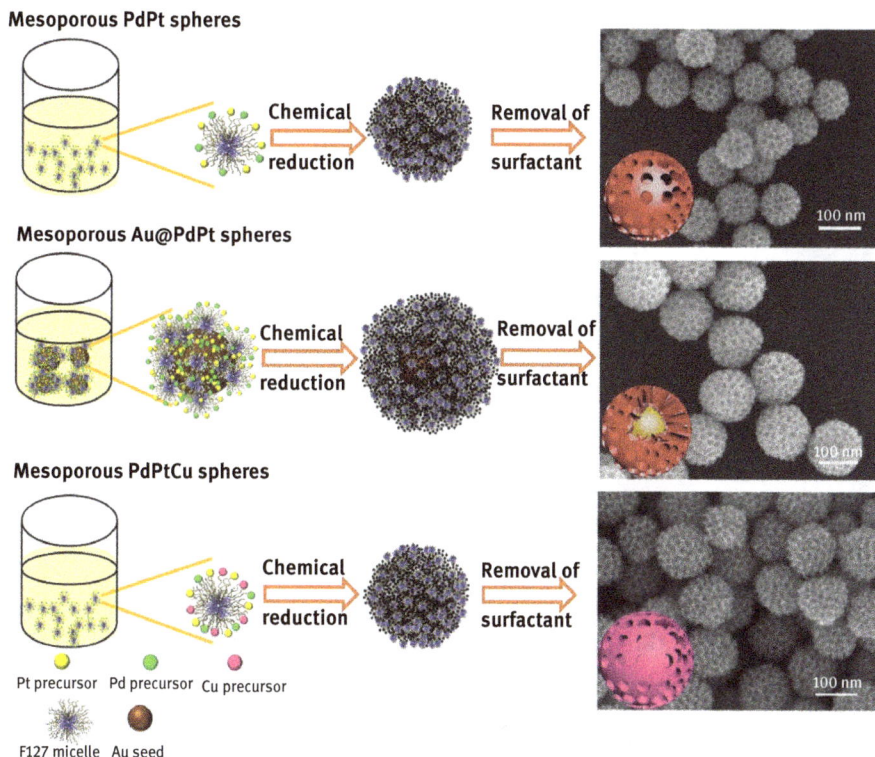

Figure 1.13: Systematic illustration of synthesis route and SEM images of mesoporous bimetallic PdPt spheres, mesoporous trimetallic Au@PdPt spheres, and mesoporous trimetallic PdPtCu spheres. Reproduced from Ref. [102] with permission from WILEY-VCH.

Different from the hard and soft template methods, sacrificial template acts as a reactant which is involved in the reaction. Cu_2O octahedral particles as sacrificed hard templates were immersed in a surfactant free oil-in-water emulsion solution, and then $HAuCl_4$ solution was added to trigger the galvanic reaction with Cu_2O. Finally, porous gold particles with octahedral shape were obtained as seen in Figure 1.14, and the octahedral shape was transcribed from template Cu_2O particles. It is interesting to notice, that if the Cu_2O octahedral particles were immersed in pure water instead of the surfactant free oil-in-water emulsion solution for the followed galvanic reaction, only solid gold particles with octahedral shape can be obtained [115].

Metallic nanosponges can be also prepared by simple solution based reduction process with metal precursors and reduction agent, and bimetallic PtNi nanosponges and porous PtIr tripods can be synthesized by that way [116, 117]. Different PtCu alloys nanosponges can be achieved by the solution based reactions with Pt $(acac)_2$ and $Cu(acac)_2$ as metal precursors and formaldehyde as a reducing agent,

Figure 1.14: SEM images of Au (a), Au-Pt (b) and Au-Pd (c) porous particles formed via interfacial nanodroplets guided galvanic reaction. Reproduced from Ref. [115] with permission from Nature publishing group.

and the PtCu alloys nanosponges demonstrated better electrocatalytic activity than Pt nanosponges [118]. Bimetallic PtRh nanopsonges have been synthesized through the co-reduction of two metallic precursors ($H_2PtCl_6 \cdot 6H_2O$ and $RhCl_3 \cdot xH_2O$) using sodium borohydride ($NaBH_4$) as reduction agent at room temperature, and demonstrated superior electrocatalytic performance for methanol oxidation reaction [119]. Trimetallic $Pt_{53}Ru_{39}Ni_8$ nanosponges were obtained by co-reduction of three metallic precursors (H_2PtCl_4, $RuCl_3$, and $NiCl_2$) using sodium borohydride ($NaBH_4$) at room temperature, and displayed excellent electrocatalytic performances for hydrogen evolution reaction and hydrazine oxidation reaction [120]. Furthermore, nanosponges can be synthesized by self-regulated reduction of sodium dodecyl sulfate without additional reduction agent, and for instance, PdAg and PdAgAu nanosponges have been synthesized via such self-regulated reduction and adding the second or third metal salt in the synthesis period [121].

In summary, different synthesis methods have been summarized in Table 1.1. Chemical dealloying-based methods and reduction reaction-based methods are applicable for synthesis of size controlled metallic and bimetallic nanosponges, while the other methods are only applicable for synthesis of bulk nanoporous metals.

1.3 Characterization methods and instrumentation

1.3.1 Structural characterization via topographical reconstruction

The structural investigation of nanoporous metals is quite challenging but necessary for the optimization of these structures and for understanding of the functional mechanisms in different applications [122, 123]. Many material characterization methods were applied to study the structural properties. SEM and TEM are usually

Table 1.1: Summary of different methods for synthesis of bulk nanoporous metals, size controlled metallic nanosponges, and size controlled bimetallic nanosponges.

Methods		Bulk nanoporous metals	Size controlled metallic nanosponges	Size controlled bimetallic nanosponges
Dealloying-based methods	Electrochemical dealloying	☺ Ref. [5, 46–52, 58]	☹	☹
	Chemical dealloying	☺ Ref. [46–52, 56, 57, 59–61]	☺ Through combination with other nanostructuring technologies (solid-state dewetting, nanoimprint lithography, E beam lithography, Laser interference lithography, or AAO templates and so on). Ref. [28–45, 73, 79, 80]	☺ Either through controlled partial dealloying or additive deposition in as-prepared nanosponges. Ref. [44, 45, 81]
	Photoelectrochemical dealloying	☹	☺ Ref. [82]	☺ Ref. [82]
	Liquid metal dealloying	☺ Ref. [62]	☹	☹
	Vapor phase dealloying	☺ Ref. [63]	☹	☹

(continued)

Table 1.1 (continued)

Methods		Bulk nanoporous metals	Size controlled metallic nanosponges	Size controlled bimetallic nanosponges
Reduction reaction-based methods	Hard template	☹	☺ Ref. [99]	☺ Ref. [99]
	Soft template	☹	☺ Ref. [100, 101]	☺ Ref. [102–113]
	Sacrificial template	☹	☺ Ref. [115]	☺ Ref. [115]
	Simple solution based method	☹	☺	☺ Ref. [116–120]
Other methods	solid-state conversion reactions	☺ Ref. [64]	☹	☹

used for the morphological characterization. The hybrid nature of the bimetallic porous structure (Figure 1.10) can be well identified with TEM. Focused ion beam (FIB) can be used to study the interior structure by milling the nanosponges (cross-sectional views of the Au–Ag hybrid nanosponges shown in Figure 1.11 i-n). EDS measurements can be applied to quantitatively analyze element contents, and for instance, the Au and Ag contents in the hybrid Au–Ag hybrid nanosponges (Figure 1.11) has been analyzed, and thereby the volume porosity can be exactly calculated [44]. The calculated porosity is very important information to the model construction for the optical simulations. EDS mapping can be also used for structural determination (Pd/Pt core/shell structure) [103]. X-ray diffraction can be used to characterize the crystalline structure of the nanosponges in an ensemble way. Selective area electron beam diffraction can be used to determine the crystalline properties of individual nanosponges, and the individual gold nanosponges (Figure 1.4) with particle size of few hundred nanometers are found single crystalline [38]. Brunauer-Emmett-Teller measurement is based on the theory of physical adsorption of gas molecules on a solid surface, and can be used as an important analysis technique for measuring the specific surface area.

In addition, the distinct 3D bi-continuous structural properties of the nanoporous metals can be further explored by using different topological reconstruction methods including with (i) TEM [124–126], (ii) FIB [127–129], (iii) synchrotron X-ray nanotomography [130–132], and (iv) atom-probe topography (APT) [133]. All these methods can be surely also applied for the investigation of nanosponges. High-angle annular dark-field scanning transmission electron microscopy (HAADF-STEM) imaging is a suitable method for electron tomography due to its good Z-contrast resulting from the scaled intensity of Rutherford scattering with atomic number and thickness [124, 134]. A series of images should be recorded and collected over a tilt range in small steps. After each tilt step, re-centering and refocusing should be done before recording the next image. The alignment of the collected individual images can then be done by iterative cross-correlations to sub-pixel accuracy. The reconstruction can be done with the series of processed images by using the iterative back-projection process. Porosity and characteristic ligament and pore sizes can be quantitatively obtained by the 3D reconstruction with TEM and stereographical analysis. The nanoporous gold obtained by chemical dealloying has been investigated with TEM reconstruction, and has average ligament size of 11±4.3 nm, pore channel size of 13±6.6 nm, and specific surface area of around 12 m^2/g [125].

X-ray microcomputed tomography is a non-destructive method, and usually has limited resolution over micron. But Transmission x-ray microscopy (TXM) using synchrotron X-ray can have a clearly better resolution around 30 nm. A series of projections can be recorded with TXM, and then the material structure can be reconstructed with the projections using a standard filtered back-projection algorithm [135]. The metallic porous structure will be coarsened when heat treatment is performed [136]. Due to the non-destructive advantage of synchrotron X-ray nanotopography, the coarsening process and the structure evolution can be well studied with this method. Interfacial shape distribution, an analysis method to display the probability of finding a patch of interface with a given pair of principle curvatures, is usually used to analyze structural evolution during the coarsening [132]. It has been revealed that the coarsening of nanoporous gold was occurring mainly via anisotropic surface diffusion [130].

APT can be applied for the 3D element analysis of materials, and provides compositional profiling and mapping with very high spatial and mass resolution. However, the investigation of nanoporous gold with APT is very challenging due the difficulty in sample preparation. APT samples are usually prepared with FIB milling and lift-out procedure, and the porous structure can be damaged during this process. Recently, it is reported that the sample of nanoporous gold can be conveniently prepared, when the pores are infiltrated with Cu, Fe, and Co [133, 137]. The recent APT profiling investigation of nanoporous gold synthesized with chemical dealloying using HClO4, revealed not only that residual Ag remains in the Au ligaments, but also that the ligaments have a core-shell type variable composition with Au-rich outer parts and Ag-rich cores [133].

FIB is a destructive tomographic 3D reconstruction method. With the dual beam FIB/SEM system, the sample can be iteratively or sequentially milled and imaged. The stack of images is further processed for the reconstruction of the 3D microstructure. Figure 1.15(a) shows a cross sectional SEM image and Figure 1.15(b) is the reconstruction of $(1\ \mu m)^3$ subvolume nanoporous gold. The ligament size and the pore size distribution can be well investigated and quantitatively analyzed by that. In addition, recent research work shows that the feature of the ligaments might not be sufficiently characterized by only one metric of ligament size [129], because two distinct morphologies can be discriminated as ligaments themselves, and tori formed by the interconnected ligaments, as shown in Figure 1.16(a) and Figure 1.16(b). Tori with dead-end parts (Figure 1.16(c) and Figure 1.16(d)) can be often observed after coarsening porous structures by thermal annealing. It is still under debate if the porous structure coarsens in a self-similar manner [132]. Nevertheless, it

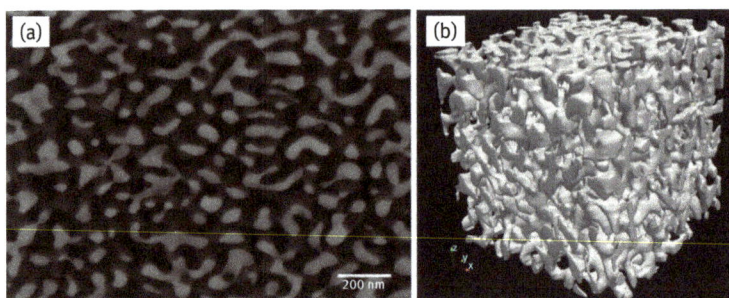

Figure 1.15: (a) a processed cross sectional SEM image of nanoporous gold infiltrated with epoxy, and (b) reconstruction of a $(1\ \mu m)^3$ subvolume. Reproduced from Ref. [127] with permission from Elsevier Ltd.

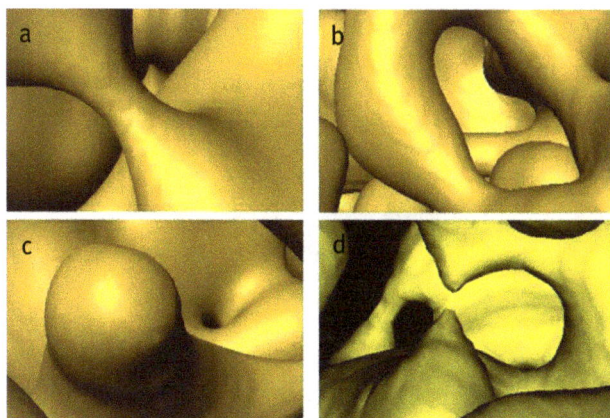

Figure 1.16: Snapshots from a 3D FIB tomography reconstruction of a nanoporous gold sample with mean ligament diameter of about 420 nm. The images reflect the typical structural features: (a) single ligaments that are connected to (b) irregular tori, which are sometimes opened with dead-end parts (c) and (d). Reproduced from Ref. [129] with permission from Elsevier Ltd.

is suggested that two metrics of both ligament and torus sizes should be characterized through the analysis of the principle curvatures, because both types of morphologies can clearly have different impact on the mechanical and other functional properties.

1.3.2 Optical characterizations and applications

In contrast to solid NPs (or nanowires), nanosponges possess a unique structural feature, which combines both 0D (as NPs), or 1D (as nanowires), and 3D (with porous structure) characteristics, and therefore, have quite complex linear and nonlinear optical properties. The optical properties of the nanosponges are highly tailorable through controlling the nanosponge size, form, pore size and porosity [38, 40]. The plasmon peak of the nanosponges is largely red-shifted compared to the solid NPs with the same particle size [38]. The individual nanosponges possess a strong polarization effect and multiple resonances behavior [36, 37]. High density of hot spots with spatially and spectrally confined modes at 10 nm scale and high quality factor (over 40), have been also experimentally probed on the surface of nanosponges [138]. High order ($n = 5 \ldots 7$) nonlinear optical behavior has been proven through light-induced electron emission [39]. Furthermore, long-lived plasmon modes have been revealed through the time dynamics of electron emission with femtosecond time-resolution [39].

The linear optical scattering and photoluminescence (PL) emission of the individual nanosponges can be measured with dark field florescence-confocal microscopy,

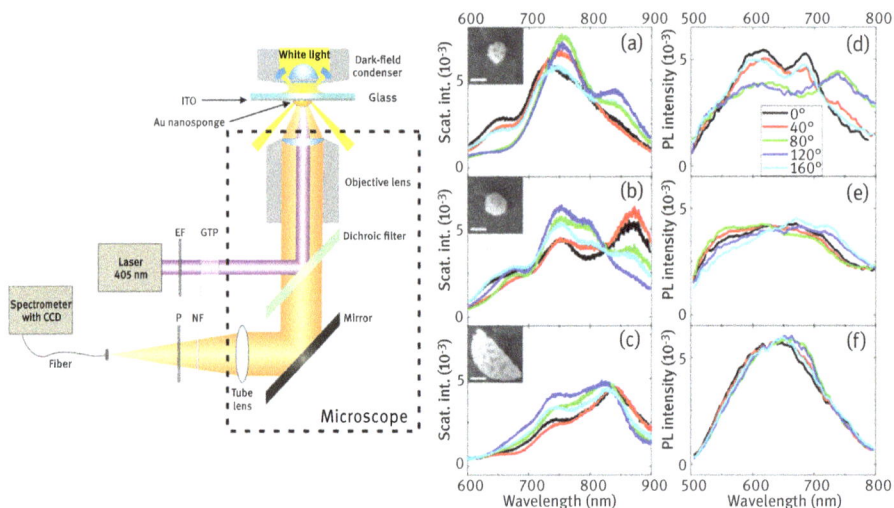

Figure 1.17: Left: Schematic of the dark-field/fluorescence-confocal setup used for the scattering and photoluminescence (PL) measurements of single nanosponges. Right: SEM images (scale bars 100 nm) and polarization-dependent scattering spectra of Au nanosponges, two almost spherical with diameters of (a) 115 and (b) 155 nm and one with lateral dimensions of (c) 410 and 195 nm. (d–f) Corresponding PL spectra. The legend in part d gives the color-coding of the polarization. Reproduced from Ref. [37] with permission from American Chemical Society.

as schematically shown in the left panel of Figure 1.17. Significant polarization anisotropy has been observed in both scattering and PL spectra (right panels of Figure 1.17) even for nearly spherical nanosponges [36, 37]. Multiple peaks in both scattering and PL spectra can be resolved when the polarized light is applied. It has been further revealed that the individual inner topography is more crucial for its optical response than its outer shape and size. However, the anisotropy ceases for both scattering and PL with larger sized nanosponges, and the dependence of the anisotropy on the sponge size is different between the scattering and PL. The PL from the recombination of electrons with d-band holes in gold NPs is usually strongly related to localized plasmon modes, but here the intrinsic PL from the gold nanosponges is substantially less polarized than the scattering spectra, suggesting a "plasmonic horizon" of appromimately 60 nm as finite coherence length of localized plasmons at the surface of gold nanosponges [37].

The significant polarization effect indicates that strongly localized plasmon modes (hot spots) were excited at the surface of the nanosponges. Scattering type scanning near-field optical microscopy (s-SNOM) was used experimentally to probe the hot spots formation on the individual gold nanosponges [138]. The s-SNOM is schematically shown in Figure 1.18(a). These measurements reveal pronounced spatio-spectral fluctuations of the local near-field intensity with local intensity maxima exceeding the average intensity by up to a factor of 10 (Figure 1.18(d) and Figure

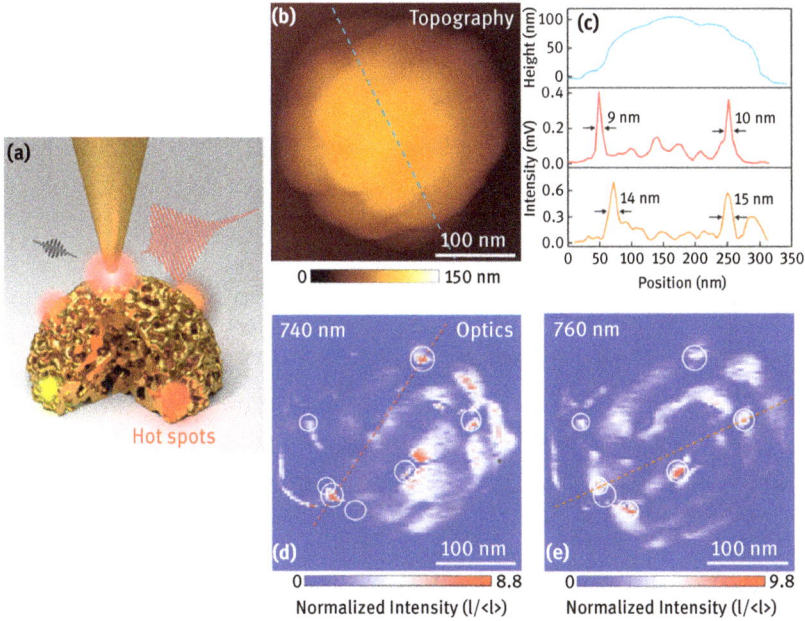

Figure 1.18: (a) Schematic of the scattering-type near-field spectroscopy experiment probing surface plasmon localization (hot-spots) in individual gold nanosponges. (b) AFM topographic image of a single nanosponge. (c) Top: Cross section of the sponge topography along the dashed line in panel (b). Bottom: Cross sections of the scattering signal along the dashed lines in panels d and e, respectively. It has been revealed that surface plasmon localization in individual hot-spots with diameters of about 10 nm. (d–e) Optical near-field scattering image of the same nanosponge for laser excitation at 740 and 760 nm, respectively. Several randomly distributed hot-spots are marked with white circles in parts d–e. Reproduced from Ref. [138] with permission from American Chemical Society.

1.18(e)) and demonstrate hot-spot localization at the scale of 10 nm (Figure 1.18(c)). These spectra reveal surprisingly large quality factors Q of the localized plasmon modes exceeding the number of 40. This suggests that the plasmonic excitations of the nanosponges are spatially strongly localized with intensity fluctuations similar to those of the strong localization regime of Anderson localization.

The mentioned strongly localized surface plasmon modes (hot spots) are highly desired for the non-linear optical applications. Measurement of the light induced electron emission can be used to study the non-linear optical response of the nanosponges [39]. Photoemission of a single gold nanosponge using ultrashort laser pulse excitation with femtosecond time resolution is schematically demonstrated in Figure 1.19(a). By comparing the photoelectron emission microscopy image (Figure 1.19(c)) with the SEM image (Figure 1.19(b)) of the same sample area with five individual nanosponges, it can be clearly seen that all nanosponges demonstrated strong electron emission. A high order (n = 5 ... 7) of optical non-linearity of the

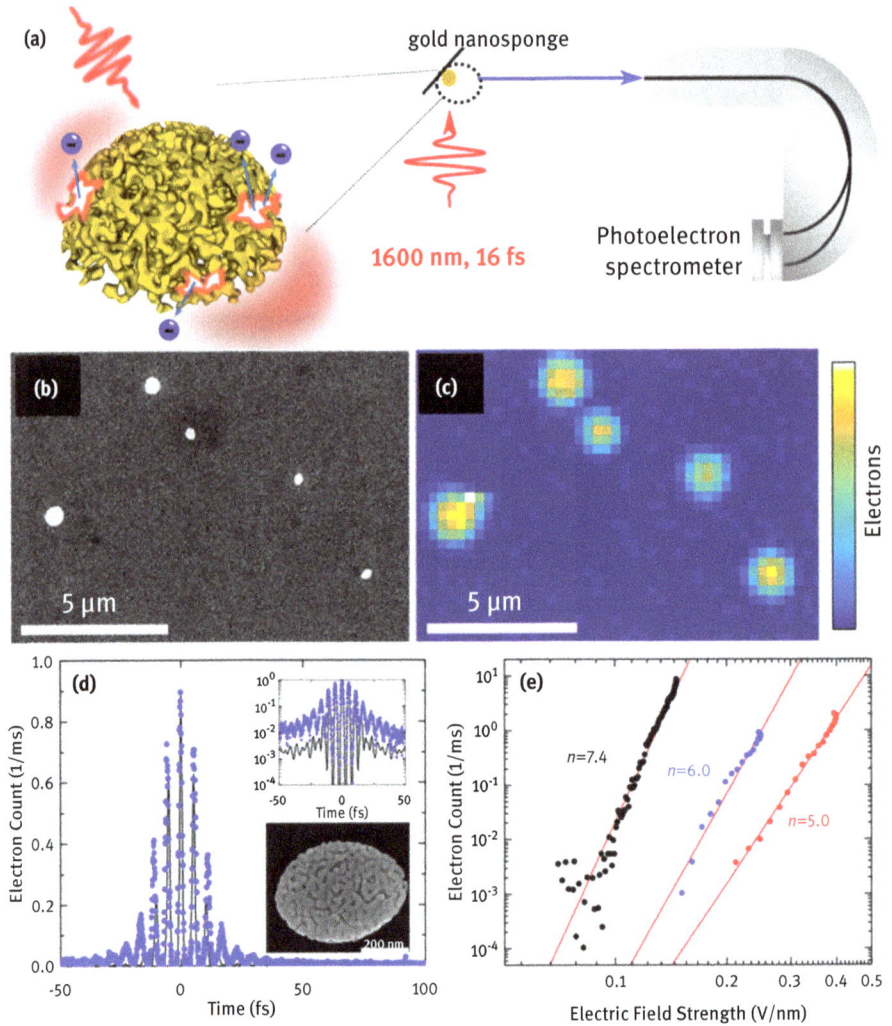

Figure 1.19: (a) Schematic of light induced electron emission from single gold nanosponge using ultrashort laser pulse excitation with femtosecond time resolution. (b) SEM image of an overview of 5 gold nanosponges, and (c) photoelectron emission microscopy (PEEM) image of the same area of the sample with visible individual sponges. (d) Interferometric autocorrelation trace $C(\Delta t)$ of the photoemission from a single nanosponge (SEM image inset). The instrument response $C_0(\Delta t)$ simulated from the measured electrical field of the laser pulses $E_L(t)$ for a 5th order optical nonlinearity is shown as a solid line. The data are also represented on a logarithmic intensity scale (inset). (e) Optical nonlinearity of the photoemission deduced by plotting the recorded photoelectron signal from three representative nanosponges as a function of the maximum electric field amplitude of the laser pulses. Optical nonlinearities of order n = 5 ... 7 are deduced. Reproduced from Ref. [39] with permission from Nature publishing group.

multiphoton emission process has been proven, as seen in Figure 1.19(e), which confirms also the highly local field enhancement in the hot spots. The time dynamics of the photoemission can be determined by performing the measurements of interferometric autocorrelation. As shown in Figure 1.19(d), the photoemission persists even clearly longer than the laser pulse duration, indicating the excited hot spots were actually long-lived localized plasmon modes.

The exciting optical properties of nanosponges have a great potential for the applications in optical sensing and spectroscopy. For instance, the intensity of the inelastic Raman scattering can be largely enhanced by the enhanced local field of the hot spots. As example, butter yellow (BY) can be better detected using SERS with Au–Ag hybrid nanosponges as support substrates [44]. BY is a possible carcinogen, and so the detection of BY is an important process for the food safety inspection. Figure 1.20 shows the Raman spectra of BY on different substrates (gold nanosponges and Au–Ag hybrid nanosponges with varied content of Ag). The Au–Ag hybrid nanosponges were prepared through the cyclic electroless deposition of Ag into the porous

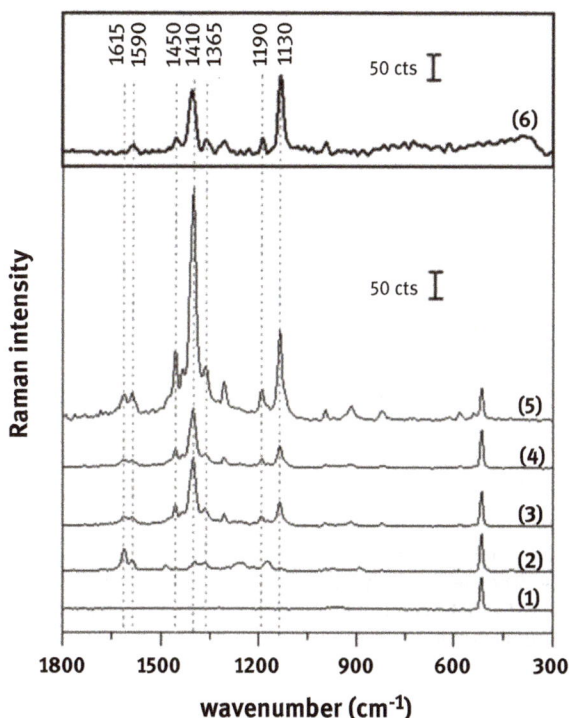

Figure 1.20: Raman spectrum of the background (1) and the SERS spectra of a 100 μM BY solution on the substrates of gold nanosponges (2), Au–Ag-1c hybrid nanosponges (3), Au–Ag-3c hybrid nanosponges (4), and Au–Ag-6c hybrid nanosponges (5). A 488 nm laser was used for excitation. A reference Raman spectrum of BY measured on powder is also presented (spectrum 6). Reproduced from Ref. [44] with permission from American Chemical Society.

structure of the as-prepared gold nanosponges (Figure 1.11). The bimetallic hybrid nanosponges show a clearly improved SERS performance than that of gold nanosponges. More research on the SERS with nanosponges or bimetallic nanosponges can be found in recent literature [45, 139–141].

Molecular and chemical information based on harmonics and combination bands of the fundamental vibrational modes can be investigated by using near-infrared (NIR) absorption spectroscopy. Unfortunately, sensitivity is limited by the generally weak absorption and the relatively poor detector performance compared to other wavelength ranges. Recently, surface enhanced near-field absorption (SENIRA) has been demonstrated with an enhancement factor of 10^4 by using nanoporous gold nanodisks as molecule supporting substrates [43]. Octadecanethiol (ODT) self-assembled monolayer has been used for the detection, and the Figure 1.21 (a) shows the spectra of ODT with and without support of nanoporous gold nanodisks. A large enhancement is demonstrated. The plasmon resonance of the nanoporous gold nanodisks is in the NIR range (Figure 1.21(b) and Figure 1.21(c)). It can be imagined that the enhancement of the near-field absorption benefits mainly from the local field enhancement of hot spots with mode wavelength in the same NIR range.

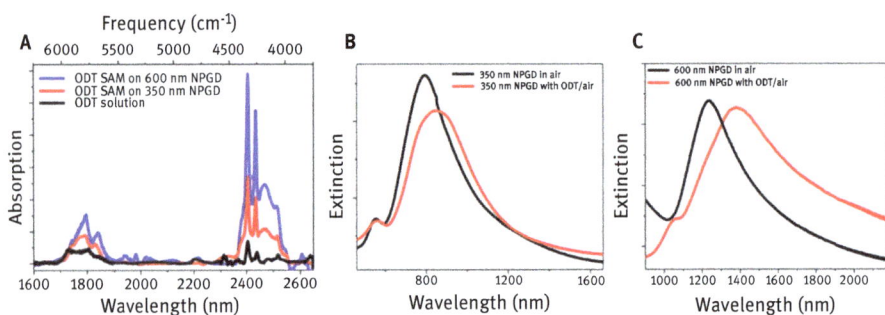

Figure 1.21: (A) SENIRA of octadecanethiol (ODT) self-assembled monolayer (SAM) on nanoporous gold nanodisks (NPGD) with two different diameters: 350 (red) and 600 (blue) nm compared against the absorption of bulk ODT solution (black). The peak represents the first C–H overtone and combination band at ~1725 and ~2400 nm, respectively. (B and C) Extinction spectra showing redshift of LSPR band of 350 and 600 nm NPG disks upon addition of ODT SAM. Reproduced from Ref. [43] with permission from American Chemical Society.

1.4 Critical safety considerations

For the synthesis of the metallic nanosponges, acids, bases, and solvents are often used. Severe burns, tissue damage, organ damage, asphyxiation, and genetic damage can be caused, if these chemicals are used improperly. Laboratory coat, glasses and hand gloves must be always taken during dealing these chemicals and synthesis of nanosponges, to avoid direct contacting and exposing to the chemicals. Volatile solutions must be treated in fume hood, and breathing mask or breathing apparatus must be worn. Standard first aid kit (for eye washing) and safety devices (fire alarm and fire extinguisher) must be equipped in the chemical lab.

Metallic nanosponges belong to the class of nanomaterials, and often show very strong catalytic activity. Therefore, the nanosponges must be avoided to entry human body through breathing, drinking, eating or other misbehaviors. Some recent research reports on safety issues with NPs can be also referred here [142–144].

1.5 Conclusions and future perspective

Nanoporous metals and bimetallic nanoporous metals have been clearly identified as a new class of novel materials for promising applications in catalysis and sensing. Dealloying-based methods are the most popular approach for the synthesis of nanoporous metals. Combining dealloying with nanostructuring technologies, the disordered nanoporous structure can be further tailored in nanosponges with defined outer forms or shapes sized in the submicron range. Bimetallic nanosponges can be obtained via either controlled partial dealloying or additive deposition of the 2nd element into the porous structure of the as-prepared nanosponges. In addition, bimetallic nanosponges can be also well synthesized via reduction reaction-based methods. Both metallic nanosponges and bimetallic nanosponges possess a largely enriched ability for tailoring the optical and catalytic properties through controlling the feature sizes of the outer form, topographical structure and element composition.

Synthesis and application of size controlled bimetallic nanosponges is a new emerging research area, and there are many aspects which still need to be explored in more details and depth. The first challenge for the future perspective is the control of the feature sizes and shapes, and the element composition towards the specific requirements for their optical and catalytic applications. For instance, high local field enhancement with eigenmodes at a specific wavelength range is required for some optical applications. The question here is how can these requirements be fulfilled by controlling the structural and compositional features during the synthesis. The second challenge is to establish a full understanding of the relation between properties and structures. So far, the recent studies have demonstrated that the optical properties of the nanosponges are very different and fascinating, but only some preliminary understanding about their optical behavior is obtained and a deep and comprehensive knowledge is still absent. Further systematic research is required in this direction.

References

[1]   Morriss RH, Collins LF. Optical properties of multilayer colloids. J Chem Phys. 1964;41:3357–63.
[2]   Sinfelt JH. Catalysis by alloys and bimetallic clusters. Acc Chem Res. 1977;10:15–20.
[3]   Gilroy KD, Ruditskiy A, Peng H-C, Qin D, Xia Y. Bimetallic nanocrystals: syntheses, properties, and applications. Chem Rev. 2016;116:10414–72.
[4]   Ding Y, Chen M, Erlebacher J. Metallic mesoporous nanocomposites for electrocatalysis. J Am Chem Soc. 2004;126:6876–7.
[5]   Wittstock A, Biener J, Baumer M. Nanoporous gold: a new material for catalytic and sensor applications. Phys Chem Chem Phys. 2010;12:12919–30.
[6]   Ding Y, Chen M. Nanoporous metals for catalytic and optical applications. MRS Bulletin. 2009;34:569–76.
[7]   Zielasek V, Jürgens B, Schulz C, Biener J, Biener MM, Hamza AV, et al. Gold catalysts: nanoporous gold foams. Angew Chem Int Ed. 2006;45:8241–4.
[8]   Luc W, Jiao F. Nanoporous metals as electrocatalysts: state-of-the-art, opportunities, and challenges. ACS Catal. 2017;7:5856–61.
[9]   Stenner C, Shao L-H, Mameka N, Weissmüller J. Piezoelectric gold: strong charge-load response in a metal-based hybrid nanomaterial. Adv Funct Mater. 2016;26:5174–81.
[10]   Forty AJ. Corrosion micromorphology of noble metal alloys and depletion gilding. Nature. 1979;282:597–8.
[11]   Erlebacher J, Aziz MJ, Karma A, Dimitrov N, Sieradzki K. Evolution of nanoporosity in dealloying. Nature. 2001;410:450–3.
[12]   Erlebacher J. An atomistic description of dealloying - Porosity evolution, the critical potential, and rate-limiting behavior. J Electrochem Soc. 2004;151:C614–26.
[13]   Ding Y, Kim YJ, Erlebacher J. Nanoporous gold leaf: "ancient technology"/advanced material. Adv Mater. 2004;16:1897–900.
[14]   Chen L, Guo H, Fujita T, Hirata A, Zhang W, Inoue A, et al. Nanoporous PdNi bimetallic catalyst with enhanced electrocatalytic performances for electro-oxidation and oxygen reduction reactions. Adv Funct Mater. 2011;21:4364–70.
[15]   Chen LY, Chen N, Hou Y, Wang ZC, Lv SH, Fujita T, et al. Geometrically controlled nanoporous pdau bimetallic catalysts with tunable Pd/Au ratio for direct ethanol fuel cells. ACS Catal. 2013;3:1220–30.
[16]   Chen LY, Fujita T, Ding Y, Chen MW. A three-dimensional gold-decorated nanoporous copper core–shell composite for electrocatalysis and nonenzymatic biosensing. Adv Funct Mater. 2010;20:2279–85.
[17]   Chen LY, Guo XW, Han JH, Liu P, Xu XD, Hirata A, et al. Nanoporous metal/oxide hybrid materials for rechargeable lithium–oxygen batteries. J Mater Chem A. 2015;3:3620–6.
[18]   Ge X, Chen L, Kang J, Fujita T, Hirata A, Zhang W, et al. A core-shell nanoporous Pt-Cu catalyst with tunable composition and high catalytic activity. Adv Funct Mater. 2013;23:4156–62.
[19]   Kang J, Hirata A, Qiu H-J, Chen L, Ge X, Fujita T, et al. Self-grown oxy-hydroxide@ nanoporous metal electrode for high-performance supercapacitors. Adv Mater. 2014;26:269–72.
[20]   Lang X, Zhang L, Fujita T, Ding Y, Chen M. Three-dimensional bicontinuous nanoporous Au/polyaniline hybrid films for high-performance electrochemical supercapacitors. J Power Sources. 2012;197:325–9.
[21]   Lang XY, Guo H, Chen LY, Kudo A, Yu JS, Zhang W, et al. Novel nanoporous Au–Pd alloy with high catalytic activity and excellent electrochemical stability. J Physl Chem C. 2010;114:2600–3.

[22]   Qiu HJ, Shen X, Wang JQ, Hirata A, Fujita T, Wang Y, et al. Aligned nanoporous Pt–cu bimetallic microwires with high catalytic activity toward methanol electrooxidation. ACS Catal. 2015;5:3779–85.

[23]   Qiu HJ, Wang JQ, Liu P, Wang Y, Chen MW. Hierarchical nanoporous metal/metal-oxide composite by dealloying metallic glass for high-performance energy storage. Corros Sci. 2015;96:196–202.

[24]   Wang K, Kobler A, Kübel C, Jelitto H, Schneider G, Weissmüller J. Nanoporous-gold-based composites: toward tensile ductility. Npg Asia Mater. 2015;7:e187.

[25]   Wang K, Weissmüller J. Composites of nanoporous gold and polymer. Adv Mater. 2013;25:1280–4.

[26]   Xu C, Wang R, Chen M, Zhang Y, Ding Y. Dealloying to nanoporous Au/Pt alloys and their structure sensitive electrocatalytic properties. Phys Chem Chem Phys. 2010;12:239–46.

[27]   Malgras V, Ataee-Esfahani H, Wang H, Jiang B, Li C, Wu KC, et al. Nanoarchitectures for mesoporous metals. Adv Mater. 2016;28:993–1010.

[28]   Wang D, Schaaf P. Nanoporous gold nanoparticles. J Mater Chem. 2012;22:5344–8.

[29]   Li X, Chen Q, McCue I, Snyder J, Crozier P, Erlebacher J, et al. Dealloying of noble-metal alloy nanoparticles. Nano Lett. 2014;14:2569–77.

[30]   Qi J, Motwani P, Gheewala M, Brennan C, Wolfe JC, Shih W-C. Surface-enhanced Raman spectroscopy with monolithic nanoporous gold disk substrates. Nanoscale. 2013;5:4105–9.

[31]   Santos GM, Zhao F, Zeng J, Shih W-C. Characterization of nanoporous gold disks for photothermal light harvesting and light-gated molecular release. Nanoscale. 2014;6:5718–24.

[32]   Ji C, Searson PC. Synthesis and characterization of nanoporous gold nanowires. J Phys Chem B. 2003;107:4494–9.

[33]   Chauvin A, Delacôte C, Molina-Luna L, Duerrschnabel M, Boujtita M, Thiry D, et al. Planar arrays of nanoporous gold nanowires: when electrochemical dealloying meets nanopatterning. ACS Appl Mater Interfaces. 2016;8:6611–20.

[34]   Shin T-Y, Yoo S-H, Park S. Gold nanotubes with a nanoporous wall: their ultrathin platinum coating and superior electrocatalytic activity toward methanol oxidation. Chem Mater. 2008;20:5682–6.

[35]   Pedireddy S, Lee HK, Koh CS, Tan JM, Tjiu WW, Ling XY. Nanoporous gold bowls: a kinetic approach to control open shell structures and size-tunable lattice strain for electrocatalytic applications. Small. 2016;12:4531–40.

[36]   Vidal C, Wang D, Schaaf P, Hrelescu C, Klar TA. Optical plasmons of individual gold nanosponges. ACS Photonics. 2015;2:1436–42.

[37]   Vidal C, Sivun D, Ziegler J, Wang D, Schaaf P, Hrelescu C, et al. Plasmonic horizon in gold nanosponges. Nano Lett. 2018;18:1269–73.

[38]   Rao W, Wang D, Kups T, Baradács E, Parditka B, Erdélyi Z, et al. Nanoporous gold nanoparticles and Au/Al2O3 hybrid nanoparticles with large tunability of plasmonic properties. ACS Appl Mater Interfaces. 2017;9:6273–81.

[39]   Hergert G, Vogelsang J, Schwarz F, Wang D, Kollmann H, Groß P, et al. Long-lived electron emission reveals localized plasmon modes in disordered nanosponge antennas. Light Sci Appl. 2017;6:e17075.

[40]   Wang D, Schaaf P. Plasmonic nanosponges. Adv Phys: X. 2018;3:1456361.

[41]   Arnob MM, Zhao F, Li J, Shih W-C. EBL-based fabrication and different modeling approaches for nanoporous gold nanodisks. ACS Photonics. 2017;4:1870–8.

[42]   Qi J, Zeng J, Zhao F, Lin SH, Raja B, Strych U, et al. Label-free, in situ SERS monitoring of individual DNA hybridization in microfluidics. Nanoscale. 2014;6:8521–6.

[43] Shih W-C, Santos GM, Zhao F, Zenasni O, Arnob MM. Simultaneous chemical and refractive index sensing in the 1–2.5 μm near-infrared wavelength range on nanoporous gold disks. Nano Lett. 2016;16:4641–7.

[44] Yan Y, Radu AI, Rao W, Wang H, Chen G, Weber K, et al. Mesoscopically bi-continuous Ag-Au hybrid nanosponges with tunable plasmon resonances as bottom-up substrates for surface-enhanced Raman spectroscopy. Chem Mater. 2016;28:7673–82.

[45] Liu K, Bai Y, Zhang L, Yang Z, Fan Q, Zheng H, et al. Porous Au–ag nanospheres with high-density and highly accessible hotspots for SERS analysis. Nano Lett. 2016;16:3675–81.

[46] Wittstock A, Biener J, Bäumer M. *Nanoporous gold: from an ancient technology to a high-tech material*. Cambridge,UK: RSC Publishing, 2012

[47] McCue I, Benn E, Gaskey B, Erlebacher J. Dealloying and dealloyed materials. Annu Rev Mater Res. 2016;46:263–86.

[48] Collinson MM. Nanoporous gold electrodes and their applications in analytical chemistry. ISRN Anal Chem. 2013;2013:21.

[49] Seker E, Reed M, Begley M. Nanoporous gold: fabrication, characterization, and applications. Materials. 2009;2:2188.

[50] Li GG, Wang H. Dealloyed nanoporous gold catalysts: from macroscopic foams to nanoparticulate architectures. ChemNanoMat. 2018;4:897–908.

[51] Wittstock A, Wichmann A, Biener J, Bäumer M. Nanoporous gold: a new gold catalyst with tunable properties. Faraday Discuss. 2011;152:87–98.

[52] Qiu HJ, Xu H-T, Liu L, Wang Y. Correlation of the structure and applications of dealloyed nanoporous metals in catalysis and energy conversion/storage. Nanoscale. 2015;7:386–400.

[53] Juarez T, Biener J, Weissmüller J, Hodge AM. Nanoporous metals with structural hierarchy: a review. Adv Eng Mater. 2017;19:1700389.

[54] Fujita T. Hierarchical nanoporous metals as a path toward the ultimate three-dimensional functionality. Sci Technol Adv Mater. 2017;18:724–40.

[55] Li H, Zhang A, Fang Z, Zeng J. Bimetallic nanoframes and nanoporous structures. In: Zhang Y-W, editor(s). *Bimetallic Nanostructures*. Weinheim: Wiley, 2018

[56] Detsi E, Vuković Z, Punzhin S, Bronsveld PM, Onck PR, Hosson JT. Fine-tuning the feature size of nanoporous silver. CrystEngComm. 2012;14:5402–6.

[57] Zhang Z, Wang Y, Qi Z, Zhang W, Qin J, Frenzel J. Generalized fabrication of nanoporous metals (Au, Pd, Pt, Ag, and Cu) through chemical dealloying. J Physl Chem C. 2009;113:12629–36.

[58] Dursun A, Pugh DV, Corcoran SG. Dealloying of Ag-Au alloys in halide-containing electrolytes: affect on critical potential and pore size. J Electrochem Soc. 2003;150: B355–60.

[59] Artymowicz DM, Erlebacher J, Newman RC. Relationship between the parting limit for de-alloying and a particular geometric high-density site percolation threshold. Philos Mag. 2009;89:1663–93.

[60] Parida S, Kramer D, Volkert CA, Ouml, Sner H, Erlebacher J, et al. Volume change during the formation of nanoporous gold by dealloying. Phys Rev Lett. 2006;97:035504.

[61] Bai Q, Wang Y, Zhang J, Ding Y, Peng Z, Zhang Z. Hierarchically nanoporous nickel-based actuators with giant reversible strain and ultrahigh work density. J Mater Chem C. 2016;4:45–52.

[62] Wada T, Yubuta K, Inoue A, Kato H. Dealloying by metallic melt. Mater Lett. 2011;65:1076–8.

[63] Lu Z, Li C, Han J, Zhang F, Liu P, Wang H, et al. Three-dimensional bicontinuous nanoporous materials by vapor phase dealloying. Nat Commun. 2018;9:276.

[64] Coaty C, Zhou H, Liu H, Liu P. A scalable synthesis pathway to nanoporous metal structures. ACS Nano. 2018;12:432–40.

[65] Kosinova A, Wang D, Schaaf P, Kovalenko O, Klinger L, Rabkin E. Fabrication of hollow gold nanoparticles by dewetting, dealloying and coarsening. Acta Mater. 2016;102:108–15.

[66] Thompson CV. Solid-state dewetting of thin films. Annu Rev Mater Res. 2012;42:399–434.

[67] Wang D, Schaaf P. Ni–Au bi-metallic nanoparticles formed via dewetting. Mater Lett. 2012;70:30–3.

[68] Herz A, Wang D, Kups T, Schaaf P. Solid-state dewetting of Au/Ni bilayers: the effect of alloying on morphology evolution. J Appl Phys. 2014;116:044307.

[69] Herz A, Friák M, Rossberg D, Hentschel M, Theska F, Wang D, et al. Facet-controlled phase separation in supersaturated Au-Ni nanoparticles upon shape equilibration. Appl Phys Lett. 2015;107:073109.

[70] Wang D, Schaaf P. Solid-state dewetting for fabrication of metallic nanoparticles and influences of nanostructured substrates and dealloying. Phys Status Solidi (A). 2013;210:1544–51.

[71] Herz A, Theska F, Rossberg D, Kups T, Wang D, Schaaf P. Solid-state dewetting of Au–ni bi-layer films mediated through individual layer thickness and stacking sequence. Appl Surf Sci. 2018;444:505–10.

[72] Herz A, Franz A, Theska F, Hentschel M, Kups T, Wang D, et al. Solid-state dewetting of single- and bilayer Au-W thin films: unraveling the role of individual layer thickness, stacking sequence and oxidation on morphology evolution. AIP Adv. 2016;6:035109.

[73] Wang D, Ji R, Albrecht A, Schaaf P. Ordered arrays of nanoporous gold nanoparticles. Beilstein J Nanotechnol. 2012;3:651–7.

[74] Wang D, Ji R, Schaaf P. Formation of precise 2D Au particle arrays via thermally induced dewetting on pre-patterned substrates. Beilstein J Nanotechnol. 2011;2:318–26.

[75] Herz A, Wang D, Schaaf P. Dewetting of Au/Ni bilayer films on prepatterned substrates and the formation of arrays of supersaturated Au-Ni nanoparticles. J Vac Sci Technol B. 2014;32:021802.

[76] Giermann AL, Thompson CV. Solid-state dewetting for ordered arrays of crystallographically oriented metal particles. Appl Phys Lett. 2005;86:121903.

[77] Wang D, Schaaf P. Thermal dewetting of thin Au films deposited onto line-patterned substrates. J Mater Sci. 2012;47:1605–8.

[78] Wang D, Schaaf P. Two-dimensional nanoparticle arrays formed by dewetting of thin gold films deposited on pre-patterned substrates. J Mater Sci: Mater Electron. 2011;22:1067–70.

[79] Khristosov MK, Bloch L, Burghammer M, Kauffmann Y, Katsman A, Pokroy B. Sponge-like nanoporous single crystals of gold. Nat Commun. 2015;6:8841.

[80] Zhao F, Zeng J, Parvez Arnob MM, Sun P, Qi J, Motwani P, et al. Monolithic NPG nanoparticles with large surface area, tunable plasmonics, and high-density internal hot-spots. Nanoscale. 2014;6:8199–207.

[81] Cai S, Jia X, Han Q, Yan X, Yang R, Wang C. Porous Pt/Ag nanoparticles with excellent multifunctional enzyme mimic activities and antibacterial effects. Nano Res. 2017;10:2056–69.

[82] Nishi H, Tatsuma T. Photoregulated nanopore formation via plasmon-induced dealloying of Au–ag alloy nanoparticles. J Physl Chem C. 2017;121:2473–80.

[83] Qian L-H, Ding Y, Fujita T, Chen M-W. Synthesis and optical properties of three-dimensional porous core–shell nanoarchitectures. Langmuir. 2008;24:4426–9.

[84] Zeis R, Mathur A, Fritz G, Lee J, Erlebacher J. Platinum-plated nanoporous gold: an efficient, low Pt loading electrocatalyst for PEM fuel cells. J Power Sources. 2007;165:65–72.

[85] Xiao S, Xiao F, Hu Y, Yuan S, Wang S, Qian L, et al. Hierarchical nanoporous gold-platinum with heterogeneous interfaces for methanol electrooxidation. Sci Rep. 2014;4:4370.

[86] Du Y, Xu J-J, Chen H-Y. Ultrathin platinum film covered high-surface-area nanoporous gold for methanol electro-oxidation. Electrochem commun. 2009;11:1717–20.

[87] Jia C, Yin H, Ma H, Wang R, Ge X, Zhou A, et al. Enhanced photoelectrocatalytic activity of methanol oxidation on TiO2-decorated nanoporous gold. J Physl Chem C. 2009;113:16138–43.

[88] Biener MM, Biener J, Wichmann A, Wittstock A, Baumann TF, Bäumer M, et al. ALD functionalized nanoporous gold: thermal stability, mechanical properties, and catalytic activity. Nano Lett. 2011;11:3085–90.

[89] Kosinova A, Wang D, Baradács E, Parditka B, Kups T, Klinger L, et al. Tuning the nanoscale morphology and optical properties of porous gold nanoparticles by surface passivation and annealing. Acta Mater. 2017;127:108–16.

[90] Lang X, Hirata A, Fujita T, Chen M. Nanoporous metal/oxide hybrid electrodes for electrochemical supercapacitors. Nat Nanotechnol. 2011;6:232.

[91] Kang J, Chen L, Hou Y, Li C, Fujita T, Lang X, et al. Electroplated thick manganese oxide films with ultrahigh capacitance. Adv Energy Mater. 2013;3:857–63.

[92] Yu Y, Gu L, Lang X, Zhu C, Fujita T, Chen M, et al. Li storage in 3D nanoporous au-supported nanocrystalline tin. Adv Mater. 2011;23:2443–7.

[93] Chen LY, Hou Y, Kang JL, Hirata A, Fujita T, Chen MW. Toward the theoretical capacitance of RuO2 Reinforced by highly conductive nanoporous gold. Adv Energy Mater. 2013;3:851–6.

[94] Ge X, Chen L, Zhang L, Wen Y, Hirata A, Chen M. Nanoporous metal enhanced catalytic activities of amorphous molybdenum sulfide for high-efficiency hydrogen production. Adv Mater. 2014;26:3100–4.

[95] Tan Y, Liu P, Chen L, Cong W, Ito Y, Han J, et al. Monolayer MoS2 films supported by 3D nanoporous metals for high-efficiency electrocatalytic hydrogen production. Adv Mater. 2014;26:8023–8.

[96] Hou Y, Chen L, Liu P, Kang J, Fujita T, Chen M. Nanoporous metal based flexible asymmetric pseudocapacitors. J Mater Chem A. 2014;2:10910–6.

[97] Meng F, Ding Y. Sub-micrometer-thick all-solid-state supercapacitors with high power and energy densities. Adv Mater. 2011;23:4098–102.

[98] Detsi E, Onck P, De Hosson JT. Metallic muscles at work: high rate actuation in nanoporous gold/polyaniline composites. ACS Nano. 2013;7:4299–306.

[99] Wang H, Jeong HY, Imura M, Wang L, Radhakrishnan L, Fujita N, et al. Shape- and size-controlled synthesis in hard templates: sophisticated chemical reduction for mesoporous monocrystalline platinum nanoparticles. J Am Chem Soc. 2011;133:14526–9.

[100] Jiang B, Li C, Tang J, Takei T, Kim JH, Ide Y, et al. Tunable-sized polymeric micelles and their assembly for the preparation of large mesoporous platinum nanoparticles. Angew Chem. 2016;128:10191–5.

[101] Jiang B, Li C, Dag Ö, Abe H, Takei T, Imai T, et al. Mesoporous metallic rhodium nanoparticles. Nat Commun. 2017;8:15581.

[102] Jiang B, Li C, Imura M, Tang J, Yamauchi Y. Multimetallic mesoporous spheres through surfactant-directed synthesis. Adv Sci. 2015;2:1500112.

[103] Shim K, Lin J, Park M-S, Shahabuddin M, Yamauchi Y, Hossain MS, et al. Tunable porosity in bimetallic core-shell structured palladium-platinum nanoparticles for electrocatalysts. Scr Mater. 2019;158:38–41.

[104] Jiang B, Li C, Henzie J, Takei T, Bando Y, Yamauchi Y. Morphosynthesis of nanoporous pseudo Pd@Pt bimetallic particles with controlled electrocatalytic activity. J Mater Chem A. 2016;4:6465–71.

[105] Li C, Xu Y, Li Y, Yu H, Yin S, Xue H, et al. Engineering porosity into trimetallic PtPdNi nanospheres for enhanced electrocatalytic oxygen reduction activity. Green Energy Environ. 2018;3:352–9.

[106] Eid K, Ahmad YH, Yu H, Li Y, Li X, AlQaradawi SY, et al. Rational one-step synthesis of porous PtPdRu nanodendrites for ethanol oxidation reaction with a superior tolerance for CO-poisoning. Nanoscale. 2017;9:18881–9.

[107] Deng K, Xu Y, Li C, Wang Z, Xue H, Li X, et al. PtPdRh mesoporous nanospheres: an efficient catalyst for methanol electro-oxidation. Langmuir. 2019;35:413–9.

[108] Yamauchi Y, Sugiyama A, Morimoto R, Takai A, Kuroda K. Mesoporous platinum with giant mesocages templated from lyotropic liquid crystals consisting of diblock copolymers. Angew Chem Int Ed. 2008;47:5371–3.

[109] Luo K, Walker CT, Edler KJ. Mesoporous silver films from dilute mixed-surfactant solutions by using dip-coating. Adv Mater. 2007;19:1506–9.

[110] Attard GS, Bartlett PN, Coleman NR, Elliott JM, Owen JR, Wang JH. Mesoporous platinum films from lyotropic liquid crystalline phases. Science. 1997;278:838–40.

[111] Jiang B, Li C, Malgras V, Imura M, Tominaka S, Yamauchi Y. Mesoporous Pt nanospheres with designed pore surface as highly active electrocatalyst. Cheml Sci. 2016;7:1575–81.

[112] Lv H, Sun L, Zou L, Xu D, Yao H, Liu B. Size-dependent synthesis and catalytic activities of trimetallic PdAgCu mesoporous nanospheres in ethanol electrooxidation. Cheml Sci. in press. 2019. DOI: 10.1039/C8SC04696D.

[113] Fu G, Liu H, You N, Wu J, Sun D, Xu L, et al. Dendritic platinum−copper bimetallic nanoassemblies with tunable composition and structure: arginine-driven self-assembly and enhanced electrocatalytic activity. Nano Res. 2016;9:755–65.

[114] Xie X-W, Lv J-J, Liu L, Wang A-J, Feng J-J, Xu Q-Q. Amino acid-assisted fabrication of uniform dendrite-like PtAu porous nanoclusters as highly efficient electrocatalyst for methanol oxidation and oxygen reduction reactions. Int J Hydrogen Energy. 2017;42:2104–15.

[115] Ma A, Xu J, Zhang X, Zhang B, Wang D, Xu H. Interfacial nanodroplets guided construction of hierarchical Au, Au-Pt, and Au-Pd particles as excellent catalysts. Sci Rep. 2014;4:4849.

[116] Huang X, Zhu E, Chen Y, Li Y, Chiu C-Y, Xu Y, et al. A facile strategy to Pt3Ni nanocrystals with highly porous features as an enhanced oxygen reduction reaction catalyst. Adv Mater. 2013;25:2974–9.

[117] Lu S, Eid K, Deng Y, Guo J, Wang L, Wang H, et al. One-pot synthesis of PtIr tripods with a dendritic surface as an efficient catalyst for the oxygen reduction reaction. J Mater Chem A. 2017;5:9107–12.

[118] Hu Y, Liu T, Li C, Yuan Q. Facile surfactant-free synthesis of composition-tunable bimetallic PtCu alloy nanosponges for direct methanol fuel cell applications. Aust J Chem. 2018;71:504–10.

[119] Lu Q, Huang J, Han C, Sun L, Yang X. Facile synthesis of composition-tunable PtRh nanosponges for methanol oxidation reaction. Electrochim Acta. 2018;266:305–11.

[120] Shi Y-C, Yuan T, Feng J-J, Yuan J, Wang A-J. Rapid fabrication of support-free trimetallic Pt53Ru39Ni8 nanosponges with enhanced electrocatalytic activity for hydrogen evolution and hydrazine oxidation reactions. J Colloid Interface Sci. 2017;505:14–22.

[121] Lee C-L, Huang Y-C, Kuo L-C, Oung J-C, Wu F-C. Preparation and characterization of Pd/Ag and Pd/Ag/Au nanosponges with network nanowires and their high electroactivities toward oxygen reduction. Nanotechnology. 2006;17:2390.

[122] Lilleodden ET, Voorhees PW. On the topological, morphological, and microstructural characterization of nanoporous metals. MRS Bulletin. 2018;43:20–6.

[123] Soyarslan C, Bargmann S, Pradas M, Weissmüller J. 3D stochastic bicontinuous microstructures: generation, topology and elasticity. Acta Mater. 2018;149:326–40.

[124] Rösner H, Parida S, Kramer D, Volkert CA, Weissmüller J. Reconstructing a nanoporous metal in three dimensions: an electron tomography study of dealloyed gold leaf. Adv Eng Mater. 2007;9:535–41.

[125] Fujita T, Qian L-H, Inoke K, Erlebacher J, Chen M-W. Three-dimensional morphology of nanoporous gold. Appl Phys Lett. 2008;92:251902.

[126] Krekeler T, Straßer AV, Graf M, Wang K, Hartig C, Ritter M, et al. Silver-rich clusters in nanoporous gold. Mater Res Lett. 2017;5:314–21.

[127] Mangipudi KR, Radisch V, Holzer L, Volkert CA. A FIB-nanotomography method for accurate 3D reconstruction of open nanoporous structures. Ultramicroscopy. 2016;163:38–47.

[128] Hu K, Ziehmer M, Wang K, Lilleodden ET. Nanoporous gold: 3D structural analyses of representative volumes and their implications on scaling relations of mechanical behaviour. Philos Mag. 2016;96:3322–35.

[129] Ziehmer M, Hu K, Wang K, Lilleodden ET. A principle curvatures analysis of the isothermal evolution of nanoporous gold: quantifying the characteristic length-scales. Acta Mater. 2016;120:24–31.

[130] Chen Y-C, Chu YS, Yi J, McNulty I, Shen Q, Voorhees PW, et al. Morphological and topological analysis of coarsened nanoporous gold by x-ray nanotomography. Appl Phys Lett. 2010;96:043122.

[131] Fam Y, Sheppard TL, Diaz A, Scherer T, Holler M, Wang W, et al. Correlative multiscale 3D imaging of a hierarchical nanoporous gold catalyst by electron, ion and X-ray nanotomography. ChemCatChem. 2018;10:2858–67.

[132] Chen-Wiegart Y-C, Wang S, Chu YS, Liu W, McNulty I, Voorhees PW, et al. Structural evolution of nanoporous gold during thermal coarsening. Acta Mater. 2012;60:4972–81.

[133] El-Zoka AA, Langelier B, Botton GA, Newman RC. Enhanced analysis of nanoporous gold by atom probe tomography. Mater Charact. 2017;128:269–77.

[134] Kübel C, Voigt A, Schoenmakers R, Otten M, Su D, Lee T-C, et al. Recent advances in electron tomography: TEM and HAADF-STEM tomography for materials science and semiconductor applications. Microsc Microanal. 2005;11:378–400.

[135] Nattere Frank. *The mathematics of computerized tomography*. Philadelphia, PA, USA: Society for Industrial and Applied Mathematics, 20010-89871-493-1.

[136] Li R, Sieradzki K. Ductile-brittle transition in random porous Au. Phys Rev Lett. 1992;68:1168–71.

[137] Pfeiffer B, Erichsen T, Epler E, Volkert CA, Trompenaars P, Nowak C. Characterization of nanoporous materials with atom probe tomography. Microsc Microanal. 2015;21:557–63.

[138] Zhong J, Chimeh A, Korte A, Schwarz F, Yi J, Wang D, et al. Strong spatial and spectral localization of surface plasmons in individual randomly disordered gold nanosponges. Nano Lett. 2018;18:4957–64.

[139] Mei H, Bai H, Bai S, Li X, Zhao X, Cheng L. Tuning Ag content in AuAg nanosponges for superior SERS detection. Mater Lett. 2018;230:24–7.

[140] Zhang T, Sun Y, Hang L, Li H, Liu G, Zhang X, et al. Periodic porous alloyed Au–ag nanosphere arrays and their highly sensitive SERS performance with good reproducibility and high density of hotspots. ACS Appl Mater Interfaces. 2018;10:9792–801.

[141] Zhang T, Zhou F, Hang L, Sun Y, Liu D, Li H, et al. Controlled synthesis of sponge-like porous Au–ag alloy nanocubes for surface-enhanced Raman scattering properties. J Mater Chem C. 2017;5:11039–45.

[142] Sufian MM, Khattak JZ, Yousaf S, Rana MS Safety issues associated with the use of nanoparticles in human body. Photodiagn Photodyn Ther. 2017;19:67–72.

[143] Berube D, Cummings C, Cacciatore M, Scheufele D, Kalin J. Characteristics and classification of nanoparticles: expert delphi survey. Nanotoxicology. 2011;5:236–43.

[144] Ray PC, Yu H, Fu PP. Toxicity and environmental risks of nanomaterials: challenges and future needs. J Environ Sci Health Part C, Environ Carcinog Ecotoxicol Rev. 2009;27:1–35.

Bionotes

Dong Wang studied chemical engineering at TU Wuhan for his B.Sc., and materials science at RWTH Aachen University for his M.Sc. He obtained his PhD from Karlsruhe Institute of Technology in 2007. He conducted his two years PostDoc research at Hannover University, and then has moved to TU Ilmenau. In 2016, he finished the Habilitation at TU Ilmenau, and currently is working as Privatdozent there. His research interest is focused on tailored nanostructures and nanomaterials for photonic and energy applications.

Peter Schaaf studied material physics at Saarland University and obtained his diploma degree there in 1988. This is followed by earning his doctoral degree (PhD) in 1991 with honors at the same university. After that, he moved to Göttingen University for a PostDoc position in 1992. In 1995, he got an assistant professorship and was promoted to associate professor there in 1999. He accomplished habilitation at Göttingen University in 1999. Since 2008, he is full professor at TU Ilmenau. Currently, he is chair of Materials of Electrical Engineering and Electronics in the Institute of Materials Science and Engineering and the Institute of Micro and Nanotechnologies MacroNano® and dean of the Department of Electrical Engineering and Information Technology of TU Ilmenau. His research interests lie in nanomaterials, electronic materials, nanotechnologies, thin films, functional materials and materials analysis.

Yunyun Zhou

2 Controllable design, synthesis and characterization of nanostructured rare earth metal oxides

Abstract: Rare earth metal oxide nanomaterials have drawn much attention in recent decades due to their unique properties and promising applications in catalysis, chemical and biological sensing, separation, and optical devices. Because of the strong structure–property correlation, controllable synthesis of nanomaterials with desired properties has long been the most important topic in nanoscience and nanotechnology and still maintains a grand challenge. A variety of methods, involving chemical, physical, and hybrid method, have been developed to precisely control nanomaterials, including size, shape, dimensionality, crystal structure, composition, and homogeneity. These nanostructural parameters play essential roles in determining the final properties of functional nanomaterials. Full understanding of nanomaterial properties through characterization is vital in elucidating the fundamental principles in synthesis and applications. It allows researchers to discover the correlations between the reaction parameters and nanomaterial properties, offers valuable insights in improving synthetic routes, and provokes new design strategies for nanostructures. In application systems, it extrapolates the structure–activity relationship and reaction mechanism and helps to establish quality model for similar reaction processes. The purpose of this chapter is to provide a comprehensive overview and a practical guide of rare earth oxide nanomaterial design and characterization, with special focus on the well-established synthetic methods and the conventional and advanced analytical techniques. This chapter addresses each synthetic method with its advantages and certain disadvantages, and specifically provides synthetic strategies, typical procedures and features of resulting nanomaterials for the widely-used chemical methods, such as hydrothermal, solvothermal, sol–gel, co-precipitation, thermal decomposition, etc. For the nanomaterial characterization, a practical guide for each technique is addressed, including working principle, applications, materials requirements, experimental design and data analysis. In particular, electron and force microscopy are illuminated for their powerful functions in determining size, shape, and crystal structure, while X-ray based techniques are discussed for crystalline, electronic, and atomic structural determination for oxide nanomaterials. Additionally, the advanced characterization methodologies of synchrotron-based techniques and *in situ* methods are included. These non-traditional methods become more and more

This article has previously been published in the journal *Physical Sciences Reviews*. Please cite as: Zhou, Y. Controllable design, synthesis and characterization of nanostructured rare earth metal oxides. *Physical Sciences Reviews* [Online] 2020, 5. DOI: 10.1515/psr-2018-0084.

https://doi.org/10.1515/9783110345001-002

popular because of their capabilities of offering unusual nanostructural information, short experiment time, and in-depth problem solution.

Graphical Abstract:

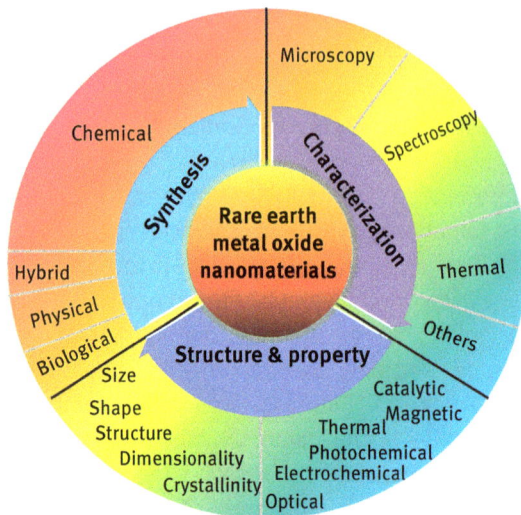

Keywords: Rare earth oxides, synthesis, characterization, structure, property, spectroscopy, microscopy, synchrotron methods

2.1 Introduction

Nanoparticles, typically in size of 1–100 nm, yield significantly distinguished size-dependent physical and chemical properties compared with their bulk counterparts. The difference is driven by the quantum confinement effect, greatly enhanced surface to volume ratio, and increased surface functional groups of nanomaterials. Nanomaterials have attracted intense interest in a wide range of fields including catalysis, electronics, energy storage, and healthcare. Their increasing applications in these fields leverage advances in nanotechnology and exert critical requirements over nanomaterial synthesis, characterization and fundamental understandings.

Nanomaterials synthesis is essential for nanoscience and nanotechnology. High-quality nanoparticles are of monodispersed size distribution, uniform shape and dimensionality, great purity and limited degree of agglomeration under operation conditions. However in reality, nanoparticles frequently exhibit non-uniform compositions, size and shapes, and lead to inconsistency in their resulting properties, and pose difficulties in reproducible fabrication. In order to avoid inhomogeneity in nanoparticles and provide desired properties of building blocks for nanodevices, a variety of methods have been developed to control the shape, size, dimensionality, and assembly of nanostructures. Generally, the methods are divided into three general techniques: chemical, physical, and biological techniques. Currently, the physical method is usually limited to the physical vapor deposition method, and mainly used for obtaining thin layers or atomic layer clusters on substrate. Chemical methods are more preferred due to their large quantities of production, low cost, and flexibility of tuning various controlling parameters. Particularly, the solution or liquid chemical methods are highly sophisticated, and extensively employed to fabricate well-defined and uniform crystallite sizes and shapes which possess extraordinary assembly properties [1–3]. The popular solution-based chemical routes include hydrothermal, solvothermal, sol–gel, co-precipitation as well as template assisted synthetic approaches. More often, these methods are combined to create certain synthesis protocols to meet specific criteria of targeted nanomaterials [4–8].

Characterization of nanomaterials is equally as important as synthetic route. Not only characterization yield information of chemical and physical properties of nanocrystals, and build relationships between structure and properties, but also it benefits the synthesis of nanomaterials-questions regarding control over size, shape, composition, and crystal structure and the undergoing reaction mechanism can be addressed. Parallel to the development of highly advanced analytical tools, characterization of nanostructures down to atomic scale resolution is enabled and hence fundamental understanding of synthesis principles is achieved. The basic physical properties of size, shape, and crystal structure can be elucidated by scanning electron microscopy (SEM), transmission electron microscopy (TEM), X-ray diffraction (XRD), dynamic light scattering (DLS), etc. To gain insights into the atomic level of nanomaterials, high resolution transmission electron microscopy (HRTEM) is able to probe the exposed facets, lattice fringes and defect features [9]. To extract information beyond the laboratory techniques, synchrotron based X-ray absorption spectroscopy (XAS) and X-ray photoelectron spectroscopy (XPS) enable the structural characterization on poorly crystalline nanostructures. They are also powerful in investigating the bulk and surface structures and local environments of atoms with extremely low concentration, and thus become increasingly important in nanomaterials analysis [10–13]. Additionally, *in situ* techniques become popular in nano research work to investigate the growth/formation mechanism of nanoparticles as well as their performances and property changes under operation conditions in application systems [14–19].

Rare earth metal oxides (REO), including lanthanides as well as Sc and Y, have been successfully applied in catalysis, coatings, polishing, microelectronic devices, and biomedicine, due to their special electronic and magnetic properties [20–25]. Many of the key properties of a rare earth oxide arise from their 4f electrons of rare earth elements [26]. Rare earth oxides exist in various forms, e.g. Ln_2O_3, LnO_2, LnO and Ln_3O_4, with sesquioxides and dioxides as the most common ones. The crystal structure varies with ionic radius and oxidation state of rare earth elements [20]. Of special interest are the defect structures of the fluorite type in dioxides. Defects, existing both on surface and in bulk material, play important roles in a variety of technological applications. For example, ceria is mainly used in three-way catalyst for the removal of toxic gases from automotive exhaust. The importance of ceria in catalysis originates from its remarkable oxygen storage capacity, which depends on its density of oxygen vacancy defects in ceria [22]. In order to increase the oxygen vacancy defects, methods such as doping and low-pressure activation have been explored and proved useful approaches both experimentally and computationally [27, 28].

Due to the unique crystal structures, special properties, and successful applications of rare earth oxides, fabrication of nanoscaled rare earth oxides has attracted huge interest in both academic research and industry over the past decade. On one hand, it is to apply the nano-oxides in previous applications to explore new features and performances. It was observed that the rare earth oxides may exhibit electron-photon interaction related spectral behavior when down to a few nanometers in size [26]. On the other hand, researchers keep seeking new applications of rare earth nano-oxides. For instance, the utilization of ceria nanoparticles has been extended to bioimaging, biosensor, drug delivery and therapies.

Impressive progress has been made in controlled synthesis of rare earth oxide nanoparticles in the past decade. A large number of research works have reported on the design of size and shape controlled nanoparticles, including nanorods, nanoplates, nanopolyhedrons and other nanostructures [4, 29–31]. Among these synthesis routes, solid state methods have been mainly used for complex nano-oxides, and required high temperatures. Solution-based methods attracted much more attention and have been further developed based on its low operating temperatures, simple apparatus, and versatile control. Hydro-(solvo-) thermal, co-precipitation, sol–gel and thermal decomposition methods are among the well-established ones. Parameters like reaction temperature, time, solvents, and precursor concentrations are usually tuned to adjust the nanomaterial properties. Templates have been frequently incorporated into these methods for generating uniform nano-oxides in nanoplates, nanodisks, hollow and tubular shapes [32–36]. However, the use of a template may inevitably increase the complexity of a synthetic procedure, and greatly limit the quantity of oxides obtained in each run of synthesis. Moreover, the residues of templates might affect the performance of nanomaterials in real reaction systems. For different rare earth elements, the synthesis procedures are usually not the same. For the same synthesis routes, different rare earth metal precursors typically result in distinguished shape and size of nano-oxides [37–39]. Additionally, when preparing the low-valence rare earth oxides containing Eu (II), Ce (III) or Yb (II), protection against oxidation from the environment is often required [40–42]. All the complexities bring uncertainties in nanomaterial synthesis. It is critical to take good control of each step and every parameter involved, as well as to follow the synthetic procedures exactly.

The book chapter herein discusses the details of the popular synthetic methods for rare earth nanostructured oxides, including the parameters and procedures, advantages and disadvantages, combination with other methods, roles of synthetic parameters, and reaction mechanisms. The solution-based routes are the main focus, including hydrothermal, solvothermal, co-precipitation, sol–gel, microemulsion, thermal decomposition synthesis, and methods assisted with microwave, sonication, and electrochemistry. Solid-based routes and vapor deposition method are also introduced. The discussion of characterization methodologies and instrumentation techniques encompasses both the conventional and advanced techniques, with a target on the microscopies, such as SEM, TEM, STM and AFM, and spectroscopies of XRD, XPS, XAS and Raman, that have been extensively used for rare earth oxides. *In situ* methods have made great contributions to the fundamental understandings of nanomaterials, which are also discussed in several characterization techniques.

2.2 Preparation methods

The preparation of rare earth oxide nanomaterials with desirable properties remains a grand challenge, which derives from the difficulties in synthesizing high-quality materials with controlled size and shape, and exploring robust pathways in different systems [43]. Synthesis of rare earth metal oxides involves a wide variety of methods, including chemical, physical, and hybrid methods. Among them, solution-based chemical methods draw most interests and have remarkable developments with different strategies. Those methods include numerous synthetic processes, such as hydrothermal, solvothermal, sol–gel, microemulsion, co-precipitation methods, and thermal decomposition. Herein this chapter will focus on the solution-based chemical methods, and briefly discuss about other chemical methods, physical methods, and some of the hybrid methods.

2.2.1 Hydrothermal synthesis

The hydrothermal method is one of the most important and well-established methods for nanomaterial synthesis. It involves with aqueous chemical reactions in a sealed heated vessel generally in an autoclave or a bomb calorimeter-type apparatus [44]. This method has several advantages: (1) producing particular shaped nanoparticles with relatively narrow size distribution (2) low operation temperature, below the melting point of reactants (3) environmentally friendly (4) different reactivity of inorganic solids at elevated temperatures and pressures in water [8, 13, 28, 45–47].

Rare earth oxides synthesized by simple hydrothermal method generally are obtained through the rare earth hydroxide gels, which are precipitated in basic solutions (NaOH, KOH or $NH_3 \bullet H_2O$) at room temperature or high temperatures. In contrast to vapor-liquid-solid or template-confined methods, a simple hydrothermal method neither can absorb particular reactant molecules nor assist nanoparticle growth in a certain direction. Therefore, hydrothermal methods have been effective in low dimension materials synthesis, such as nanorods, nanowires, nanotubes, and nanoparticles [13, 46–49]. One of the very successful examples is the synthesis of lanthanide hydroxide (RE = Y, La, Nd, Sm-Tm) single-crystal nanowires series reported by Wang and Li [50]. Xu et al. also showed a facile hydrothermal synthesis method for dysprosium and holmium hydroxide nanotubes from bulky rare earth oxide powder or colloidal hydroxide precipitation [51]. To obtain the final rare earth oxides, these hydroxides have to be separated, washed, dried, and calcined at high temperatures of 300–600 °C [28, 49, 50, 52–54]. Ceria nanostructures have also been synthesized successfully using template-free hydrothermal methods reported by Yan's and Tang's groups in early 2000s [53, 54].

Despite nanorods and nanowires being the most popular and feasible morphology of hydrothermal methods, other nanocrystal shapes have also been obtained by modifying metal precursors, reaction time and temperature, and precipitators [13, 28, 46, 47, 54]. Mai et al. reported a hydrothermal method for synthesizing ceria nanorods, nanocubes and nanopolyhedra by tuning temperatures in the range of 100–180 °C and NaOH concentrations from 0.01 to 9 mol/L (see details in Table 2.1). Ceria nanorods were synthesized at high base concentration (6–9 mol/L) at low temperature of 100 °C, while ceria nanopolyhedra were achieved at extremely low NaOH solution (0.01 mol/L) at 100–180 °C. Ceria cubes were generated at high NaOH concentration and high temperatures (140–180 °C). The very representative hydrothermal reactor, stainless steel autoclave with a Teflon liner, was used for hydrothermal treatment for 24 h. The white precipitates were collected by centrifugation and washed with deionized water and ethanol. Yellow powders were obtained after drying in air at 60 °C overnight. The last calcination step was fulfilled at 1000 °C for 4 h. The different nanostructures are elucidated in Figure 2.1 [53].

Table 2.1: Cyrstal structures, shapes and sizes of CeO$_2$ samples. All samples are synthesized under [Ce^{3+}] = 0.05 mol/L. (Reprinted with permission from Ref. [53]. Copyright (2005) American Chemical Society) [53].

No.	[NaOH] mol/L	T (°C)	t (h)	Structure	Shape	Size (nm)
1	0.01	100	24	Cubic	Polyhedra	11.5 ± 1.8
2	0.01	180	24	Cubic	Polyhedra	9–25
3	1	100	24	Cubic	Polyhedra; rods	
4	3	100	24	Cubic	Polyhedra; rods	
5	6	100	24	Cubic	Rods	(9.6 ± 1.2) × (50–200)
6	6	140	24	Cubic	Rods; cubes	
7	6	180	24	Cubic	Cubes	36.1 ± 7.1
8	9	100	24	Cubic	Rods	(13.3 ± 2.8) × (100–400)

Figure 2.1: (a) TEM and (b) HRTEM images of CeO$_2$ nanopolyhedra. (c) TEM and (d) HRTEM images of CeO$_2$ nanorods, inset is a fast Fourier transform (FFT) analysis. (e) TEM and (f) HRTEM images of CeO$_2$ nanocubes, inset is a fast Fourier transform (FFT) analysis. (Reprinted with permission from Ref. [53]. Copyright (2005) American Chemical Society.) [53].

Yan and co-workers later reported using Na$_3$PO$_4$ as a precipitator/mineralizer to obtain single crystalline ceria nanooctahedrons and nanorods by tuning the hydrothermal treatment time [55]. The formation of multinanostructures and morphology evolution from nano-octahedron to nanorod were proposed to follow a nucleation-dissolution-recrystallization process (see Figure 2.2) [55, 56]. In terms of controlling

nano-octahedron   multi-nanostructure   nanorods

Figure 2.2: TEM images of (left) ceria nano-octahedron (middle) ceria nanooctahedron with nanorods multi-structure and (right) ceria nanorods. (Reprinted with permission from Ref. [55]. Copyright (2008) American Chemical Society.) [55].

nanoparticle morphology, the anions from metal precursor and base solution in the solvents were found of great significance. For example, halogen (Cl^-, Br^-, and I^-) and SO_4^{2-} solutions typically formed ceria nanorods or nanopolyhedra, while NO_3^- based precursors were generally useful in generating ceria nanorods and nano-cubes [57]. Modifying the pH values using different bases NaOH, KOH and $NH_3 \bullet H_2O$ could also result in different composition and morphology. Zhang and co-workers reported that weak base $NH_3 \bullet H_2O$ with pH = 10–12 formed metastable γ-ScOOH loz-enge-like plates. Strong base NaOH or KOH resulted in α-ScOOH nanorods, nano-sized hexagonal-like plates, and cubic $Sc(OH)_3$ cubes/cuboids by varying the pH from 10 to [OH^-] = 5 mol/L [58].

In these template-free hydrothermal methods, it is very challenging to obtain high-yield, monodisperse, well shape- and size-controlled nanoparticles. For example, in the hydrothermal synthesis of ceria nanocubes reported by Mai et al., nanoparticles easily formed and the nanocubes exhibited various sizes [53]. It is also difficult to avoid the octahedron and rod multi-nanostructural composites when synthesizing ceria nano-octahedrons, even though exact synthetic procedures were followed based on our experience. Furthermore, the nano-octahedron yield is very low because of very small amount of metal precursor (1 mmol) and precipitator (0.01 mmol) [59]. To solve these problems, various modifications like template-assisted, microwave-as-sisted, and supercritical hydrothermal methods have been explored, among which the template assisted hydrothermal method is the most common strategy [8].

Surfactants acting as soft templates are very popular in preparing nanoparticles with specific shapes [30]. For example, sheets [60], flowers [61, 62], octahedrons [63, 64], hollow structures [32, 64–67], and particles exhibit certain sizes or exposed facets [63, 64]. The surfactants typically serve as a facilitator of arranging certain oriented nanoparticles to aggregate into desired shapes. The common surfactants include polyvinylpyrrolidone (PVP), cetyltrimethylammonium bromide (CTAB), polyethylene glycol (PEG), copolymer P123, glucose or fructose, and so on [30, 32, 62–64, 67, 68]. Yang and co-workers reported a PVP assisted simple hydrothermal method for fabricating monodispersed CeO_2 hollow spheres assembled from nano-octahedrons building blocks [64]. The synthesis procedure involved mixing 0.099 g of $CeCl_3 \cdot 7H_2O$ with 0.178 g of PVP in 19 mL of deionized water under vigorous magnetic stirring, followed by adding 1 ml of formamide and 0.1 mL of H_2O_2 and stirring for 30 min. The obtained yellow solution was transferred into a Teflon-lined autoclave and heated at 180 °C for 24 h. After cooling to room temperature, the collected light brown products were washed with deionized water and ethanol, and finally dried in oven at 70 °C for 6 h to get the CeO_2 hollow spheres. The hollow spheres were in diameters of 120–140 nm, with a pore size distribution from 3 to 15 nm. The nano-octahedrons composite with an average edge length of 20 nm underwent an Oswald ripening process to form novel structured hollow spheres. The inner small crystallites in the core slowly dissolved, whereas the outer larger ones acted as the growth seed and kept growing. With prolonged reaction time and continuous mass transfer, the hollow spheres finally formed [64].

CTAB and Pluronic P123 copolymer were used to synthesize ceria-zirconia-yttria (CZY) solid solutions in nanorod, microspherical, microbowknot, and micro-octahedron shapes [63]. $Ce(NO_3)_3 \cdot 6H_2O$, $ZrO(NO_3)_2 \cdot 2H_2O$, and $Y(NO_3)_3 \cdot 6H_2O$ were used as the metal sources, and urea as the precipitation agent. Stoichiometric amounts of $Ce(NO_3)_3 \cdot 6H_2O$, $ZrO(NO_3)_2 \cdot 2H_2O$, and $Y(NO_3)_3 \cdot 6H_2O$ were mixed and dissolved in water, followed by adding to the surfactant-containing solutions. Urea precipitate was then dropwise added to the mixture, and stirred for 2 h. The coprecipitation mixture solution was transferred to a glass vessel, and heated at 80 °C for 72 h in an oven. The obtained product was then ready for hydrothermal treatment. CTAB assisted hydrothermal treatment at 120 °C for 72 h resulted in CZY nanorods after calcination at 550 °C, while P123 involved hydrothermal process at 100, 120, and 240 °C for 48 h fabricated microsphere, micorbow-knot and micro-octahedron CZY particles, respectively. Apparently the reaction temperature, time, and surfactants were crucial for the morphologies. Ce, Zr and Y hydroxide nanoparticles were formed after hydrothermal treatment and assembled to rod nanoentities. The CZY nanorod precursor and CTAB enabled the formation of CZY nanorods during calcination. The nanorod CZY precursors in the presence of P123 generated an array of spherical, bow-knot-like or octahedral structure at different temperatures, and thus resulted in microsphere, microbow-knot and micro-octahedron CZY particles (see the formation mechanism in Figure 2.3).

Hard template assisted synthesis is another important route to generate novel nanostructure materials, such as wires, nanotubes and hollow spheres [30]. Compared with the soft templates, the hard templates can control the nanostructures more easily and exhibit more uniform size. Carbon nanotubes (CNT), carbon fiber, porous anodic alumina, silica spheres, and polystyrene spheres are the common hard templates [33, 36, 69–72]. Du et al. reported an *in situ* synthesis

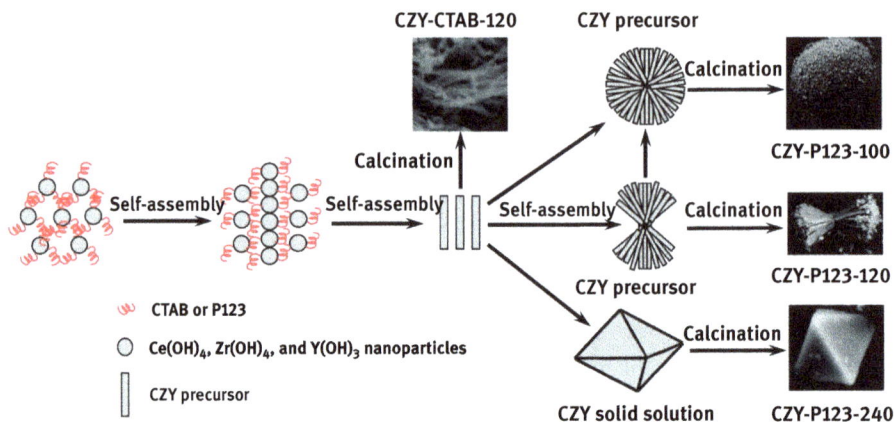

CZY-CTAB-120 **CZY precursor**

Calcination

CZY-P123-100

Calcination

Self-assembly Self-assembly Self-assembly Calcination

CZY precursor **CZY-P123-120**

CTAB or P123

Ce(OH)$_4$, Zr(OH)$_4$, and Y(OH)$_3$ nanoparticles

CZY precursor

Calcination

CZY solid solution **CZY-P123-240**

Figure 2.3: The schematic illustration of formation mechanisms of the CZY precursors and CZY solid solutions under surfactant-assisted hydrothermal and calcination conditions. (Reprinted with permission from Ref. [63]. Copyright (2009) American Chemical Society.) [63].

of Nb$_2$O$_5$ nanowires on carbon fibers using a hydrothermal method. The amorphous Nb$_2$O$_5$•nH$_2$O precursor mixed with carbon fiber and other assisting ingredients hydrofluoric acid, ammonium oxalate, and SDS were heated in a Teflon-lined stainless steel autoclave at 160 °C for 8–14 h. After filtration, washing, and drying steps, Nb$_2$O$_5$/Carbon fiber were obtained with high surface areas up to 130 m^2/g. The sample with 14 h hydrothermal treatment exhibited high Cr (VI) adsorption capacity (115 mg/g), excellent photocatalytic performance for Cr (VI) reduction (99 %) and good stability (10 cycles) [69].

Unlike the hybrid structure of Nb$_2$O$_5$/carbon fiber directly used for reaction systems, most templates need to be removed or sacrificed. The reason is because the hard templates could influence nanomaterial physicochemical properties, and undesirably interact with chemicals involved in systems, particularly in catalytic systems [72, 73]. For example, Titirici et al. reported a generalized hydrothermal approach for metal oxide hollow spheres synthesis assisted by carbon spheres [71, 73]. Carbon spheres were formed *in situ* during hydrothermal treatment, with metal ions incorporated into their hydrophilic shell. The carbon spheres were removed via calcination, and the metal oxide spheres were left behind as the final product. Other hard template such as SiO$_2$ spheres has to be removed using chemical etching method using NaOH [72, 74]. However, the template removal process often causes structure distortion or dissolves material composites. It is also very challenging to remove the template completely, and the residues might have negative effect on nanomaterial properties and their performances [30, 71, 72]. Thus extra caution is needed in nanoparticles preparation using template-assisted methods. Exploring controllable synthesis of nanomaterials is still crucial, and inspiring ideas are demanded.

The addition of templates in assisting morphology and size control of nanomaterials is broadly used not only in hydrothermal method, but also in other methods like sol–gel, solvothermal, precipitation, etc. The selection of template materials, the removal process, and their specific roles in nanoparticle nucleation and growth follow similar criteria in different methods [21, 29, 44, 75]. Since the template-assisted method is reviewed in the hydrothermal synthesis, only brief discussions will be stated in the following methods.

2.2.2 Solvothermal synthesis

The solvothermal method, a nanomaterial synthetic route quite similar to the hydrothermal method, has been extensively applied in producing precisely controlled metal oxide nanostructures and microstructures. Solvothermal processes are mainly controlled by the nature of the reagents and of the solvents, temperature, and pressure [76]. The difference between the hydrothermal and solvothermal methods is the reaction medium. Solvothermal synthesis is conducted in organic solvents, such as ethanol, CCl_4 or mixed solvents [48, 77–79], whereas hydrothermal reactions are taken in aqueous solutions [44]. Due to their similar synthetic policies and principles, many researchers liked to combine these two methods together and make it as hydro/solvothermal method for discussion [21, 30]. However, hydrothermal methods are mainly applied for synthesizing hydroxides, oxides or oxyhydroxides, whereas solvothermal methods can also be used for the preparation of non-oxide materials, such as nitrides, chalcogenides, etc [76].

Solvothermal synthesis is very important for novel materials development, particularly for materials with specific structures and properties. Lin's group reported a large-scale and facile solvothermal method for producing homogeneous Y_2O_3:Eu^{3+} microspheres [48]. The spheres were composed of randomly aggregated nanoparticles. Their formation was believed to follow an isotropic growth mechanism. Temperature, ethylene glycol and CH_3COONa played important roles in such structure formation. Ethylene glycol acted as a capping agent to control the growth rate of nanoparticles in different directions, rather than a solvent. No solid product was obtained without the presence of CH_3COONa.

Solvothermal synthesis for single-crystalline-like hollow nanostructures has also been reported by Chen and co-workers, shown in Figure 2.4 [77, 78]. In the synthesis of CeO_2 hollow nanocubes, cerium chloride heptahydrate was dissolved in anhydrous ethanol, with proxyacetic acid subsequently added as the oxidant. The obtained slurry was transferred into a Teflon-lined steel autoclave and heated at 160 °C for 9 h in oven. After cooling, the white-brown products were collected and washed. The final product ceria hollow nanocubes were achieved under a vacuum dry at 60 °C for 24 h [77]. In the CeO_2 hollow nanocrystals synthesis, the same

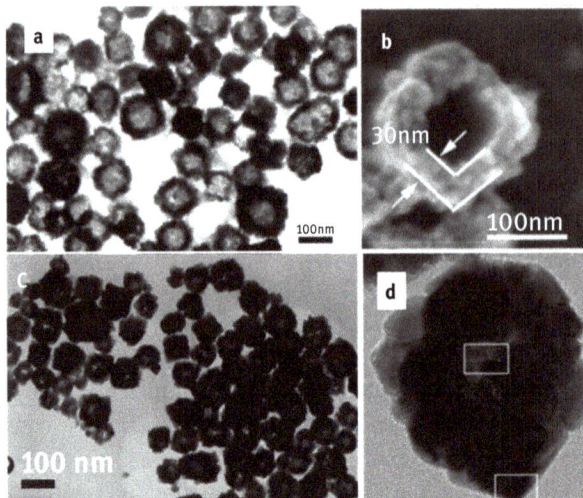

Figure 2.4: (a) TEM and (b) FE-SEM images of CeO_2 hollow nanocubes. (c) and (d) TEM images of CeO_2 hollow nanocrystals. Reprinted with permission from Ref. [77], Copyright (2008) American Chemical Society, and Ref. [78] from Royal Society of Chemistry.) [77, 78].

cerium precursor $CeCl_3 \cdot 7H_2O$ was applied. PVP was selected as the capping agent, water-ethanol mixture as the solvent and no oxidant was used. The obtained mixture was also sealed in a 20 ml of Teflon-lined steel autoclave but heated at 160 °C for 24 h. The rest of washing and drying steps were the same as the ceria hollow nanocubes [78]. Both the final CeO_2 hollow nanocubes and nanostructures exhibited higher catalytic activity towards CO oxidation than commercial CeO_2 powder. The formation of hollow nanocubes was proposed to follow a combined mechanism of oriented attachment and Ostwald ripening [77]. The latter nanostructure formation mechanism was proposed as a dissolution followed by a recrystallization process [78].

2.2.3 Co-precipitation synthesis

The co-precipitation method has a long history for various materials synthesis. This method typically involves with two salt precursors, such as oxalates, carbonates, and hydroxides dissolving in a common medium and precipitating out with a precipitant by pH adjustment [80–82]. The precipitates are subsequently calcined at elevated temperatures to yield the final product. Typical advantages of this method include (1) low temperature requirement, low energy consumption, (2) well-controlled homogeneity and adjustable particle size and (3) direct and relatively simple synthetic procedures [44]. Co-precipitation method is hasty and relatively simple. However, in order to obtain quality nanoparticles, the main parameters including feed solution and precipitant concentration, temperature, mixing time and pH of the medium have to be well controlled [81–83]. To achieve good homogeneity, the solubility products of the precipitates of metal cations must be close. Therefore, the disadvantages of this method are also obvious: (1) difficult to prepare samples with high purity and accurate phases, (2) time-consuming due to slow precipitation, and (3) not always capable to find a suitable precursor with similar solubility and precipitation rates [44]. The vast majority of rare earth nano-oxides synthesized by co-precipitation method are nanoparticles, and there are very few reported for the synthesis of other particular shaped nanostructures.

Rare-earth-doped ceria powders with a composition of $Ce_{0.8}RE_{0.2}O_{1.9}$ (RE = Yb, Y, Gd, Sm, Nd, and La) were prepared using co-precipitation method by mixing rare earth/cerium nitrate solution and an oxalic solution [81]. Taking the $Ce_{0.8}Y_{0.2}O_3$ sample for example, the oxalate solid solutions were obtained by dropping the mixed 0.20 M nitrate solution into the oxalic acid solutions ranging from 0.025 to 0.75 M. The formed oxalate was in the form of platelike particles with an average aspect ratio of 2.2–2.4. After heating at 600 °C for 1 h, the oxalate transformed into fine polycrystalline oxides with smaller particle size 10 nm. Fine pores were also formed due to the production of CO_2, H_2O and CO gases [81].

A similar series of nanocrystalline powders of ceria doped with 20 at. % of various rare-earth cations ($Ce_{0.8}RE_{0.2}O_{1.9}$, RE = La, Nd, Sm, Gd, Dy, Y, Ho, Er, and Yb) have also been synthesized via co-precipitation method but using ammonium carbonate as the precipitant (see Figure 2.5) [82]. The starting salts are rare-earth nitrate hexahydrates $RE(NO_3)_3 \bullet 6H_2O$ with high purities ≥ 99.99 %. In the synthesis, each starting salt was then mixed with cerium nitrate hexahydrate with the correct Ce^{3+}/RE^{3+} ratio to obtain the stock solution. The final concentration of each stock solution was 0.15 M of Ce^{3+}. A solution of 1.5 M ammonium carbonate in water was then used as the precipitant. 300 ml of the mixed salt solution was added at a speed of 5 ml/min into the precipitant solution kept at 70 °C under mild stirring. The precipitation took about 1 h to complete, and the precipitate was subsequently filtered and washed with distilled water and anhydrous alcohol. It was then dried under N_2 gas flow at room temperature for 24 h or over. The obtained products were ammonium rare-earth carbonate hydrates, which decomposed to rare earth oxide products upon heating for 2 h at various temperatures. The advantage of using ammonia carbonates was that the precursors showed less agglomeration after drying, and the oxides exhibited good dispersion and excellent sinterability. The RE^{3+} dopants smaller than Sm^{3+} exhibited primary particles with spherical shape, while those with larger cations (La^{3+}, Nd^{3+}, and Sm^{3+}) mainly consisted of secondary agglomerates of fibrous primary particles [82].

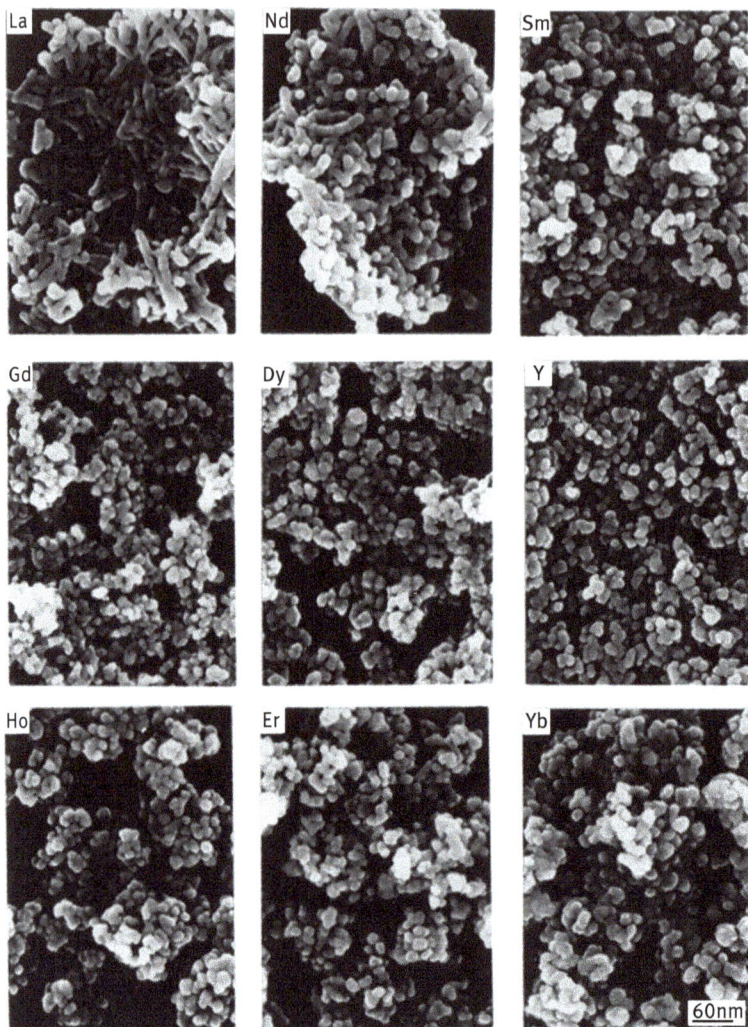

Figure 2.5: HRSEM micrographs showing particle morphologies of the $Ce_{0.8}R_{0.2}O_{1.9}$. The corresponding dopants are indicated in the pictures, and the scale bar for Yb sample is applicable to other materials. (Reprinted with permission from Ref. [82]. Copyright (2001) American Chemical Society.) [82].

2.2.4 Sol–gel process

The sol–gel method is another popular soft chemical route for fabricating solid materials such as metal oxide from solution state precursors. It typically involves converting the liquid precursors to a colloidal suspension defined as a sol and then to a multiphase network structure called a gel at low temperatures [21, 84, 85]. The size of a sol particle is dependent upon the solution composition, temperature, pH, etc., while the quality of gel is mainly correlated to the hydrolysis and condensation processes, which are influenced by the electronegativity of metal ions, precursor types, pH, solvent and temperature [6, 44, 84]. In order to obtain high-purity and high-crystalline nanostructures, calcinations at high temperature is usually required [21, 36, 84, 86]. Actually, depending on the properties of final products, sol–gel chemistry can go through different processing techniques to generate materials, i.e. films, powder, dense ceramics, bulk and nanoparticles, etc. (see Figure 2.6). For example, different drying conditions can result in different materials. A normal drying process or evaporation leads to xerogel formation, which has smaller surface area and pore size compared with its parent gel structures. However, when dealt with supercritical drying method, the resulting sample ends up with aerogels. The aerogels do not go through structure collapse or shrinkage and thus maintain highly open networks or porous structures [6, 44]. Aerogels are usually applied for thermal insulation materials, which created about $400 million USD market in 2016, and is expected to reach US$3.29 billion by 2025 [87].

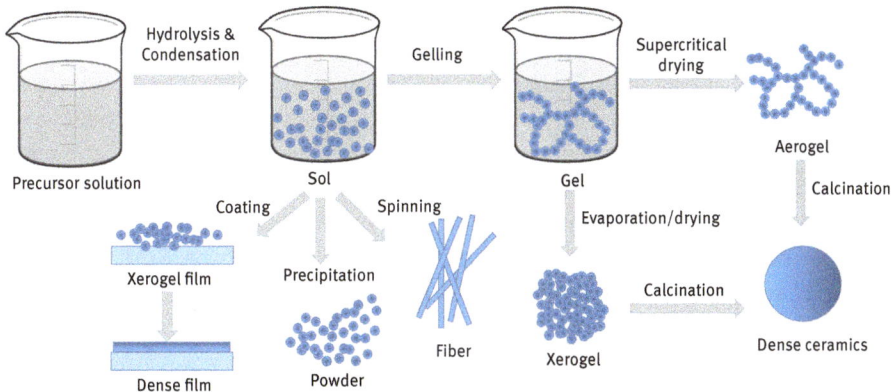

Figure 2.6: Sol–gel process with various processing techniques for different morphological materials [6, 44].

The sol–gel synthesis is categorized into aqueous sol–gel route and non-aqueous route based on the solvents [6]. In the aqueous sol–gel process, water is both the solvent and ligand [6, 84]. Reaction parameters such as metal precursors, pH, and temperature should be strictly controlled to achieve good reproducibility. Metal alkoxide and inorganic metal salts are the common precursors due to their high solubility and relatively more predictable behavior in water, and easy removal from the final metal oxide product [6, 21, 36, 84, 86, 88]. Cerium (IV) t-butoxide has been used for CeO_2 nanoparticle synthesis with a combined sol–gel and solvothermal method [88]. High surface area up to 277 m^2/g of ceria was obtained prior to calcination treatment. The high surface area together with the surface groups created by post-synthesis treatment enabled such ceria exhibit high catalytic activity towards CO oxidation [88]. Compared with metal alkoixdes, rare earth inorganic salts are more desirable in sol–gel synthesis due to the stability, low toxicity, and abundance [36, 86, 89, 90]. Hu et al. reported a sol–gel method for large scale fabrication of ceria hollow spheres derived from cerium nitrate on polymeric templates. Poly (styrene-co-acrylic acid)(PSAA) colloids were dispersed evenly in water to adsorb Ce^3+ ions onto the surface by electrostatic interaction. Ce(OH)$_3$ was precipitated out by adding NaOH, and then it was calcinied at 773 K for 3 h to obtain cerium hollow spheres [86].

In the previously reported rare earth oxide synthesis with sol–gel methods, templates have been usually applied to control the hydrolysis and condensation rate of metal precursors to further control its resulting morphology and size [36, 86]. Actually, one major challenge of aqueous sol–gel method is the control over hydrolysis and condensation rates in aqueous solutions, especially for mixed oxides synthesis with different metal precursors. Therefore, nonaqueous sol–gel chemistry using organic solvents has been explored as a promising alternative to fabricate metal oxide nanoparticles with controllable size and shape [6]. Recent developments on the preparation of rare earth mesoporous structures involved the nonaqueous sol–gel process. Yuan and coworkers synthesized ordered mesoporous ceria-zirconia solid solutions based on a sol–gel process combined with evaporation-induced self-assembly in ethanol with Pluronic P123 as the template and ceric nitrate and zirconium oxide chloride as the precursors. A series of mesoporous Ce$_{1-x}$Zr$_x$O$_2$ with different Ce/Zr ratio were obtained under optimized temperature and humidity conditions [90]. The large area of Ce$_{1-x}$Zr$_x$O$_2$ from the TEM images in Figure 2.7 suggested a long-range ordered mesostructure. The Ce$_{1-x}$Zr$_x$O$_2$ structure

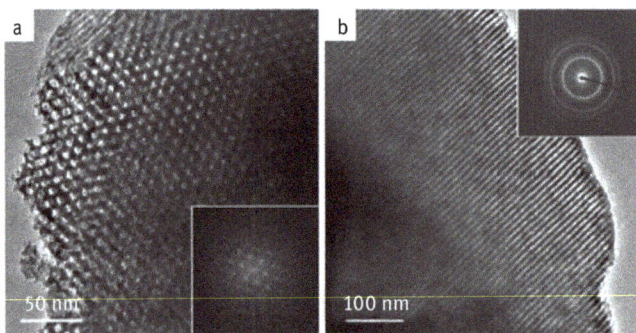

Figure 2.7: TEM images of the mesoporous Ce$_{1-x}$Zr$_x$O$_2$ (x = 0.5) imaged from (a) [001] and (b) [110] orientations. The inset in (a) is the corresponding FFT (fast Fourier transform) diffraction image, and the one in (b) is the corresponding SAED pattern. (Reprinted with permission from Ref. [90]. Copyright (2007) American Chemical Society.) [90].

exhibited high-crystalline pore walls composed of several nanocrystallites with well-defined lattice planes. The nanocrystallite size was about 3–4 nm. The highly ordered and porous $Ce_{1-x}Zr_xO_2$ solid solutions have proven to be ideal catalyst supports for Pt in CO oxidation and cyclohexene hydrogenation [90].

Compared with the furnace-based techniques like hydrothermal, co-precipitation or solvothermal methods, which are primarily based on inhomogeneous reacting materials, sol–gel chemistry offers the advantage of producing metal oxides from homogenous precursors, and thus enables good control of complex inorganic metal oxides involved three or four metal precursors. However, producing a homogeneous precursor at room temperature does not guarantee homogenous end product. Additionally, many sol–gel routes suffer phase segregation during synthesis [84]. To produce high-quality metal oxide nanoparticles, other techniques have been combined with sol–gel processes, for instance, the emulsion involved sol–gel routes for rare earth elements doped Y_2SiO_5 mixed oxide nanocrystals and TiO_2 nanoparticles [91–93]. All these extra steps result in complex synthetic procedures. Moreover, the utilization of metal alkoxides, various organic solvents, and templates gives rise to high cost, toxicity and unavailability issues [21].

2.2.5 Thermal decomposition synthesis

Rare earth oxides achieved through pyrolyzing oxysalt precipitates, such as oxalate and hydroxide at high temperatures, are usually not well controlled regarding the size, shape and dispersion. It is also difficult to achieve high order dimension nanomaterials with good quality [94]. Thermolysis or thermal decomposition of rare earth complexes such as acetate, oleate, carbonate, and acetonate have been shown as an effective way to produce monodisperse, single crystalline, and high-dimension rare earth oxide nanocrystals [21, 43, 95, 96]. It is also the most efficient way to synthesize high-quality nanoparticles less than 10 nm and is beneficial for generating core-shell structures. The use of organic surfactants was believed to play important roles in shape and size controlling. They act as capping agent to prevent nanoparticles growth and agglomeration, and control the nanocrystal growth by a selective absorption effect [21, 39, 43, 95, 97]. This method is relatively new compared with the previously discussed methods, but becoming more and more popular because of its good control over nanocrystalline properties [21, 39, 43, 95, 97]. Despite the various advantages, thermal decomposition method exhibit obvious drawbacks, including high operation temperature, typically from 250 to 330 °C, high cost in metal precursors, various surfactants and sometimes toxic byproduct [21].

Cao pioneered the rare earth nano-oxides synthesis using thermal decomposition method. He reported a colloidal synthesis of squire Gd_2O_3 nanoplates through thermolysis of gadolinium acetate in oleylamine, oleic acid, and octadecene solvents [95]. In a typical synthesis, 0.75 mmol gadolinium nitrate hydrate dissolved in a solution with oleylamine, oleic acid and octadecene at 100 °C with stirring under ~20 mTorr. The resulting solution was then heated to 320 °C for ~5 min and cooled down to room temperature after 1 h under Ar flow. A mixture of hexane and acetone (1:4) was used as precipitant. The nanocrystals precipitated out and dried under an Ar flow. These nanoplates were single crystalline with uniform shape and size, and could form superlattice structures via a self-assembly process (see Figure 2.8) [95].

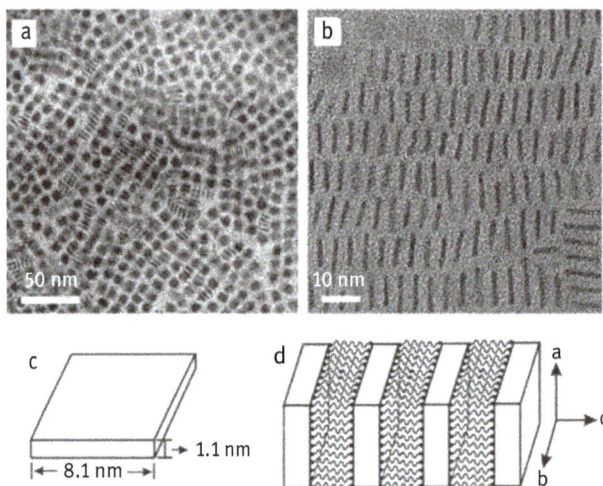

Figure 2.8: (a and b) TEM images of Gd_2O_3 nanoplates. (c and d) Proposed model for the nanoplates and assembly of nanoplate stacks, respectively. The *c*-axis of cubic Gd_2O_3 crystals is assigned as the thickness direction of the nanoplates. (Reprinted with permission from Ref. [95]. Copyright 2004 American Chemical Society) [95].

With similar synthetic ideas, Yan's group explored and fulfilled a systematic synthesis of rare earth oxide nanopolyhedra/nanoplates/nanodisks via thermal decomposition of rare earth benzoylacetonate (BA) complexes [37]. Rare-earth complexes were obtained by dissolving REO precursors in HNO_3, and then mixing with HBA and $NH_3 \bullet H_2O$ and aging for 12 h at room temperature. The resulting $Ln(BA)_3(H_2O)_2$ was mixed with oleic acid (OA)/oleylamine (OM). After the solvents were removed, the mixture was heated at around 300 °C (310–330 °C) for 1 h under Ar. For CeO_2 nanoparticles fabrication, $Ce(BA)_4$ was mixed with OM and heated at 250 °C for 20 min. After the reaction was complete, the resultant mixture was separated through centrifuge. The precipitates were then collected, washed and dried to obtain the final REO nanocrystals. The details of OA/OM ratio, temperature, reaction time and resulting structure for each REO are listed in Table 2.2. The nanocrystals exhibited high-crystalline, uniform size and shape, and self-assembly behavior. The capping ligands oleic acid and oleylamine played critical roles in determining the phases and morphologies of

Table 2.2: Crystal structures of the as-obtained rare-earth oxides synthesized by thermolysis of Ln (BA)$_3$(H$_2$O)$_2$ (Ln = La–Y) or Ce(BA)$_4$ in oleic acid (OA)/oleylamine (OM) at 250–330 °C for 20–60 min.[a] (Reprinted from Ref. [37] with permission from John Wiley and Sons, Inc.) [37].

Sample	OA/OM	T (°C)	t (min)	Structure	Morphology
La$_2$O$_3$	1:7	330	60	Ia3	7 nm nanoplate
CeO$_2$	0	250	20	Fm3m	2.6 nm nanopolyhedron
Pr$_2$O$_3$	3:5	310	60	Ia3	20 nm nanoplate
Nd$_2$O$_3$	3:5	310	60	Ia3	11 nm nanoplate
Sm$_2$O$_3$	3:5	310	60	Ia3	11 nm nanoplate 26 nm nanodisk
Eu$_2$O$_3$	3:5	310	60	Ia3	12 nm nanoplate 32 nm nanodisk
Gd$_2$O$_3$	3:5	310	60	Ia3	30 nm nanodisk
Tb$_2$O$_3$	3:5	310	60	Ia3	34 nm nanodisk
Er$_2$O$_3$	3:5	310	60	Ia3	43 nm nanodisk
Y$_2$O$_3$	3:5	310	60	Ia3	65 nm nanodisk

[a] Determined by powder X-ray diffraction (Rigaku D/MAX-2000, Cu$_{K\alpha}$ radiation) and transmission electron microscopy (Philips Tecnai F30, 300 kV).

these nanocrystals. For example, oleic acid bound more strongly than oleylamine to the surface atoms due to its higher oxophilicity. Due to the diverse adsorption behaviors of capping agent on different facets, CeO$_2$ exhibited a polyhedron shape enclosed by (111) and (200) planes, while the other rare earth oxides either showed formation of square nanoplates (La$_2$O$_3$-Eu$_2$O$_3$) or nanodisks (Sm$_2$O$_3$-Y$_2$O$_3$) crystals (see Figure 2.9). Tuning the capping agent concentration, synthetic time, and temperature may also result in different crystalline shapes and sizes [37].

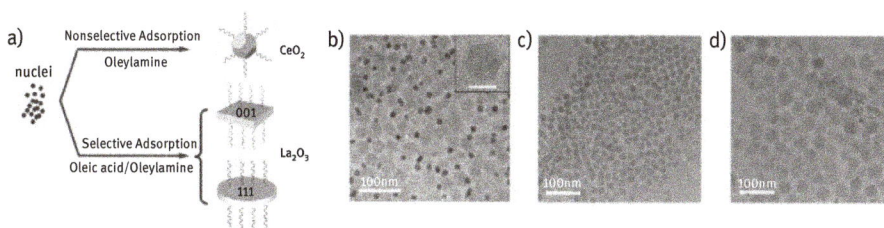

Figure 2.9: (a) Formation of rare-earth oxide nanopolyhedra, nanoplates, and nanodisks. (b)- (d) TEM images of the as-obtained Eu$_2$O$_3$: b) OA/OM = 1:7, 310 °C, 1 h (inset: HRTEM image of an Eu$_2$O$_3$ nanoparticle; scale bar: 10 nm); c) OA/OM = 3:5, 310 °C, 20 min; d) OA/OM = 3:5, 330 °C, 1 h. (Reprinted from Ref. [37] with permission from John Wiley and Sons, Inc.) [37].

Based on the pioneer work, various materials such as rare earth oxysulfide, sodium rare earth fluoride and other transition metal oxides, were synthesized and later applied in biomedical imaging, water treatment, and catalysis [21, 30, 39, 98–101]. Modifications of the method were also made to synthesize particular shaped oxides with better shape-control and increased yield [34, 39, 102–104]. For example, Murray's group has explored a mineralizer-assisted thermodecomposition method to synthesize rare earth oxide nanoplates with promising catalytic and optical property [39, 104]. Sodium oleate, sodium phosphate, and sodium nitrate were used as mineralizers to facilitate the formation of oxide nanoplates and control the final morphology. In a typical La_2O_3 nanoplate synthesis, 0.1 g of $La(Ac)_3$ and 0.53 g of sodium diphosphate were dissolved in a mixture of 2 ml of oleic acid (OA), 4.5 ml of oleylamine (OM), and 2 ml of 1-octadecene (ODE). The mixture was then heated to 120 °C under stirring and vacuum for 20–30 min. Subsequently, the temperature was brought up to 320 °C and maintained at this level for 1 h. The resulting product was then cooled and washed with ethanol, and finally separated by centrifuge to obtain the nanoplates. It showed that La_2O_3 and CeO_2 nanoplates could not be achieved without mineralizer. By varying the mineralizer and its concentration, capping agent concentration, and the precursor concentration, the resulting oxide morphology and yield were significantly tailored (see Table 2.3 and Figure 2.10) [39].

Table 2.3: Reaction parameters for synthesis of nanoplates. OA-oleic acid, OM-oleylamine, ODE- 1-octadecene (Reprinted with permission from Ref. [39]. Copyrigh (2014) Americal Chemical Society.) [39].

Product	RE precursor	Mineralizer	OA (ml)	OM (ml)	ODE (ml)	T (°C)	t (min)
Gd_2O_3 round nanoplates	0.1 g $Gd(Ac)_3$		0.2	8		360	25
Gd_2O_3 branched tripod nanoplates	0.4 g $Gd(Ac)_3$	0.17 g $NaNO_3$	1.5	1.5	7	310	40
Gd_2O_3 triangular nanoplates	0.4 g $Gd(Ac)_3$	0.17 g $NaNO_3$	1.5	2	6.5	310	40
Y_2O_3 round nanoplates	0.68 g $Y(Ac)_3$		3	3	14	310	60
Y_2O_3 tripod nanoplates	0.68 g $Y(Ac)_3$	0.34 g $NaNO_3$	2	1	17	310	30
Y_2O_3 branched tripod nanoplates	0.68 g $Y(Ac)_3$	0.34 g $NaNO_3$	3	2	15	310	30
Y_2O_3 triangular nanoplates	0.68 g $Y(Ac)_3$	0.34 g $NaNO_3$	8	8	5	310	30

2.2.6 Microemulsion method

The microemulsion method is another low-temperature route to generate monodisperse rare earth oxide nanoparticles in the range of 1 to 100 nm. Microemulsions are

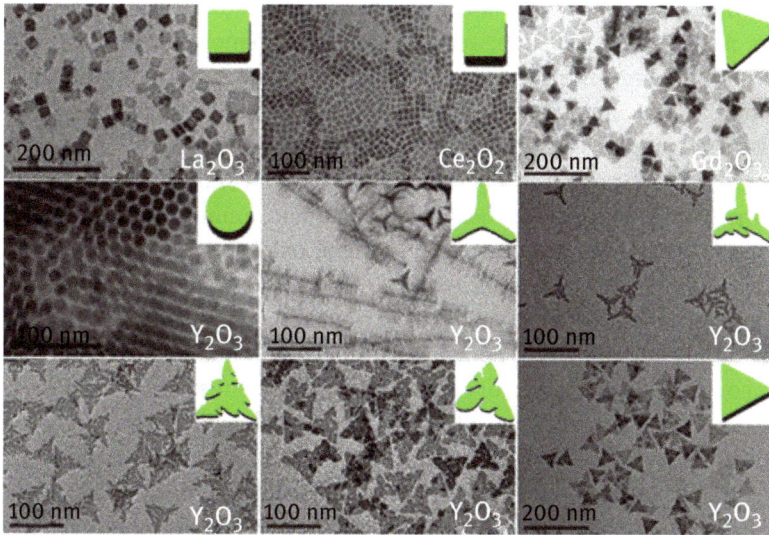

Figure 2.10: TEM images of (a) 25.2 nm (σ = 7 %) La$_2$O$_3$ nanoplates; (b) 11.9 nm (σ = 7 %) CeO$_2$ nanoplates; (c) 39.4 nm (σ = 6 %) Gd$_2$O$_3$ nanoplates; and (d) 13.6 nm (σ < 5 %), (e) 36.7 nm (σ = 6 %), (f) 43.7 nm (σ = 8 %), (g) 58.6 nm (σ = 7 %), (h) 62.7 nm (σ = 7 %), and (i) 56.9 nm (σ = 9 %) Y$_2$O$_3$ nanoplates. σ - product yield. (Reprinted with permission from Ref. [39]. Copyright (2014) American Chemical Society.) [39].

homogeneous, clear, and thermodynamically stable liquid systems composed of water, oil and surfactant. Oil and water are immiscible, but surfactant molecules are able to form an interfacial film between water phase and oil phase, and result in spontaneous mixing of one phase in another. Cosurfactant is a frequent addition to ensure the flexibility of interfacial layer [21, 44]. A promising advantage of this method is to produce microemulsion carriers that can improve drug loading capacity, solubility and bioavailability, and therefore could be potentially influential in pharmaceutical industry [105, 106].

There are three basic types of microemulsions: oil in water, water in oil, also known as reverse microemulsion, and bicontinuous microemulsion. Another well-known classification of microemulsions is from Winsor, who identified four general types of phase equilibria [107]. The traditional synthetic route for nanoparticle fabrication is almost always reverse microemulsion method. The small water droplets dispersed in oil-rich continuous phase dissolve reactants and constantly exchange aqueous phase among micelles. The dynamics of the exchange of solute between micelles and the continuous phase is controlled by diffusion [108]. Over-mixing microemulsion containing different reactants or by adding another reactant, the micelles act as nanoreactors and enable initial nucleation, nanoparticles formation, and further grain growth. The nanoparticle size or grain size is controlled by the micelle size [21, 109]. During the synthesis, parameters such as the composition of reverse microemulsion, water/surfactant ratio, aging time, and cosurfactants can tailor the size, morphology, structure, and surface area of nanoparticles [108, 110]. The selection of surfactant is critical in microemulsion synthesis since surfactant films determine the static and dynamic properties of microemulsion, and thus play essential rules in controlling particle size and material uniformity [108, 111]. Surfactants possess both polar and non-polar moieties, and should be chemically inert to the other components in microemulsions. Cationic, anionic, non-ionic, and amphiphilic surfactants have been reported for metal oxide nanoparticles synthesis. Sodium bis (2-ethylhexyl) sulfosuccinate (AOT), CTAB, and didodecyldimethyl ammonium bromide (DDAB) are among the most frequently used ones [21, 44, 112].

Bumajdad et al. reported ceria nanopowder generation with heptane-microemulsified aqueous solutions of $CeCl_3$ or $Ce(NO_3)_3$ using different surfactants of AOT, DDAB, DDAB, and BRrij 35 [113]. They observed almost-agglomerate-free nanosized ceria crystallites with a size range of 6–13 nm in the presence of surfactants. Batches with AOT or (DDAB + Brij 35)-stabilized microemulsions exhibited ceria with large specific areas $ca.$ 250 m^2/g, but they agglomerated heavily when calcined at 800 °C (13–28 m^2/g). The double calcined DDAB was able to produce ceria with initial high surface area (144 m^2/g), and relatively high stability at enhanced temperatures (45–55 m^2/g at 800 °C).

A similar size range of 5–16 nm ceria nanoparticles were reported by Kaskel's group using an inverse microemulsion based method [114]. A system consisting of $Ce(NO_3)_3$, n-heptane, and surfactant Marlophen NP 5 was operated to generate microemulsion. The microemulsion was stirred for 1 h in a closed flask at 28 °C. Diluted ammonia was then added and stirred for 1 h to precipitate $Ce(OH)_3$ nanoparticles inside the micelles. After a series of steps of microemulsion destabilization by acetone, filtration, washing, drying, and calcination, CeO_2 nanoparticles were obtained. The synthetic steps are illustrated in Figure 2.11. Molar water to surfactant ratio Rw was controlled to fine tune CeO_2 particle size. By changing the Rw from 5.8 to 16.4, CeO_2 nanoparticle size increased from 2.5 to 9.5 nm from DLS data. Micelle size also exhibited an increasing trend with Rw, but their sizes were slightly larger compared with the precipitated particle size (see Figure 2.12). Drying and calcination at 400 °C had negligible effect on the particle size. However, increasing the annealing temperature to 600 °C resulted in strong crystal aggregation, and the crystallite size increased from 6.1 nm to 16.4 nm (Rw = 5.8).

2.2.7 Combustion method

The soft chemical synthesis methods discussed above usually are applied with high pressure, long reaction time, surface capping-agents or surfactants, and complex synthetic steps. Unlike these soft chemical techniques, combustion synthesis is a

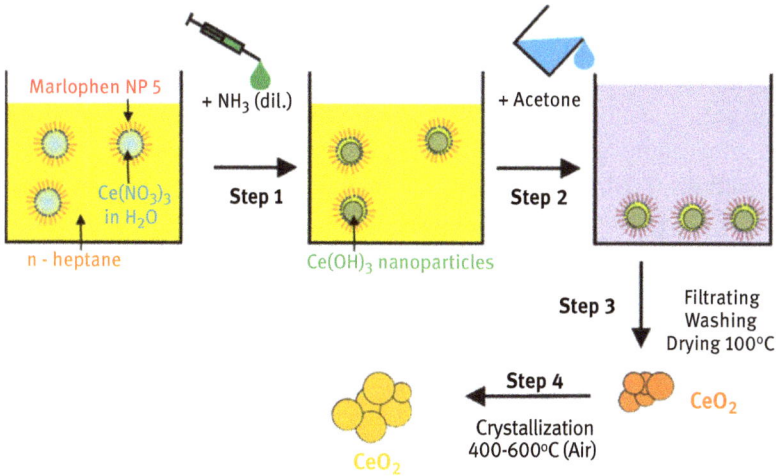

Figure 2.11: Schematic synthesis strategy for preparation of ceria nanoparticles with microemulsion and precipitation method. (Reprint from Ref. [114] with permission from Elsevier) [114].

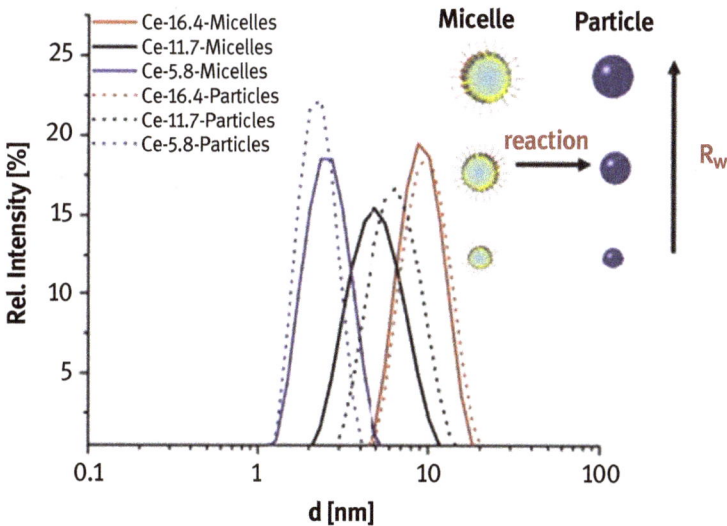

Figure 2.12: Dynamic light scattering (DLS) curves of inverse microemulsions and particle dispersion. (Reprint from Ref. [114] with permission from Elsevier) [114].

nonconventional method which involves propagation of self-sustaining exothermic reactions. This method is characterized with fast reaction rates, short reaction times, high temperatures, simple procedures, and low cost, and attracts much attention in scale-up synthesis for metal oxides for catalysis, energy conversion, and storage [59, 115, 116].

According to the initial reaction medium, combustion synthesis is classified into three categories, including (1) conventional combustion synthesis based on solid state initial reactant, also known as condensed phase combustion, (2) solution combustion synthesis (SCS), which happens in aqueous solutions and (3) gas phase combustion. The conventional combustion method is usually difficult to produce nanoscale particles with high surface area without coupling post-synthesis treatment such as chemical dispersion, milling or mechanical activation [59]. SCS is a self-sustained exothermic process which starts at the ignition temperature and generates certain amount of heat that is manifested in the maximum temperature or temperature of combustion. The SCS process takes place in homogeneous solutions of oxidizers and fuels mixed at molecular level, and produces at least one solid product and a large amount of gas. It is able to yield nanosized materials with uniform size and morphology in a time-, energy-, and cost-effective manner. Therefore, this method has received much more attention in recent years for nanoparticles synthesis [115, 116].

Metal nitrate is a good oxidizer for SCS due to its high solubility in water and relatively low decomposition temperature [59, 116–118]. Some other oxidizers are metal sulfates, metal carbonates, and ammonium nitrates. The fuels or reducing agents are glycine, urea, sucrose, hydrazine, etc. Besides providing the carbon and hydrogen source, the fuel also serves as complexing agent facilitating the mixing of metal ions to form homogenous solutions [115, 116]. The ratio of oxidant to fuel is considered one of the most important parameters in determining the final properties of nanoparticles through SCS. Properties such as morphology, surface area, porosity, phase, and degree of agglomeration can be controlled by adjusting the oxidant-fuel ratio [59, 117, 118]. Other important parameters are metal precursor, type of flame, temperature, and generated gases. Because of the large amounts of gas generated during combustion, the resulting materials are usually porous and exhibit high surface area, which is very beneficial in catalysis, adsorption and separation, and sensing applications [115].

Recently, high surface area CeO_2 nanopowders reported by Varma's group was prepared by SCS method. Cerium nitrate hexahydrate and cerium ammonium nitrate were applied as the oxidizer. Glycine and hydrous hydrazine were the fuels. In a typical experiment, a cerium precursor, a fuel, and a gas-generating agent NH_4NO_3 were dissolved in minimum amount of water with desired fuel to oxidizer ratio and NH_4NO_3/metal nitrate ratio. The mixture was subsequently placed on a hot spot to evaporate water and induce the self-sustained combustion. After 5–10 min, the solution started to boil with frothing and foaming, and then ignited. A significant amount of gases was produced while the temperature skyrocketed. Ignition temperature and maximum combustion temperature were both monitored using a K-type thermocouple during combustion. It was found that using the hydrous hydrazine fuel, with a fuel-to-oxidizer ratio of 2 and ammonium nitrate/metal nitrate ratio of 4, allowed to produce CeO_2 with ~ 8 nm crystallite size, 24.3 defect concentration, and 88 m^2/g sur-face area, which has been considered as the highest value compared with the prior reported SCS- derived CeO_2 powders. Figure 2.13 shows the size distribution, shape and crystal structure of CeO_2 powder prepared under the above conditions [118].

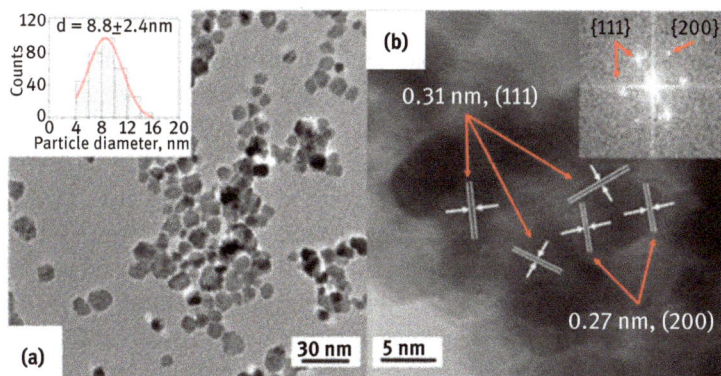

Figure 2.13: (a) TEM image, inset is the size distribution, and (b) HRTEM image of CeO_2 powder prepared from solution combustion synthesis with ammonium nitrate to metal nitrate ratio of 4. (Reprinted with permission from Ref. [118]. Copyright (2018) American Chemical Society.)[118].

Combustion synthesis is advantageous in making mixed oxides with high purity, even for ternary or quaternary oxides [115, 117, 119]. A binary oxide CeO_2/Y_2O_3 powder was synthesized by a nitrate-glycine gel-combustion route. Fuel/oxidizer ratio, mea-sured by glycine/metal nitrate ratio, was investigated to achieve particles with smaller crystallite size and higher surface area. The binary oxides with crystallite size of 4.5–7 nm and specific surface area of 25–40 m^2/g were obtained from slow com-bustion processes with glycine/metal nitrate ratios of 1.5–2 [117]. Combustion synthe-sis is also powerful in synthesizing ternary or quaternary oxides with perovskite, spinal or scheelite structures for solid oxide fuel cells applications. Generally, higher formation temperature and longer reaction time are required to prepare oxides with a more complex structure, good crystallinity and superior performance [115, 116, 119].

2.2.8 Microwave assisted synthesis

The primary advantages of microwave heating include time saving, low cost, and en-vironmentally benign [120–122]. Microwave-assisted synthesis has drawn significant attention to synthetic chemists and researchers because of such benefits. Microwave irradiation interacts selectively with chemicals based on their microwave absorbing capability, which enables microwave energy transfer on particular molecules directly without suffering thermal gradient effects, and induces superheating of solvents and supersaturation of reactants [38, 122]. The rapid heating affects the molecule movement and collision, and could lead to an increase in chemical reaction rate, and thus shorten reaction time [44, 120–123]. It has been combined with various conventional methods, such as hydrothermal, sol–gel, solvothermal, and combustion method, to accelerate reaction and save time for preparing inorganic nanomaterials [38, 44, 45, 119, 123, 124].

El-Shall's group developed a simple, rapid, and versatile microwave irradiation method for the synthesis of single crystalline and uniform rare earth oxide nanorods and square nanoplates (RE_2O_3, RE = Pr, Nd, Sm, Eu, Gd, Tb, Dy) [38]. The metal acetate or acetylacetonate mixed with capping agents oleic acid and oleylamine were preheated at 110 °C in an oil bath under vigorous stirring for 5 min. The mixture was subsequently heated for 10–15 min using a conventional microwave oven with a power set to 70 % of 650 W. The product was washed with ethanol, and then centrifuged, and re-dispersed in toluene or dichloromethane. Nanorods and nanowires were generally obtained, and they can self-assemble to form highly ordered, crystalline 2D superstructures. By tuning the microwave time, precursor concentration, and capping agent ratio, high-quality squire nanoplates could be achieved (see Figure 2.14).

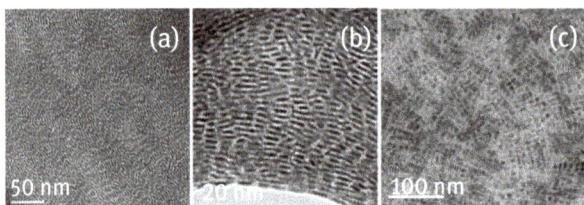

Figure 2.14: TEM images of (a) Sm_2O_3 nanorods, (b) Sm_2O_3 nanorods forming 2D supercrystalline structure, and (c) Sm_2O_3 nanoplates mixed with nanorods synthesized using a higher mole ratio 3: 1 of oleylamine/oleic acid. (Reprinted with permission from Ref. [38]. Copyright (2007) American Chemical Society) [38].

Cao et al. reported a template-free microwave assisted hydrothermal method for ceria hollow spheres synthesis with a programmable microwave oven [45]. Compared with the conventional microwave, the programmable microwave is more advanced in terms of temperature control, reaction time and experiment feasibility and repeatability. The mixture of $Ce(NO_3)_3 \cdot 6H_2O$, urea and water was heated to 170 °C in 2 min and kept at 170 °C for 30 min to obtain the ceria precursor. The precursor was then cooled, separated and washed with ethanol. The final calcination at 500 °C for 2 h converted the nanomaterial precursor to desired ceria nanostructures. The microwave heating was also believed to uniformly distribute the heat and resulted in its homogeneous shapes and sizes of ceria nanoparticles. A self-templated, self-assembly process coupled with Ostwald ripening growth mechanism was proposed for the formation of hollow structures (see Figure 2.15).

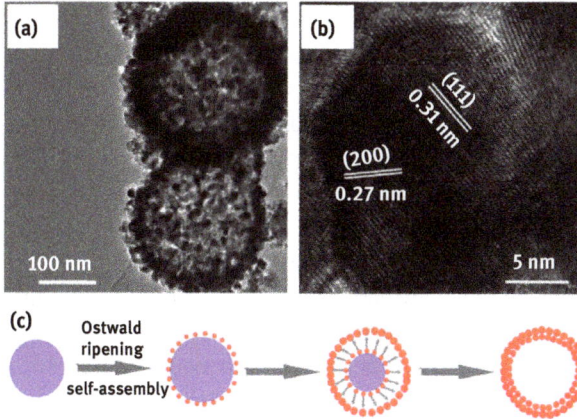

Figure 2.15: (a) TEM and (b) HRTEM image of ceria hollow nanospheres. (c) Schematic illustration of ceria precursor hollow structure growth with an Ostwald ripening coupled with self-templated, self-assembly process.(Reprinted with permission from Ref. [45]. Copyright (2010) American Chemical Society) [45].

2.2.9 Sonochemical method

Sonochemical synthesis is a synthetic process involving application of ultrasonic radiation (20 KHz-10 MHz) [125]. This method provides a route to generate novel materials at room temperature, low pressure, and short reaction time. In this method, the powerful ultrasound interacts with molecules to make chemical changes. Because of the smaller frequencies, ultrasound irradiation cannot interact with chemical bonds directly and break them. Instead, a physical phenomenon called acoustic cavitation related to the formation, growth, and implosive collapse of bubbles is responsible for the sonochemistry [5, 126]. The extreme, transient conditions produced by acoustic cavitation allow chemical reactions to take place and enable synthesis for new materials.

Sonochemical method has been used for producing nanomaterials, including metal nanoparticles, metal sulfides, metal alloys, and metal oxides, etc [126, 127]. Microjets and shock waves are considered important phenomena for nanomaterial synthesis. When bubbles collapse near an extended surface, microjets form with high speed and create pitting and erosion on the surfaces, leading to surface modification and nanostructures formation. Shock wave is a type of propagating disturbance. Rebounding bubbles compresses the surrounding liquid and makes it propagate as shock wave. Shock waves move faster than local sound wave, which can reach 4 km/s in water [128]. Because of the high energy carried by shock wave, it can increase the mass transport, accelerate solid particles suspension in the liquid, and result in modifications in particle size, morphology, phase, and composition [126].

The controlling parameters in sonochemical synthesis are sonication time, pH, temperature, ultrasound power, solvent, and pressure of gas [5, 68, 129]. Zhong et al. developed a simple sonochemical procedure to synthesize the flowerlike yttria Y_2O_3, and found various parameters could affect the particular structure development [129]. In the synthesis, commercial Y_2O_3 was dissolved in diluted HNO_3 (10 wt%) to form a solution. After evaporation, mixing with H_2O and agitation, the mixed solution was sonicated for 30 min at 35–50 °C. Subsequently the solution was filtered, washed and dried, and the obtained product was the flowerlike $Y_2(OH)_5NO_3 \cdot 1.5H_2O$. Y_2O_3 flowerlike product was yielded through calcinations of $Y_2(OH)_5NO_3 \cdot 1.5H_2O$ at 600 °C (see Figure 2.16). The flowerlike structure was controlled by various critical parameters, including time, precursor concentration, pH and application of ultrasonication. The time-dependent experiment indicated that the flowerlike structure was retained in 2 h reaction time. Beyond 2 h, the flower structure completely changed to flakes. Flowerlike $Y_2(OH)_5NO_3 \cdot 1.5H_2O$ could only be achieved when < 0.5 mmol of commercial Y_2O_3 was used. The optimal pH value to maintain the flowerlike structure is 6–7. Ultrasonication is a must step for this material synthesis. Without sonication, it either formed no solid or the hierarchical structure agglomerated. The hypothesis was that the cavitation generated by ultrasonication carried high enough energy to overcome agglomeration of the particles [129].

Figure 2.16: (a) FESEM image of the as-prepared Y_2O_3, (b and c) HRTEM images of the as-prepared Y_2O_3 (Reprinted with permission from Ref. [129]. Copyright (2009) American Chemical Society.) [129].

Sonoelectrochemistry, a method dating back to 1934, has attracted more and more attention in the last several decades and become a very important synthesis method for nanoparticles synthesis [5, 127]. Combining ultrasound and electrochemistry provides powerful induced effects for chemical reactions. The ultrasound in electrochemistry can enhance mass transport, accelerate reaction rate, increase surface area, and reduce diffusion layer [127]. The experimental variables controlling crystal size and process efficiency include sonic intensity, current pulse time, bath temperature, and stabilizer [5, 127]. Sonoelectrochemistry methods have been mainly applied to synthesize metal, semiconductor, and conducting polymer nanoparticles [127].

2.2.10 Electrochemical method

Electrochemical synthesis is another popular route for preparing nanopowders, thin films, and coatings in a simple and inexpensive way [130]. In this method, electric current is applied between electrodes which are separated by an electrolyte. The electrochemical reaction or synthesis takes place close to the electrode within the electric double layer. The product is electrodeposited on the electrode in the form of a thin film or a coating. One of the biggest advantages of this method is the low processing temperature, which typically is room temperature. Another advantage of this method is the control of driving force-the continuous tuning of applied cell potential. This is hardly obtained from other chemical synthesis. The major disadvantages of this method are the poorly-ordered products and amorphous impurities [130, 131].

Rare earth oxide fabrication from the electrochemical method has been majorly focused on the cerium oxide coatings and powders [131]. Switzer firstly described one of the electrodeposition methods–electrogeneration of base by cathodic reduction to prepare oriented ceramic films and polycrystalline powders in 1987 [132]. His group further utilized this cathodic electrodeposition method and successfully synthesized nanocrystalline CeO_2 powders from a cerium nitrate solution. The CeO_2 crystallites were randomly oriented, and the powder consisted of nonaggregated, faceted particles. The average crystallite size could be controlled by reaction temperature, which was 10 nm at 29 °C and increased to 14 nm at 80 °C [133]. Later, the reaction mechanism was revealed by Aldykiewicz and Li with different probe techniques [16, 134]. H_2O_2 was generated from dissolved O_2 in the electrolyte, and then oxidized Ce(III) to Ce(IV), which finally formed CeO_2 film on the electrode. Aldykiewicz proposed that cerium oxide film formation was achieved through a four or two-electron process to a hydroxide intermediate, and the hydroxide intermediate was only stable above pH = 8.7 [134]. Li found the CeO_2 deposition occurred through a nucleation and growth mechanism, in which CeO_2 nuclei crystallized out from a hydroxide intermediate gel through *in situ* atomic force microscopy studies [16]. The overall reaction is given as follows:

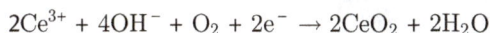

$$2Ce^{3+} + 4OH^- + O_2 + 2e^- \rightarrow 2CeO_2 + 2H_2O$$

Nanocrystals with particular shapes and orientations from electrochemical synthesis have also been reported in recent years. For example, Lu and co-workers reported a simple electrochemical process for large scale fabrication of CeO_2 octahedrons and nanospheres on F-doped SnO_2 coated (FTO) glass substrate. A conventional three-electrode cell was used, with an FTO glass as the working electrode, a 4.0 cm^2 graphite rod as the auxiliary electrode, and an Ag/AgCl reference elec-trode. The FTO glass was cleaned thoroughly by ultrasonication in water, ethanol and acetone, and rinsed in distilled water before electrodeposition. 10 mM Ce(NO$_3$)$_3$ and DMSO were dissolved in the electrolytic solution. The electrodeposition was con-ducted at 70–90 °C. The octahedrons exhibited structural defects and have a size ranging from 200–300 nm. The nanospheres were highly crystalline and formed through an oriented attachment mechanism. The morphology of CeO$_2$ could be changed by controlling the DMSO concentration and the temperature [135]. Lu et al. also synthesized hexagonal CeO$_2$ nanorods with (110) plane as the predomi-nant exposed plane via a template-free electrochemical method. A similar conven-tional three-electrode cell was used. The working electrode was a 1.5 cm × 3 cm Ti foil. A graphite rod and a saturated Ag/AgCl electrode were employed as the auxil-iary electrode and reference electrode, respectively. The same cleaning steps were also required before carrying out electrodeposition. Subsequently, CeO$_2$ nanorods were electrodeposited on Ti substrates in a solution containing 0.01 M Ce (NO$_3$)$_3$•6H$_2$O, 0.1 M NH$_4$Cl and 0.05 M KCl at 70 °C. It was found that these CeO$_2$ nanorods showed relatively smooth surface, with diameters ~ 240 nm and lengths up to 820 nm (see Figure 2.17). They also exhibited great photocatalytic activity for

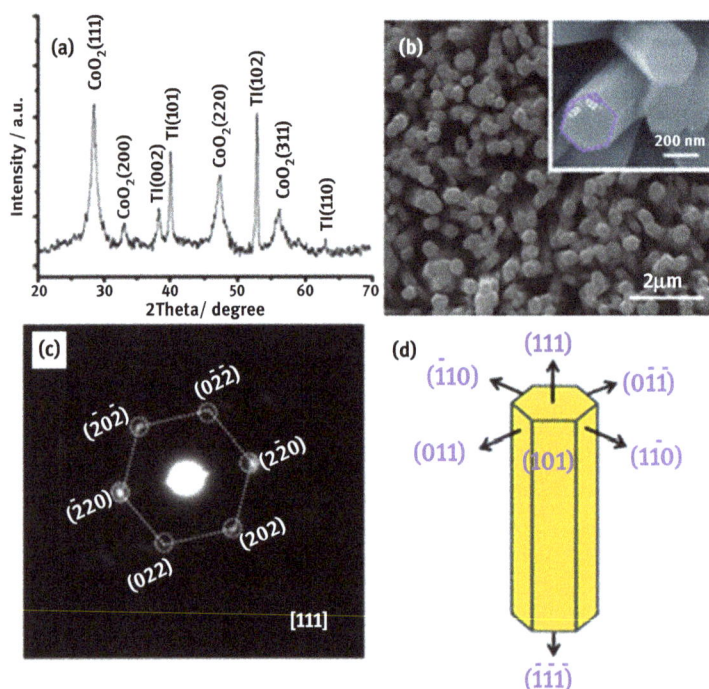

Figure 2.17: (a) XRD pattern, (b) SEM images, (c) SAED pattern, and (d) schematic diagram of as synthesized CeO$_2$ NRs. (Reprinted from Ref. [136] with permission from Royal Chemical Society) [136].

hydrogen evolution with Na_2S-Na_2SO_3 as sacrificial agents. The special redox capacity of CeO_2 ($E^0_{Ce^{4+}/Ce^{3+}} = 1.44$ eV) was considered as the main reason for their application in photocatalysis [136].

Other rare earth oxides such as La_2O_3, PrO_2, Sm_2O_3, and TbO_2 have also been synthesized through electrochemical deposition [131]. In all of the synthesis, the oxides were obtained through adherent coatings. Current density, temperature, pH, electrolyte concentration, surfactants and other parameters have been adjusted to produce these nanomaterials with desired size, morphology, and thickness [130].

2.2.11 Other methods

While the above discussed synthetics methods are very popular in the lab-scale and industrial scale synthesis, there are also plenty of other traditional and non-traditional methods for preparing rare earth oxide nanomaterials that deserve our attention. Among these are ball milling, spray pyrolysis, chemical vapor deposition (CVD), pulsed laser deposition, and so on.

Ball milling is an ancient idea, and has been largely applied in ceramics, mineral, and paints industry. A ball mill is a type of grinder, which uses the ball mill as a media to reduce the materials to fine powders through an internal cascade effect. The conventional ball milling is mainly used for fracturing the particles and reducing the size. The new high energy ball milling employs a magnet to create strong magnetic pulling force on the magnetic milling balls, which provides much higher impact energy on the powders. This technique has successfully produced fine, uniform oxide particles and many new meta-stable materials [137, 138]. Lei et al. have reported preparation of rare earth oxides La_2O_3 and CeO_2 ultrafine nanosuspensions in pure water using high energy ball milling technique. The rare earth oxides suspensions exhibited excellent stability, high surface tension, and low viscosity. Increasing the ball-milling time and rotation speed assisted in improving the suspension stability [137].

Spray pyrolysis is a chemical deposition process of depositing a film obtained from a heated solution. Unlike many other expensive film deposition techniques, spray pyrolysis is much simpler and relatively cost-effective. It doesn't require either expensive instrument, nor high-quality substrate or chemicals. It is also capable of producing multilayered films. Therefore this method has been used to deposit a wide variety of films or coatings, which are used in glass industry, solar cells, sensors, and solid oxide fuel cells [139, 140]. Hao and co-workers used spray pyrolysis to prepare polycrystalline thin films of undoped and Eu^{3+}, Tb^{3+} and Tm^{3+} doped Ga_2O_3 [141]. $Ga(NO_3)_3$, $Eu(NO_3)_3$, $Tb(NO_3)_3$ and $Tm(NO_3)_3$ were used as the precursors. The spray was developed by an ultrasonic nebulizer and directed to the chamber by humid air. The deposition temperature was 400 °C. The annealing temperature was varied and up to 900 °C. Polycrystalline thin films with preferred peaks of (111) and (-310) in XRD were observed after annealing treatment. The presence of oxygen vacancies affected the cathodoluminescence (CL) of undoped Ga_2O_3 films. Higher density of oxygen vacancies of films showed higher CL intensity, evidenced by the annealing studies. Eu^{3+} and Tb^{3+} doped Ga_2O_3 films exhibited red and green cathodoluminescence, while Tm doped Ga_2O_3 film showed a broad band in the blue-green region from the emissions of both Ga_2O_3 and Tm^{3+}. Elidrissi et al. employed $CeCl_3 \bullet 7H_2O$ and $Ce(NO_3)_3$ as precursors to prepare thin films of CeO_2 on glass substrate using this method [142]. The CeO_2 thin film was oriented along the (100) direction. The surface morphology, grain size, and porosity of the films were dependent upon the initial cerium chloride and nitrate solutions. All the thin films were highly transparent, and > 80 % in the visible and near-infrared region.

CVD is often used in semiconductor industry to prepare thin films and coatings. The products are usually in high-quality and exhibit high performance. In CVD, one or multiple precursor gases flow into the reaction chamber, where heated objects are placed and ready to be coated. Due to the high temperature in chamber, chemical reactions occur or precursors decompose and generate thin films on the surface, while the volatile by-products are removed by the gas flow [143, 144]. Besides the advantages of producing high-quality and high-performance films and coatings, CVD also has its own disadvantages, such as the non-volatile impurities from precursors, high operating temperatures, toxic by-products, etc [144]. CVD is practiced in different formats, and have incorporated various elements *i.e.* plasma, aerosol, photons and combustion reactions to increase deposition rates and lower deposition temperatures [144]. For example, Jiang and co-workers reported an aerosol-assisted metal-organic CVD method to prepare Sm doped ceria thin films from $Sm(DPM)_3$(DPM = 2,2,6,6-tetramethyl-3,5-heptanedionato) and $Ce(DPM)_4$ [145]. The metal precursors were mixed with a molar ratio of Sm: Ce = 1:4. The solution was ultrasonically nebulized to generate a mixed precursor mist, which was subsequently sprayed over the substrates. The substrate temperature was maintained between 400 and 650 °C. It was observed that the Sm doped ceria film showed single crystal structure above 450 °C. At lower deposition temperatures, the films were not uniform and composed of clusters due to the faster nucleation rate. The Sm: Ce ratio in the final films increased from 400 to 450 °C and decreased from 450 to 500 °C because of different deposition mechanisms of Sm and Ce.

2.3 Characterization methodologies and instrumentation techniques

2.3.1 Electron microscopy

Electron microscopy is one of the most powerful techniques to provide structural, phase, and crystallographic information, and can pin down the processing–structure–

properties links to atomistic levels. It is a type of microscope applying accelerated electron beam as the probing source. Scanning electromicroscopy (SEM) and transmission electron microscopy (TEM) are the two most common techniques. Conventionally, they have been used for initial morphology and size distribution analysis, however, high resolution electron microscopies allow for the direct imaging of atomic structures in nanoparticles, like lattice fringes, displayed facets and planes, defects, etc [9, 146].

2.3.1.1 Scanning electron microscopy

Scanning electromicroscopy (SEM) provides scanning images of sample surfaces at acceleration voltages of 0.1–30 keV. It is typically used for relatively low magnification image of rough 3-dimension objects. SEM has been extensively applied in characterization of both organic and inorganic nanomaterials and bulk materials [147]. Unlike light microscopes, the signal detected by SEM is from the interactions of the electron beam with atoms at various depths in the sample. The imaging resolution is limited by the electron beam spot size, and the intensity of signal is determined by the amount of electron current at the final probe. Other parameters such as the conical angle with specimen and the applied accelerating voltage also contribute to the image resolution by governing the depth-of-focus and penetration depth. Some of these parameters are not independent due to particular SEM settings. To obtain a faithful and good image, operators must master these controls and follow standard operation procedures [147].

The most common SEM imaging mode is to collect the secondary electrons emitted from sample surface excited by inelastic scattering interactions with beam electrons. These secondary electrons are typically with low energy < 50 eV, and thus with this mode, SEM can provide excellent topographical information of sample surface. Even though quotations for the state-of-the-art SEM resolution reach 1 nm or subnanometer by instrumental companies, SEM gains its popularity in the characterization of nanosized and bulk materials with a magnification range of 10–10,000x [147]. Takaya et al. investigated Y_2O_3, La_2O_3, CeO_2, and Nd_2O_3 rare earth oxides using SEM for the information of morphology and particle size distributions. Y_2O_3, La_2O_3 and Nd_2O_3 showed similar particle size ~ 2 um while CeO_2 was in 200 nm in average and demonstrated more spherical structures [148]. Menendez and co-workers characterized their Pt/Pt: CeO_2-x nanorods mixed with Fumion "ink paste" on glassy carbon electrode for short chain alcohol electrooxidation. Agglomeration of Pt:CeO_2 nanorods promoted by Fumion polymer were clearly observed under SEM, and the electrodeposited Pt nanoparticles exhibited diameters of 100 to 600 nm [149]. Park and co-workers reported utilizing SEM for morphological characterization of cerium oxide nanoparticles for oxidative stress studies in BEAS-2B cells. The samples were prepared by sprinkling the powder oxides onto double-sided sticky tape and coated with sputtered gold before imaging.

Figure 2.18: SEM characterization of different sized cerium oxide nanoparticles. Four different sizes of cerium nanoparticles show different morphologies: (a) 15 nm; (b) 25 nm; (c) 35 nm; and (d) 40 nm size of nanoparticles. (Reprint with permission from Ref. [150] from Elsevier) [150].

Figure 2.18 shows the morphology and crystallite size of studied cerium oxide nanoparticles [150].

Backscattered electron (BSE) is another important imaging signal with great interest. These electrons are deflected or backscattered by elastic scattering from a larger interaction volume with deeper penetration and reduced electron energy. In targets of high-atomic number materials, the electrons undergo more elastic scattering and thus increase the backscattering and result in brighter display in the image. Backscattering increases with increasing atomic number indicated by Monte Carlo trajectory plots [151]. Therefore BSE is powerful in characterizing phase difference and atomic composition [151]. Zanfoni and co-workers has studied their porous platinum doped CeO_2 thin films on Si substrate using backscattering SEM. Due to the higher atomic number of Pt than Ce, Pt nanoparticles appeared much brighter than CeO_2 on the film, as indicated in Figure 2.19. It also showed that the surface density of Pt clusters increased with the Pt precursor flow rate, both on the film surface and the interface [152].

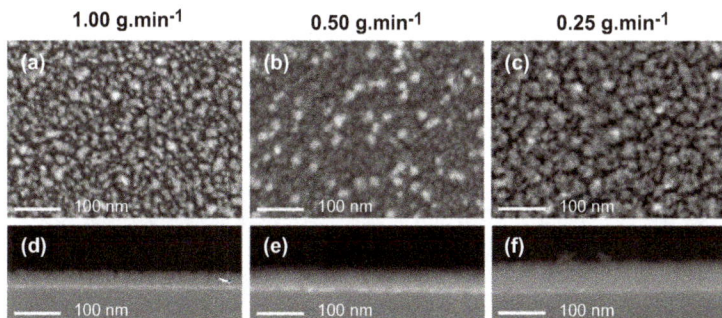

Figure 2.19: Backscattered electrons SEM images of Pt and CeO$_2$ co-deposited thin films on Si substrates with platinum precursor flow rate of (a) 1.00 (b) 0.50 and (c) 0.25 g/min.(Reprinted from Ref. [152] with permission from Elsevier) [152].

2.3.1.2 Energy dispersive X-ray spectroscopy

The interaction between electron beam and sample specimen is complex. Except the previously discussed secondary electrons and BSEs, it can also generate characteristic X-rays. These X-rays have specific energies and are equal to the energy difference between the excited and ground state of electron orbitals in the atoms. The energy dispersive X-ray spectroscopy (EDX), typically mounted onto an electron probe microanalyzer, utilizes the characteristic X-rays produced in the interaction for compositional mapping and element abundance analysis [153, 154].

EDX usually measures the mid-energy X-rays between 0.2 and 12 keV for SEM. This implies that only X-rays of the K series are observed for the light elements; K series plus L series for the intermediate elements, and L series plus the M series for the heavy elements [153]. Silicon Si (Li) and high-purity Ge solid detectors are the most common EDX detectors, and are cooled with liquid nitrogen. Newer systems might be equipped with silicon drift detectors with Peltier cooling systems [154, 155]. Due to the overlapping energy of X-ray emissions from different elements, the accuracy of EDX is strongly affected by the nature of samples. Samples with close atomic number elements cannot be accurately applied for compositional analysis without quantitative correction procedures [154].

Very different composite samples are frequently investigated by EDX for qualitative and quantitative analysis. Cheung's and Carlos's groups studied a complex Pt/Pt: CeO$_{2-x}$ electrocatalyst system using EDX mapping to identify the location and quantity of different composites. The presence and distribution of Pt clusters, CeO$_{2-x}$ support were clearly elucidated. The S and F composites were also verified because of the usage of Fumion as a binding agent of catalyst and electrode (see Figure 2.20) [149]. Moreover, Men and co-workers indicated that EDX revealed a more accurate Cu/Ce ratio (~0.1) in their Cu/CeO$_2$/γ-Al$_2$O$_3$ catalyst for steam reforming of methanol

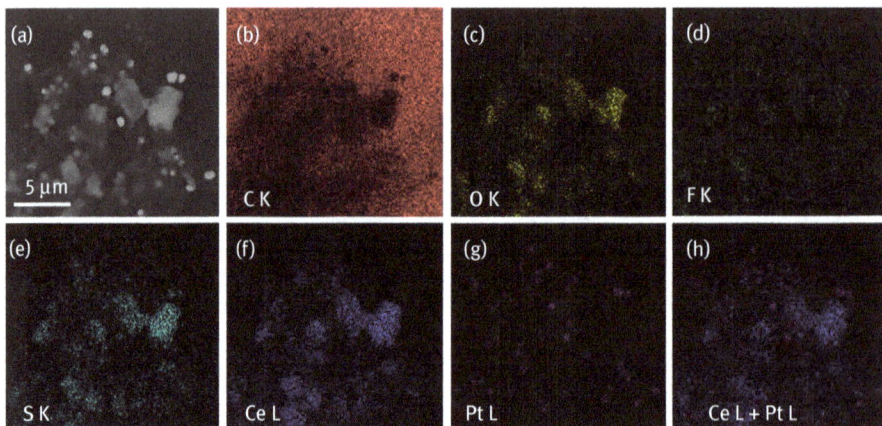

Figure 2.20: (a) SEM image of Pt/Pt:CeO$_{2-x}$ electrocatalyst with Fumion® coating on a glassy carbon electrode (thin arrows: Pt:CeO$_{2-x}$ catalyst; thick arrows: Pt nanoparticles). EDX elemental mapping of (b) carbon, (c) oxygen, (d) fluorine, (e) sulfur, (f) cerium and (g) platinum of the catalyst coating in (a). (h) Combined mapping image of Ce and Pt. (Reproduced from Ref. [149] with permission from The Royal Society of Chemistry.) [149].

than their XPS analysis (~0.3). EDX has been considered as a relatively bulk sensitive technique, and thus their result was in good agreement with the theoretical value Cu/Ce = 0.1 of the catalysts [156].

2.3.1.3 Transmission electron microscopy

2.3.1.3.1 Conventional and high resolution TEM

Transmission electron microscopy (TEM) might be the most efficient and versatile technique for the characterization of materials over spatial ranges from the atomic scale to the nano regime and up to the micrometer level. Generally it has much higher resolution than SEM, and provides various sample information such as morphology, chemical identity, crystal orientation, electronic structure, atomic defects, etc [9]. TEM utilizes the electrons transmitted through a sample to form an image, and thus the TEM samples are mostly ultra thin (< 100 nm) or a thin layer of suspension on a grid. The beam energies of TEM are in the range from 100 keV to 400 keV, which makes TEM a very expensive instrument. Unlike SEM producing 3D-like images of the surface topology, TEM is topology independent, and produces 2D images averaging through the thickness of sample. This is because TEM exhibits large depth of field, and TEM specimen is usually in focus from the top to the bottom. Therefore one should always be careful when interpreting TEM results [9].

TEM utilizes electrons going through the specimen. It is important to note that these electrons interact with samples and scatter in various directions. The forward scattering, including direct beam, elastic/inelastic scattering, and diffraction, forms most of the signals in TEM images [9]. The scattered and diffracted electrons can change both its amplitude and phase as it traverses the sample, and give rises to the amplitude and phase contrast in TEM [157, 158]. The amplitude contrast has two types: mass-thickness and diffraction contrast. Mass-thickness contrast is more important for investigating non-crystalline samples, such as polymers and biological systems. Diffraction contrast is a special form of amplitude contrast where the scattering occurs at Bragg angles. The diffraction contrast is controlled by crystal structure and orientation of the sample [157]. The phase contrast correlates to "fringes" of crystalline samples, which is essential for constructing TEM images of rare earth metal oxide nanoparticles. Phase contrast is very sensitive to many factors, *i.e.* thickness, orientation or scattering factor of the specimen, and variation in the focus or astigmatism of the objective lens. All these sensitivities make phase contrast capable of investigating thin specimens at atomic scale [158].

Zhou et al. has studied the morphology and structure of their regenerative Pd/CeO$_{2-x}$ catalyst for CO oxidation with conventional TEM and HRTEM (see Figure 2.21a and Figure 2.21b). The conventional TEM image demonstrates nanorod structure with lengths between 300 and 800 and diameter ranges from 7 to 15 nm. The HRTEM image of the Pt/CeO$_{2-x}$ nanorod shows the 3.1 Å fringes corresponding to the (111) lattice of cubic fluorite structure ceria [47]. Yan's group work for ceria nanorods in Figure 2.1(c) and Figure 2.1(d) showed three kinds of lattice fringes on a single nanorod attributed to (111), (002) and (220) planes with a preferred growth direction of [110] [53].

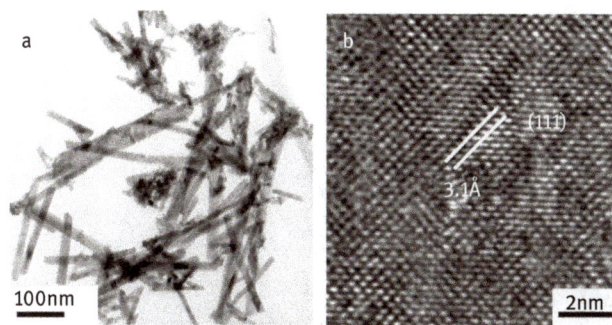

Figure 2.21: (a) TEM image of 1 at% Pd/CeO$_{2-x}$ nanorods. (b) HRTEM image of 1 at% Pd/CeO$_{2-x}$ nanorods, showing a lattice fringe spacing of 3.1 Å corresponding to the (111) planes of ceria. (Reprinted with permission from Ref. [47] from John Wiley and Sons, Inc.) [47].

The major limit of TEM resolution is controlled by electron lenses, which suffer severe spherical and chromatic aberrations due to their imperfections. Spherical aberration is the primary one, and results in a general image blurring and structure dislocation. Recent development in aberration corrected TEM by commercial companies like Thermo Scientific, FEI, JEOL, Hitachi and Nion enabled major advances in TEM, making the instrument achieve an image with 0.05 nm resolution [9, 75]. The advanced TEM technique has been employed to probe fundamental questions and principle understandings of nanomaterials in catalysis, sensing, and biomedical systems [75].

Kirkland et al. utilized a combined direct spherical aberration-corrected TEM and computational exit wavefunction restoration to elucidate surface termination of cerium oxide nanoparticles during an electron-beam-induced structure transformation. It was found that a representative ceria nanoparticle viewed close to a <110> direction consisted of {100} and {111} surfaces, and the average profile for a {111} surface demonstrated a clear oxygen peak at the surface, while this peak was absent in the profile extracted from {100} surface (see Figure 2.22) [159]. Lin and coworkers determined the atomic structures of the (100), (110) and (111) surfaces of CeO_2 nanocubes by observing O and Ce atoms using chromatic and spherical aberration-corrected HRTEM [160]. The predominantly exposed (100) surface terminated with mixed Ce, O and CeO on the outermost surface, as well as the partially occupied atoms existing in the near surface region. The (110) surface showed "sawtooth-like" (111) nanofacets and CeO_{2-x} terminations. The (111) surface was truncated with an O termination. This study indicated that the imperfect structures in nanocrystals should be taken into account, because these surface reconstruction and surface vacancies could contribute to the catalytic properties and provide insights into face-dependent catalysis (see Figure 2.23).

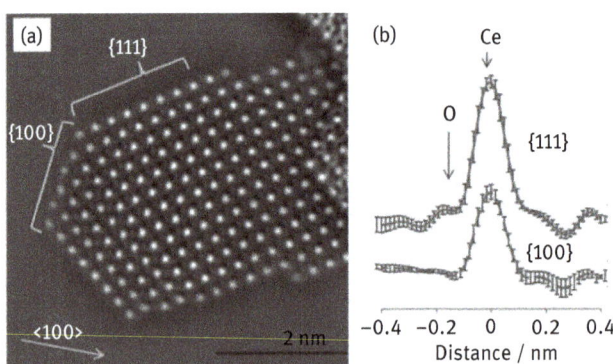

Figure 2.22: (a) Exit wave phase of a CeO_2 nanoparticle imaged close to a <110> direction. (b) Average of seven line profiles obtained from {111} and {100} surfaces along a <100> direction [as indicated in (a)] with the vacuum region on the left and the particle bulk on the right. (Reprinted from Ref. [159] with permission from John Wiley and Sons, Inc.) [159].

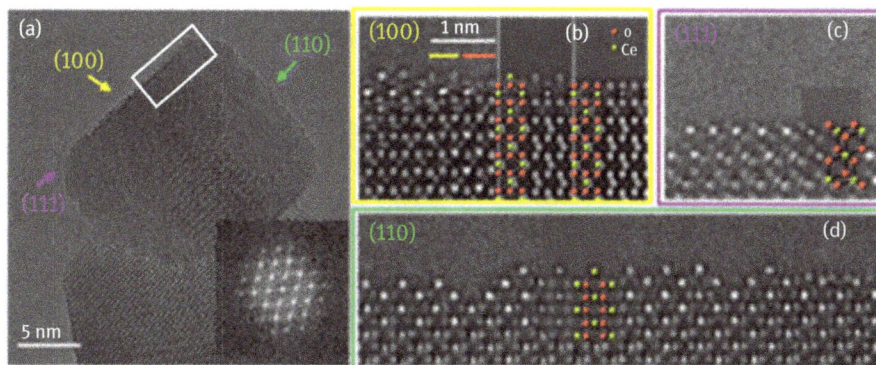

Figure 2.23: (a) An HRTEM image on a typical CeO$_2$ nanocube at the [110] zone axis with the FFT of the region of interest (highlighted with white box) shown in the inset. Magnified HRTEM images of (b) the (100) surface, (c) the (111) surface and (d) the (110) surface of (a). (Reprinted with permission from Ref. [160]. Copyright (2014) American Chemical Society) [160].

2.3.1.3.2 Environmental transmission electron microscopy

Although TEM is powerful in characterizing nanomaterial structures at atomic level, the major criticism of TEM observations lies in the fact that TEM experiments are conducted under ultra-high vacuum and don't represent the bulk world. Environmental transmission electron microscopy (ETEM), which is almost as old as TEM, has been developed in demonstrating the real-time changes in materials. It is important in observing structural changes under chemical reactions and synthesis, and obtains great interests from the catalysis and semiconductor communities [14, 161–163]. The development of ETEM started back in 1950s when Hashimoto and Naiki constructed their high-temperature reaction specimen chamber compatible with TEM based on differential pumping [164]. Nowadays, an environmental cell is integrated into the microscope column to increase the stability and improve the resolution [162, 165]. Furthermore, due to the application of field emission gun and aberration correctors, the ETEM resolution is much more improved, and gains more attention from various fields in the last recent decade [18, 162].

Crozier and co-workers applied *in situ* ETEM to study the dynamic changes during redox reactions in ceria nanoparticles in a hydrogen atmosphere from room temperature to 800 °C. Figure 2.24 shows a combination of HRTEM, selected area diffraction and electron energy-loss spectra (EELS) data of a ceria nanoparticle in 0.55 Torr of H$_2$ at 600 and 730 °C. The EELS data indicates that Ce underwent a transition from Ce^{4+} to Ce^{3+}. This reduction in ceria was associated with the appearance of superlattice reflections observed in both HRTEM images and electron diffraction patterns. Ceria surfaces also suffered significant structural reconstruction in reducing atmosphere. The ceria (110) surface displayed a saw-tooth profile composed of (111) nanofacets at 270 °C. At 730 °C, the surface formed a much smoother profile, with the

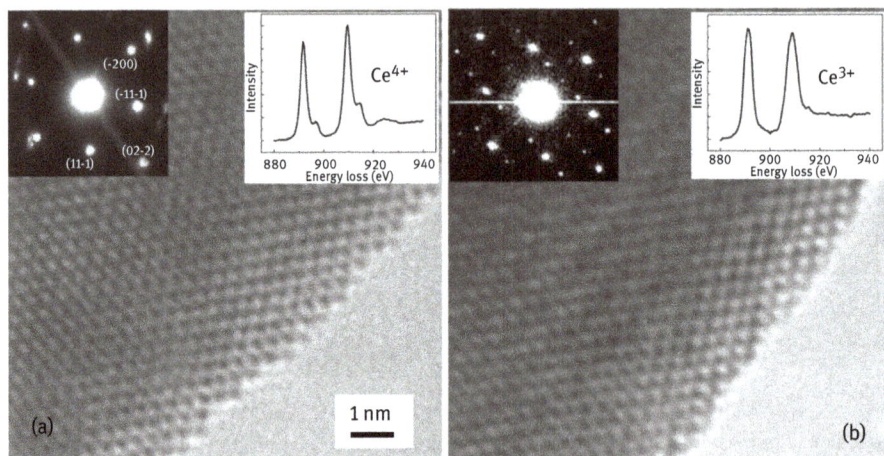

Figure 2.24: *In situ* HRTEM, electron diffraction pattern and energy-loss spectrum from a ceria crystal recorded at (a) 600 °C and (b) 730 °C in 0.5 Torr of H_2. Images and diffraction patterns are recorded from (110) fluorite projection. (Reprinted from Ref. [14] with permission from Elsevier) [14].

elimination of saw-tooth points. By cooling down to 600 °C and maintained at 600 °C for 1 h in H_2, the surface flattened and became more smooth, indicating a major composition of (110) terraces with very little (111) component (see Figure 2.25) [14].

While ETEM gives the real-time information of nanostructural transformation, one has to be very careful on the data interpretation. This is because the *in situ* or environmental conditions are still not as the same as the bulk reaction conditions. In particular, under the high keV electron flux, it brings more uncertainties to almost all in-situ experiments, especially for the kinetic studies [166].

2.3.1.3.3 Scanning transmission electron microscopy

A scanning transmission electron microscope (STEM) is a type of TEM. The illumination system in TEM is operated in two principle modes: parallel beam and convergent beam. Conventional TEM utilizes the parallel beam, while the convergent beam is mainly for STEM mode, where the electron beam is focused into a fine spot and then scanned over the sample in a roster illumination [167]. Generally, TEM images exhibit higher resolution and lower noise than STEM images, and STEM imaging process is longer than TEM imaging: it is a serial recording rather than a parallel recording. However, STEM images show higher contrast than analog TEM images. This makes STEM images useful for thick and beam-sensitive samples [157, 167].

STEM is categorized into two imaging modes, bright field mode and angular dark field mode (ADF). The bright field detector is located in the path of transmitted electron beam, and collects the direct-beam electrons with variable intensity depending on the specific point on the sample. The dark field imaging is conducted

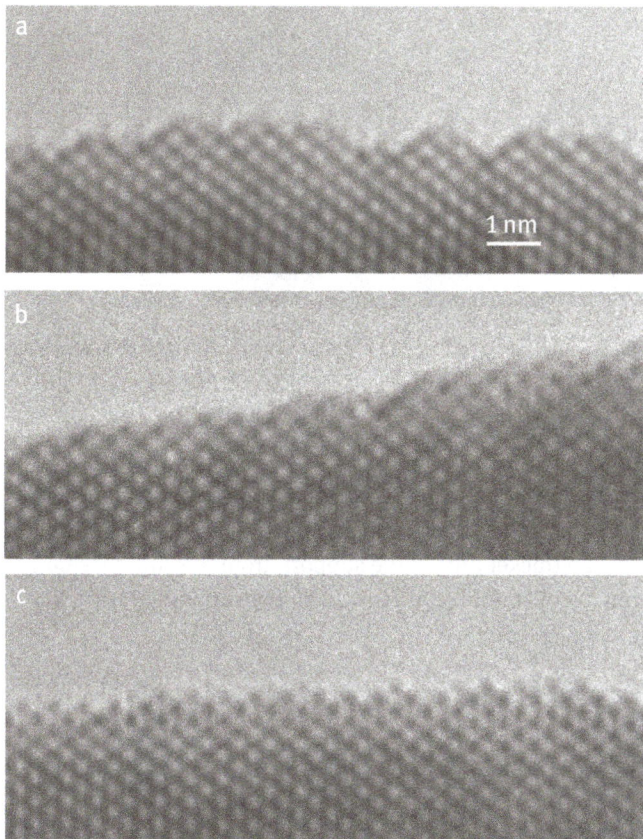

Figure 2.25: *In situ* profile images showing evolution of an identical region of a (110) surface of ceria during heating in 0.5 Torr of H_2 recorded at (a) 270 °C, (b) 730 °C and (c) 600 °C. (Reprinted from Ref. [14] with permission from Elsevier) [14].

by annular detectors, which sit around the bright field detector and collect the scattered electrons. The high-angle annular dark field (HAADF) detector collects electron scattered out to even higher angles, where Bragg reflections are excluded and Rutherford scattering is strengthened. It has been shown to provide incoherent images of crystalline materials with strong compositional sensitivity. The contrast of a component is directly related to the atomic number or Z contrast [167–169]. Figure 2.26 shows HAADF-STEM images of Pt/ceria nanorods and nanoparticles indicating uniform dispersion of Pt nanoparticles on both ceria supports. The chemical contrast between [58]Ce and [78]Pt (brighter) allows for locating the Pt nanoparticles and analyzing the Pt particle size distributions [170]. In contrast to the Pt NPs with several nanometers, for the Pt NPs with rather small size such as 1 nm or smaller, HAADF-STEM has difficulties to achieve a clear and unambiguous contrast image. Krumeich and co-workers showed that Pt particles located on two stacked CeO_2 crystals

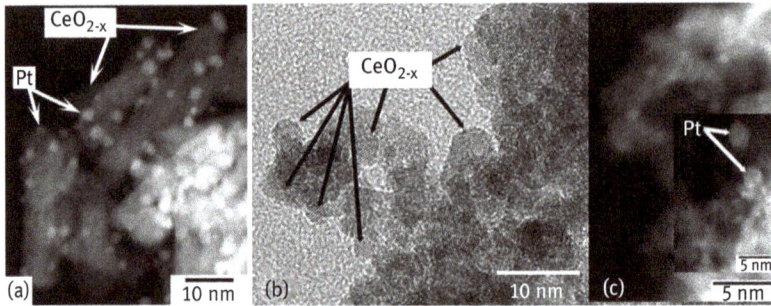

Figure 2.26: (a) HAADF-STEM image of Pt/ceria nanorods, (b) TEM image and (c) HAADF-STEM image of Pt/ceria nanoparticles. (Reproduced from Ref. [170] with permission from The Royal Society of Chemistry) [170].

(indicated by arrows in Figure 2.27) were scarcely recognizable due to their small contribution to the overall contrast. It is almost impossible to determine the brighter small patches originated from a thicker area of the support, from the superposition of two support crystals or from a looked-for metal particle on the support [171].

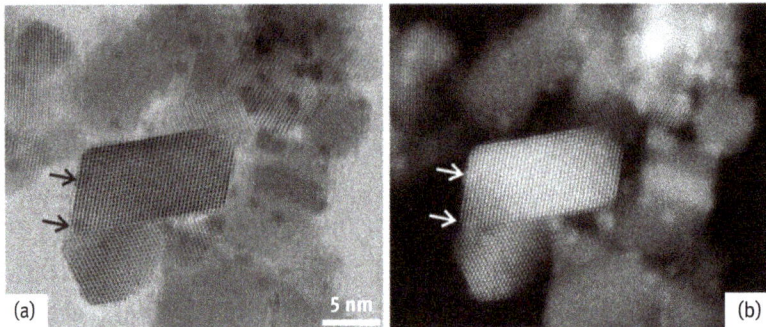

Figure 2.27: (a) Bright field- and (b) HAADF-STEM images of two areas with Pt nanoparticles located on CeO$_2$ crystallites. The arrows indicate the same two Pt particles that are hardly visible in both images as they are lying on CeO$_2$ crystals oriented along [110]. (Reprinted with permission from Ref. [171]. Copyright (2011) American Chemical Society) [171].

2.3.1.3.4 Selected area diffraction

In addition to the image mode, TEM can also be operated under diffraction mode to see the diffraction patterns by adjusting the imaging lenses. By inserting a selected area aperture into the image plane of objective lenses, one can obtain a diffraction pattern with a parallel of electron beams. This operation is called selected area (electron) diffraction (SAD or SEAD) [167]. Due to the diffraction principle, the SAD image is composed of a series of spots. Each spot corresponds to a satisfied diffraction condition of the sample's crystal structure. The SAD patterns are position dependent, which means when the sample is tilted, the resulting diffractions are different. Some new patterns might be activated and some old patterns might disappear. Single spots can only be obtained from single crystals. In multicrystalline samples, ring patterns are usually observed [167].

SAD has been employed in characterizing crystalline materials to identify the crystal structures, orientations, and defects at specific sites. It is similar to XRD, but SAD can select areas with only several hundred nanometers in size, while XRD usually collects signal from a centimeter-sized sample [167]. Figure 2.28 shows TEM images and a SAD pattern of a single CeO_2 octahedron synthesized by Xing's group. The SAD patterns corresponded to a face-centered cubic structure. It also indicated the octahedron as single crystalline, with its [110] orientation parallel to the electron beam [55].

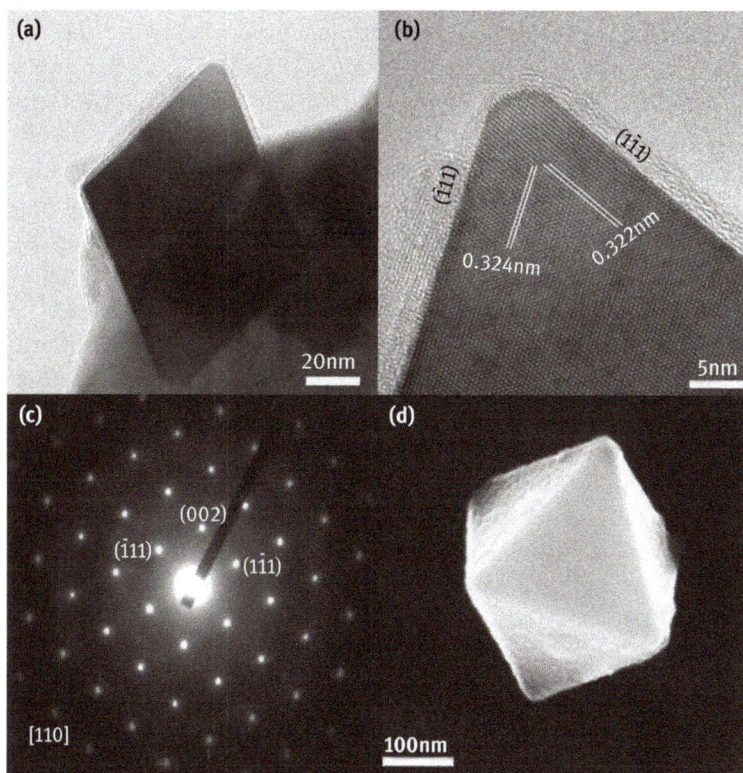

Figure 2.28: (a) TEM image, (b) HRTEM image, (c) SAED pattern, and (d) FE-SEM image of a typical individual octahedral CeO_2 nanooctahedron. (Reprinted with permission from Ref. [55]. Copyright (2008) American Chemical Society.) [55].

2.3.1.4 Electron energy-loss spectrometry

Electron energy-loss spectrometry (EELS) analyzes the energy distribution of electrons passing through the samples. These electrons may lose no energy or suffer inelastic collisions with electrons in the sample. The energy-loss spectrum has low-loss and high-loss regions. The 1–50 eV region is considered the low energy-loss spectrum, and contains peaks arising from inelastic scattering by outer shell electrons. Its energy corresponds to valence electron density and its width reflects the damping effect of single-electron transitions. At higher energy loss, the electron intensity decreases rapidly. The sharp peaks superimposing on the smoothly decreasing intensity lines represent the inner-shell excitation. The energy-loss coordinate is approximately the binding energy of the corresponding atomic shell [172, 173]. Therefore, EELS can be used to reveal tremendous amount of information, including the details of bonding/valence state, the nearest-neighbor atomic structure, the electron density, the band gap, and the thickness of sample. One may notice that EELS is also good for compositional analysis and elemental mapping, which is like EDX. However, EELS is more powerful for analyzing low atomic elements, provides higher spatial resolution and analytical sensitivity, and can detect and qualify the elements in the periodic table. EELS and EDX are complementary techniques, where EELS is more suitable for thin samples and EDX for thick samples [172–174].

Ce-$M_{4,5}$ edge is frequently used in EELS experiment for obtaining information from cerium oxide nanoparticles. Figure 2.29 shows an example of Ce-$M_{4,5}$ edge EELS data with high energy loss. It was used to monitor the change of cerium oxidation state as a function of distance from a {111} surface during beam induced-reduction. The M_4/M_5 peak ratio accompanied by a small energy shift revealed the $Ce^{4+} \rightarrow Ce^{3+}$ valency change at the imaged sample area. The EELS data indicated a transition from Ce^{3+} to Ce^{4+} from the surface to the subsurface and the bulk of CeO_2 sample [159]. In contrast, Mullin et al. has utilized valence band EELS to identify the Ce^{4+} and Ce^{3+} oxidation states on oxidized and sputtered single crystal CeO_2 films and oxidized Ce foil (see Figure 2.30). The oxidized single crystal surface CeO_2 (001) showed very different EELS spectra features compared with the sputtered CeO_2 (001) and oxidized Ce foil- Ce(III) oxide. It demonstrated an onset at 3 eV with a plateau from 4 to 6 eV and then a rise to a peak at 14 eV in the EELS spectrum. Sputtered CeO_2 (001) shared a lot of spectra similarities with oxidized Ce foil. Together with the peak at 3 eV assigned to Ce^{3+}, it confirmed a reduced surface on sputtered CeO_2 (001) [175].

2.3.2 Scanning tunneling microscopy

Scanning Tunneling microscopy (STM) is a type of scanning probe microscopy. It is another important instrument that can image sample surfaces at atomic level. A

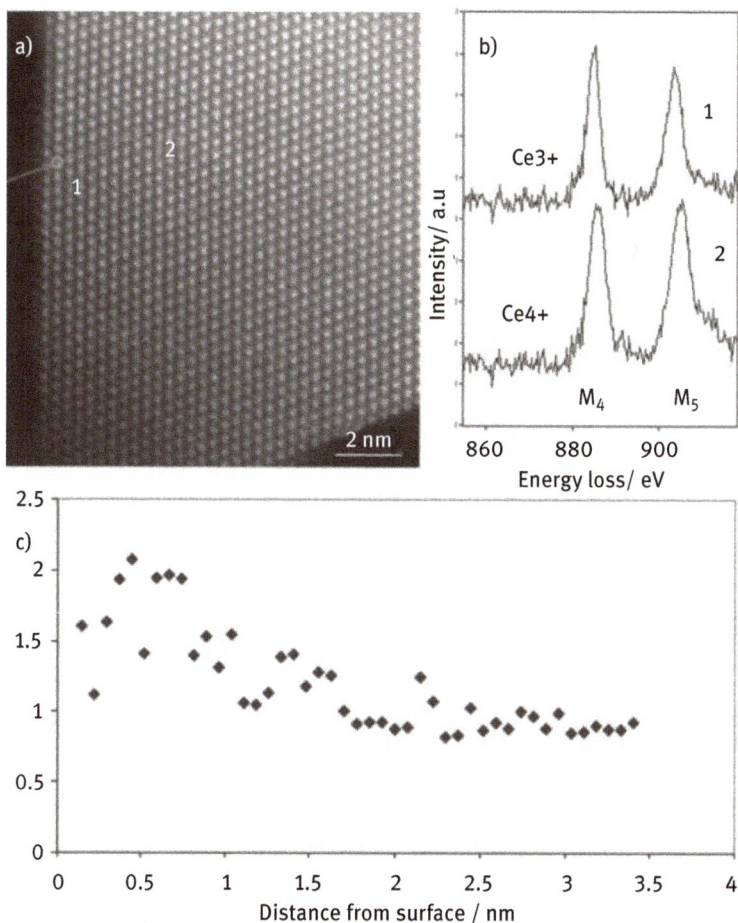

Figure 2.29: (a) HAADF-STEM image of a CeO_2 nanoparticle. (b) EEL spectra recorded at the Ce-$M_{4,5}$ edge at the positions marked in (a). Spectra were acquired sequentially along a <111> direction across a {111} surface as marked in (a). (c) Ratio of the intensity of the Ce-M_4 and Ce-M_5 transitions as a function of the distance from the surface showing a transition from Ce^{3+} to Ce^{4+} beginning at the surface. (Reprinted from Ref. [159] with permission from John Wiley and Sons) [159].

good resolution of STM is considered to be 0.1 nm in lateral and 0.01 nm in depth. It can be operated under ultra-high vacuum, moderate vacuum, air, water, and various other liquid and gas ambients. Therefore it is extensively employed in chemistry, physics, materials, and biology [176]. The STM was invented by Binnig and Rohrer, and was developed further by Binnig, Rohrer, Gerber, Weibel, and their collaborators at IBM Zurich Laboratories. The pioneering work of STM later won Binnig and Rohrer the Nobel Prize in physics in 1986 [176–178].

The STM works by scanning a probe metal tip (one electrode of the tunnel junction) over a surface (second electrode). The tip is attached to a piezodrive, consisting

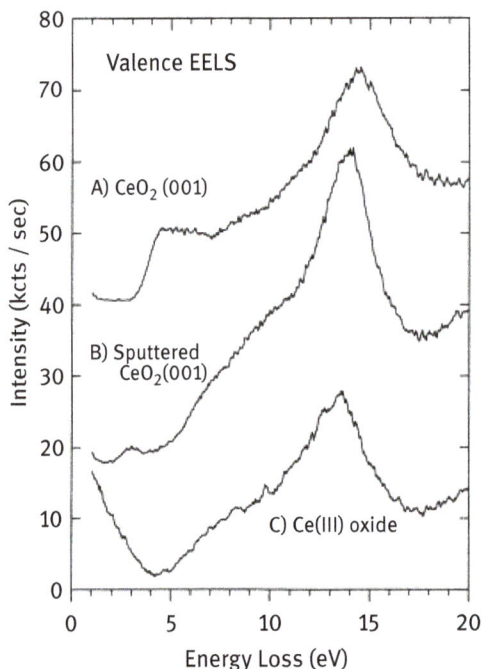

Figure 2.30: EELS spectra of (A) CeO$_2$ (001), (B) sputtered CeO$_2$ (001) and (C) Ce(III) oxide. The primary electron energy was 1000 eV. (Reprinted from Ref. [175] with permission from Elsevier) [175].

of three dimensional piezos. When the tip and sample surface is very close, by applying a bias voltage (0.01 ~ 2 V) between the tip and the sample, a tunneling current is generated. Tunneling is a quantum mechanical effect due to electrons' wavelike properties. The tunneling current is dependent upon the tip position, applied voltage, and the local density states of sample. In STM, the tunneling current is converted to a voltage by a current amplifier and compared with a reference value. A feedback loop constantly monitors the differences and makes adjustments to the tip to establish an equilibrium of the z position. As the tip scans over the xy plane, a two dimensional array of equilibrium z positions is obtained and a constant current image is displayed on the computer as a contour plot, which is very close to the actual topography of sample surface. Such a setup is referred as constant current mode. Another mode, called the constant height mode, involves scanning the surface at nearly constant height and constant voltage, and monitoring the current. It allows for much faster imaging and is mostly applied for very flat surfaces [176–179].

Esch et al. reported a high resolution STM study on the defects formation on single crystal CeO$_2$ (111) surface, assisted by DFT calculations. The sample was sputtered and annealed at 900 °C to increase the electron conductivity, suppress contaminant adsorption, and enable STM to obtain images at large-scale. STM was conducted at −3.0 V (sample with respect to tip) and a constant current of 0.3 nA at 300 °C. On slightly reduced surfaces, single vacancies were dominant, with similar coverages of 1.5 % and 1.3 % by surface and subsurface vacancies, respectively. On higher reduced surfaces, almost all vacancies became linear surface oxygen vacancies clusters (LSVCs) and occupied 92 % of all vacancies. It indicated that the electron localized on cerium ions and formed Ce^{3+} by removing oxygen. With the DFT simulations, they found that subsurface vacancies played an essential role in vacancy cluster formation. One subsurface vacancy was required to trigger the defect growth and form shorter or longer LSVCs, where Ce^{3+} ions were exposed exclusively [27].

2.3.3 Atomic force microscopy

Atomic force microscopy (AFM) is a technique developed upon the STM. It was invented by Binnig, Guate, and Gerber at IBM in 1986, and became commercially available in 1989. It has been applied in a wide range of disciplines of natural sciences, such as solid-state physics, semiconductors, polymers, and cell biology [179–181]. Unlike the STM measuring the tunneling current, AFM measures the force caused by the interaction between the tip and the sample surface. The tip is mounted on the end of a cantilever, which serves as a force sensor. The cantilever is in the scale of micrometers, typically made out of silicon or silicon nitride. The tip radius of curvature is usually a few nanometers to a few tens of nanometers. As the cantilever probes the sample surface by moving its tip along its contours, forces between the tip and the sample lead to a deflection of the cantilever. The forces measured in AFM are short-range chemical forces, Van der Walls forces, electrostatic forces, magnetic forces, capillary forces, etc.

Several operational modes have been introduced in AFM, and these modes are usually categorized into either static and dynamic modes, or contact and non-contact modes. The most important static mode is the contact mode, where the tip-sample distance is manipulated to keep the cantilever bending constant, and thus a topographic surface of constant force is recorded. Like STM, constant height above the surface can also be controlled to record the maps of force. This mode is mainly used for long-range magnetic forces. In the dynamic modes, vibration changes of cantilever induced by the tip-sample interactions are measured. Depending on the feedback parameters for the distance control, the dynamic mode has been divided into non-contact AFM (or dynamic force microscopy) and tapping mode (or intermittent contact mode). The non-contact mode exhibits high-quality factors of cantilever oscillations, and is generally used in vacuum environments. In tapping mode, the cantilever oscillation of several nanometers amplitude is excited close to its resonance frequency. This mode is the most popular AFM mode in ambient conditions and liquids nowadays [181].

Torbrügge and Reichling reported a study of surface and subsurface oxygen vacancies on reduced CeO_2 (111) surface by atomic resolution dynamic force microscopy at

80 K[183]. The surface vacancies and subsurface vacancies were clearly identified by the AFM, and their results corroborated the topography predicted by previously reported first-principles calculations done by Fabris et al. [182]. With the topography and dissipation, they have unambiguously located the subsurface oxygen vacancies buried at the third surface atomic layer. On a ceria (111) surface with high density of subsurface vacancies, the defects tended to form linear patterns and created defect free region in between (see Figure 2.31). The attractive and repulsive interactions among the subsurface vacancies might contribute to the final linear ordered arrays [183].

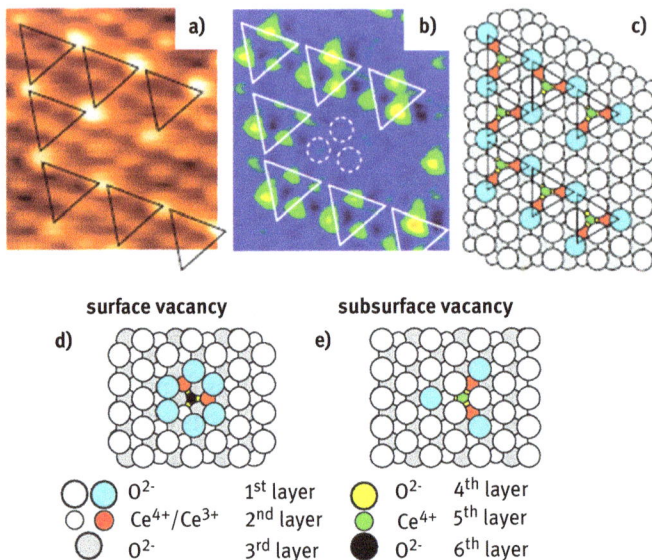

Figure 2.31: (a) Topography and (b) dissipation images together with the corresponding schematic model (c) of the typical local ordering of subsurface oxygen vacancies observed in our images. The subsurface oxygen vacancy structures are highlighted by triangles. The dashed circles indicate a defect free surface area. Structural modes of (d) surface vacancy and (e) subsurface vacancy. The image set point was -2.9 fN$\sqrt{m}$. (Reprint from Ref. [183] with permission from American Physical Society) [183].

2.3.4 X-ray diffraction

XRD is a basic technique to study the phase, composition, crystal structure, microstrains, and orientation of nanomaterials [184]. It is considered as a non-destructive, fast, and easy technique for single crystal and polycrystalline materials characterization.

Bragg's law is the physics principle of XRD [185]. William Lawrence Bragg proposed a model in which the incoming X-rays were scattered from each plane with a specific scattering angle 2θ, and the constructive interference only occurred when the path difference was an integer number multiple n of the X-ray wavelength λ. The d is the spacing between planes :

$$n\lambda = 2d sin\theta$$

The atoms in a crystal sample are a periodic array of coherent scatterers and the distance between atoms is similar to the wavelength of X-rays. These enable crystalline samples produce concerted constructive interferences at specific angles, leading to characteristic diffraction phenomena which can be studied to investigate the crystal structure of materials [185]. The peak position can provide information of lattice parameters, space group, chemical composition, macrostresses, etc. With the peak intensity, information about crystal structure, texture and quantitative phase can be obtained. The microstrains and crystallite size of samples contribute to peak broadening [184]. The concept of Bragg diffraction also applies equally to neutron diffraction and electron diffraction.

The XRD discussed here is the commonly used angle-dispersive measurements, in which a monochromatic beam is used and the angular position of the diffraction peaks is measured. The instrument in labs is composed of an X-ray tube, primary and secondary optics, a goniometer, a sample holder, and a detector [184]. Goniometer is the central part of XRD. It controls the X-ray source, the sample, and the detector moving in a precise manner. It generally uses the Bragg-Brentano geometry. In a θ-2θ scan, the X-ray tube is fixed, and the detector moves twice as fast in θ as the source, so only where the angle of incidence beam is equal to the angle of reflection, the intensity of the reflect wave can be measured. In a θ-θ mode, the sample is fixed, and the X-ray tube and detector rotate at the same speed θ/min in opposite directions. For residual stress and texture measurements, additional rotation axes are needed so that the sample can be positioned for investigation. Several types of detectors are commercially available, including gas and solid detectors, and these detectors can be divided into point, one-dimensional, and two-dimensional detectors. 1D and 2D detectors are more time-saving. 2D detectors can capture large portion of diffraction rings and hence allow for analyzing texture or residual stresses. They are very useful for materials characterization with very large grains or in microdiffraction [184, 187].

One of the most important uses of XRD is for phase identification. One can compare their data with the known standards in the Int. Center for Diffraction Data (ICDD), (previously named Joint Committee on Powder Diffraction Standards (JCPDS)), or the Inorganic Crystal Structure Database/CrystalWorks, where patterns are determined by either experiments or computational calculations [188, 189]. Using XRD patterns to confirm the synthesized materials is almost always the first step for metal oxide materials characterization. For examples, Yan's group has prepared a series of colloidal rare earth oxide nanocrystals and characterized the crystal structure with XRD for phase identification. Table 2.2 shows the crystal structures of as-obtained rare earth oxides determined by powder XRD with a Cu Kα radiation source. All the dry, trivalent rare earth oxides had a centered cubic structure with a space group of Ia3, whereas CeO_2 exhibited a face centered structure with a space group of Fm3m. The XRD patterns of Tb_2O_3, Eu_2O_3, Pr_2O_3 and CeO_2 are demonstrated in Figure 2.32. The calculated lattice constant is 5.4516, 11.36, 10.97, and 10.90 Å for CeO_2 (JCPDS: 34–0394), Pr_2O_3 [190], Eu_2O_3 (JCPDS: 34–0392), and Tb_2O_3(JCPDS: 23–1418), respectively [37].

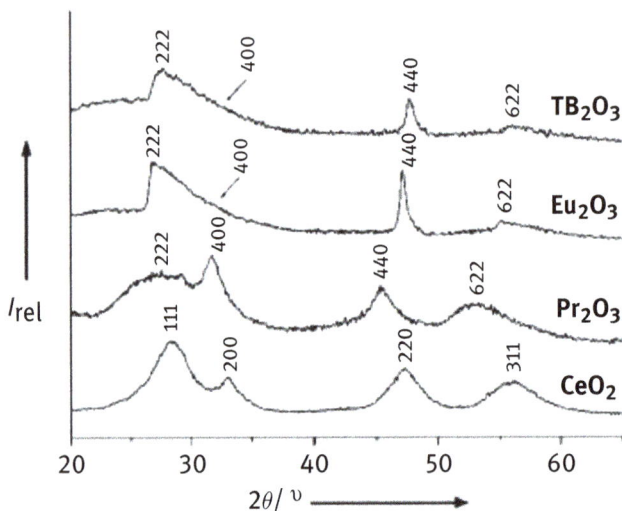

Figure 2.32: XRD patterns of dry CeO_2, Pr_2O_3, Eu_2O_3, and Tb_2O_3. (Reprinted from Ref. [37] with permission from John Wiley and Sons, Inc.) [37].

The Scherrer formula is frequently used to evaluate crystallite sizes of nanomaterial powders from their XRD patterns. It was discovered by Paul Scherrer [191]. The formula is written as follows:

$$\tau = \frac{K\lambda}{\beta cos\theta}$$

where τ is the mean size of the crystalline domains and K is a constant dependent on crystallite shape, with a typical value of 0.89 for integral breadth of spherical crystals with cubic symmetry. λ is the X-ray wavelength. β is the full width at half max (FWHM) of diffraction peak. θ is the Bragg angle. It is important to know that the Scherrer formula is based on the peak profile, however, both instrumental broadening and specimen broadening account for the peak profile. For the broadening contributions from samples, the crystallite size, microstrain, inhomogeneity, and temperature factors could all get involved and make the data interpretation more complicated. Thus one has to be very careful when using Scherrer formula for crystallite size analysis. Generally, most applications of Scherrer formula assume spherical crystallite shapes, and it only provides a lower bound on the crystallite size [192].

Bourja and co-workers reported a study on the effect of thermal treatment temperatures on the average crystallite sizes of pure ceria and bismuth doped ceria solid solutions using the Scherrer formula. The XRD patters in Figure 2.33(a) indicates a fluorite structure of CeO_2 (JCPDS: 34–0394) for pure ceria treated from 300 to 600 °C. The XRD peak width gradually decreases with increasing temperature and the average crystallite size increased from 5 to 11 nm (Figure 2.33(b)), implying a faster calcination process at higher temperature. For the bismuth doped ceria solid solutions, the average crystallite size decreased with the Bi fraction and confirmed the relationship between crystallite size and solid solution composition. The smaller crystallites obtained from substitution of Ce^{4+} by Bi^{3+} might be interesting for the catalysis applications. The thermal effect also exhibited on the crystallite sizes of Bi doped ceria samples. Figure 2.33(d) shows that the crystallite sizes of $Ce_{0.85} Bi_{0.15} O_{2-\delta}$ and $Ce_{0.8} Bi_{0.2} O_{2-\delta}$ calculated by Scherrer formula increased with calcination temperature. The variation of grain size was largely derived from sintering [193].

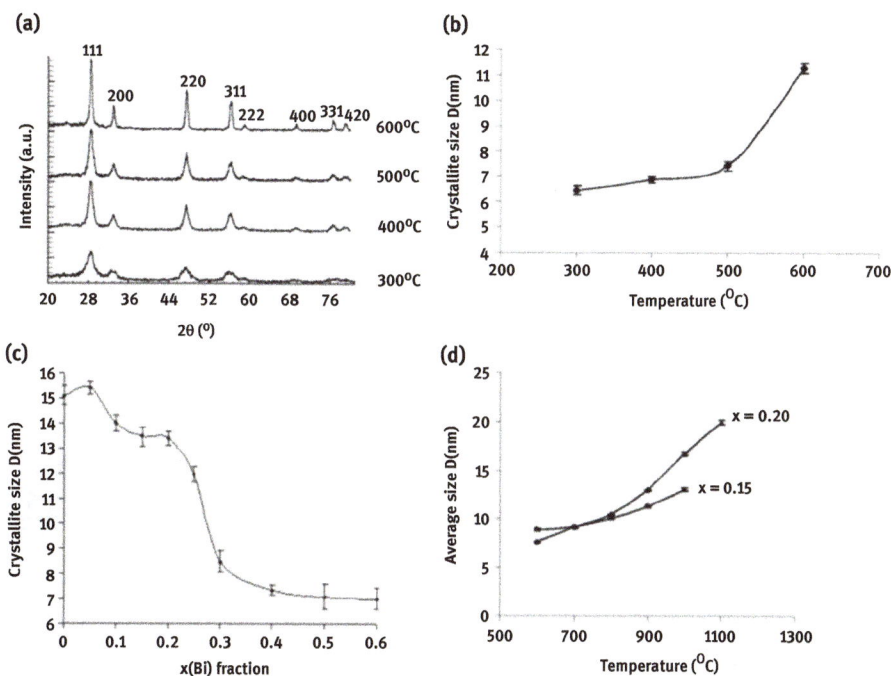

Figure 2.33: (a) X-Ray Diffraction patterns of pure ceria samples obtained at different calcination temperatures, (b) Effect of calcination temperature on the crystallite sizes of CeO_2 nanoparticles. (c) Variation of cubic crystallite size as a function of Bi fraction, and (d) Variation of crystallite average size of the bismuth doped ceria according to the calcination temperature. (Reprinted with permission from Ref. [193] with a Creative Common license) [193].

XRD can also be used for quantitative phase analysis once all present phases are identified. However, this is rather complicated in the multiphase compounds, due to the different mass absorption coefficients of different phases, and the non-linear intensity evolution. Based on the properties of sample and the purpose of analysis, four methods can be applied: methods with external and internal standards, method of intensity ratio, and Rietveld method [184]. In particular, Rietveld method is very powerful for determining the respective phases in multiphase materials, and is able to analyze samples with strongly overlapping reflections. It also can be used for evaluating crystallite size, texture, strains, and microstrains. This method uses a least squares approach, and its criterion is to minimize the difference between the calculated profile and the observed data. Although very useful this method is for diffraction peak analysis, one has to be careful about refinement parameters because many parameters without physical meaning can give good refinements and lead erroneous interpretation.

Lussier and co-workers applied powder X-ray and neutron diffraction Rietveld refinements for the structure studies of $Y_xPr_{2-x}O_3$ samples. $Y_xPr_{2-x}O_3$ has three phases, trigonal A-type, monoclinic B-type, and cubic C-type phases. Figure 2.34(a) and Figure 2.34(b) shows the Rietveld plots for the refinement of trigonal $Y_{0.05}Pr_{1.95}O_3$ and monoclinic $Y_{0.20}Pr_{1.8}O_3$ after annealing at 1200 °C. In the trigonal phase, the Y^{3+} and Pr^{3+} were fully disordered on the 2d cation site and the anions fully occupied the 2d and 1a sites. Table 2.4 demonstrates the structural details refined for the trigonal phase $Y_{0.05}Pr_{1.95}O_3$. They utilized the *ex situ* XRD refinement data collected after heating treatment to determine the phase transitions. It indicated that the average cation size was the primary factor in determining the phase of material (see Figure 2.34(c)), which would allow structure prediction based on cation sizes and design of desired materials. The *in situ* XRD experiment was also applied to understand the oxidation

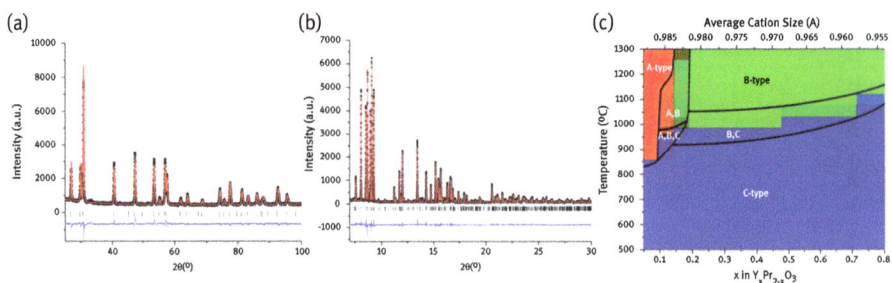

Figure 2.34: Rietveld plots for the refinement of (a) the trigonal structure $Y_{0.05}Pr_{1.95}O_{3.00}$ after annealing at 1200 °C Laboratory X-ray data (b) the monoclinic structure $Y_{0.20}Pr_{1.80}O_{3.00}$ after annealing at 1200 °C synchrotron X-ray data. Black symbols = observed data, red line = fit, blue line = difference, tick marks = Bragg positions. (c) RGB color map of the phases found during ex-situ heating experiments. Red = trigonal (A), green = monoclinic (B), and blue = cubic bixbyite (C). The black lines do not represent a fit to the data and are a guide to the eye only. (Reprinted with permission from Ref. [194]. Copyright (2018) American Chemical Society.) [194].

Table 2.4: Structural parameters of trigonal $Y_{0.05}Pr_{1.95}O_{3.00}$ after annealing at 1200 °C.[a] (Reprinted with permission from Ref. [194]. Copyright (2018) American Chemical Society.) [194].

Composition		$Y_{0.05}Pr_{1.95}O_3$
Space group		P3m1(164)
Unit cell	a (Å)	3.84974 (4)
	c (Å)	6.01129(9)
	V (Å³)	77.154(2)
	Z	1
Y/Pr (2d) (1/3, 2/3, z)	z/c	0.2467(2)
	B_{iso} (Å²)	0.0294(94)
O1 (2d) (1/3, 2/3, z)	z/c	0.6485(2)
	Biso (Å²)	0.69(2)
O2 (1a) (0, 0, 0)	Biso (Å²)	0.76(3)
Bond lengths (Å)	Y/Pr-O1	3 × 2.3102(4)
	Y/Pr-O1	1 × 2.4155(14)
	Y/Pr-O2	3 × 2.6719(6)
	Y/Pr-O average	2.480
x^2	Overall	1.56
	Neutron (T.O.F)[b]	1.52
	X-ray (Cu- Kα)[c]	1.58

[a] Values were obtained from a simultaneous Rietvel refinement against powder XRD and time-of-light powder neutron data measured at room temperature.
[b] Neutron T.O.F range: 12,000–116,000 μs, 5672 data points
[c] X-ray wavelength λ(Kα1) = 1.540598 Å, λ(Kα2) = 1.544426 Å, 25° ≤ 2θ ≤ 145°, Δ2θ = 0.0167°, 7180 data points.

pathways of all three phases. All compositions were found to oxidize to the fluorite structure at low temperatures. Additionally, only the C-type $Y_xPr_{2-x}O_3$ could accommodate extra oxide anions. The A- and B-type structures did not allow for topotactic oxygen uptake, and phase transition would occur instead [194].

2.3.5 X-ray photoelectron spectroscopy

X-ray photoelectron spectroscopy (XPS), also called electron spectroscopy for chemical analysis (ESCA), is the most frequently used surface-sensitive technique. It can conduct both quantitative and qualitative analysis of elements within a material. Particularly, it is powerful in chemical state analysis. It has been used to analyze various materials, eg. inorganic compounds, metals, semiconductors, polymers, etc. Because of its surface sensitivity and high detection limits (parts per thousand, or even ppm), it is routinely used to analyze the elemental composition and its corresponding surface chemistry in nanomaterials and thin films [195].

XPS is based on the photoelectric effect. In XPS, the sample surface is typically bombarded by a beam of X-rays causing photoelectrons to be emitted from the sample surface. The kinetic energies of these photoelectrons are then measured by an electron analyzer. With the known values of exciting X-ray energy ($E_{h\nu}$) and spectrometer wavefunction (Φ), the core-electron binding energies (E_b) relative to the Fermi level can be calculated as follows [195]:

$$E_b = E_{h\nu} - E_k - \Phi$$

From the binding energy and intensity of XPS peaks, the elemental identity, chemical state, and quantity of a detected element can be determined. The photoelectron binding energy is a result of chemical environment of the targeting atom, which is related to its oxidation state, ligand electronegativity, coordination, and spin-coupling effect. All these parameters are essential in determining the electron density of atoms. To obtain the quantity of each element, XPS fitting analysis must be conducted. Relative quantification is more common than absolute quantification, which is also more challenging. The quantitative accuracy depends upon many parameters like signal/noise ratio, peak intensity, sensitivity factors, sample homogeneity, correction for electron transmission function, etc. 5% is the generally accepted error range for quantitative precision [195–197].

As we mentioned earlier that XPS is a surface analysis technique. The photoelectrons at this top layer (0–10 nm) can escape and be captured by the analyzer, indicating the general measurement depth of XPS. XPS requires high vacuum or ultra-high vacuum conditions to minimize the interferences caused by the gases or contaminants in the chamber. For some quasi *in situ* or *in situ* experiment, the ambient pressure XPS has been developed and used, where the pressure is of a few tens of milibar [198].

The basic components of XPS instrumentation vary greatly in terms of sophistication. In a typical modern XPS, the spectrometer includes a monoenergetic X-ray source, a sample, an electron energy analyzer, a detector, and a scan and readout system. Al and Mg are the two most popular X-ray anode source materials for lab XPS. The Al Kα has a characteristic energy of 1486.6 eV with a line width of 0.85 eV. The Mg source produces a Kα line with 1253.6 eV in energy and 0.7 eV in width. Synchrotron radiation from electron storage rings is another attractive source for XPS. The most significant advantages of synchrotron radiation are the high intensity and tunable energy, which leads to dramatically reduced experimental time and allows for depth-dependent chemical analysis of a material. The biggest disadvantage of synchrotron XPS is its limited access for routine sample analysis, which is true for all the other synchrotron based techniques. The electron energy analyzer is used to measure the energy of the emitted photoelectrons. In modern XPS, a hemispherical sector analyzer (HSA) is commonly used. The HSA is designed to have a constant and as high as possible energy resolution for the detection of photoelectrons. The reported best energy resolution is 0.4 eV, corresponding to the line width of monochromator. HSA is usually operated in a fixed analyzer transmission mode with a constant pass energy of electrons, which is fulfilled by applying a constant voltage across the hemispheres. The most significant feature of this mode is the constant energy resolution in the spectrum as a function of the energy, which is unlike the cylindrical mirror analyzer usually used in Auger electron spectroscopy (AES) [197, 199, 200].

Cheung's group applied XPS for investigating the surface chemistry of ceria nanorods activated at ambient and low pressures (0.1 Torr). The changes in the oxidation state of the surface cerium and the relative density of oxygen vacancies defects of cerium oxide were reported. The Ce^{3+} fraction was used as an indicator of the density of oxygen defects. It was found that the ceria nanorods activated under ambient and low pressures exhibited very different XPS spectral features. The spectrum of ceria nanorods activated at 0.1 Torr at 400 °C showed relatively higher intensity of v^0, v', v''' and u' peaks, corresponding to the Ce^{3+} state in the sample. The spectral fitting analysis indicated a 39 % of Ce^{3+} fraction in the low-pressure activated sample, while the nanorods activated under ambient pressure demonstrated a 27 % of Ce^{3+} state (see Figure 2.35). The higher fraction of Ce^{3+}, hence the higher density of oxygen vacancy defects, was proved to enhance the catalytic activity for CO oxidation catalyzed by ceria nanomaterials. The defect engineering accomplished by a low-pressure thermal treatment is a new and effective approach to improve catalysts performances. The traditional ways target on controlling the shape and size of nanomaterials or inserting dopants [28].

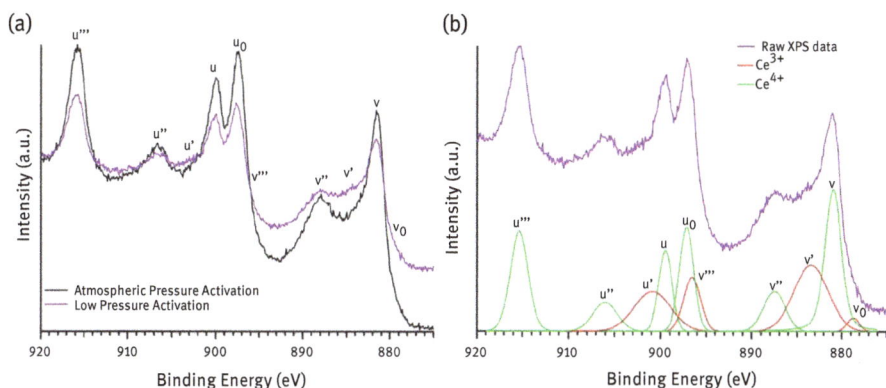

Figure 2.35: (a) Comparison of XPS spectra of ceria nanorods after (purple) low-pressure activation and (black) atmospheric pressure activation. The change in relative intensity of peaks v0, v′, v‴ and u′ demonstrate a change towards the Ce^{3+} state. (b) XPS fitting analysis of ceria nanorods after low-pressure activation. (Reprinted with permission from Ref. [28]. Copyright (2011) American Chemical Society) [28].

Ambient pressure XPS (AP- XPS) offers a promising route to study the surface chemistry of materials in or near real conditions. It is quite useful in studying the dynamic processes on a material surface, and has been widely employed in catalysis, corrosion, film growth, and electrochemistry [198]. Liu and co-workers reported using AP-XPS to probe the surface species and active phases associated with ethanol stream reforming reaction over Ni-CeO$_2$(111) catalyst under steady state reaction conditions (40–240 Torr) (see Figure 2.36). It revealed that ceria surface was highly reduced and hydroxylated under reaction conditions, illuminated by the disappearance of Ce 4d spectrum at 700 K. The active species of supported Ni nanoparticles was Ni metal, which was responsible for C-C and C-H bond cleavage in ethanol and coke. Compared with the bare ceria film, ~ 90 % reduction at the top layers of Ni-CeO$_2$ (111) film was reduced to Ce^{3+} under steam reforming conditions. Further study indicated the active species of CeO$_2$ was Ce^{3+}(OH)$_x$ compound, formed due to the reduction by ethanol and the efficient dissociation of water. The synergistic effect of Ni^0 and Ce^{3+}(OH)$_x$ through a metal-support interaction contributed to oxygen transfer, activation of ethanol and water, and the coke removal [17].

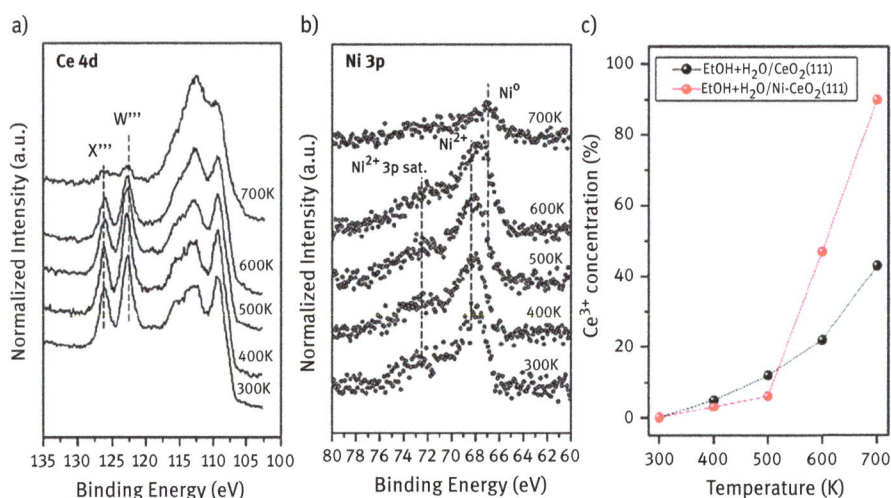

Figure 2.36: (a and b) Ce 4d and Ni 3 p spectra of the Ni-CeO$_2$(111) surface under ethanol steam reforming conditions (40 mTorr EtOH + 200 mTorr H$_2$O), (c) Surface Ce^{3+} concentration comparison between CeO$_2$ (111) and Ni-CeO$_2$ (111) catalysts. (Reproduced from Ref. [17] with permission from the Royal Society of Chemistry) [17].

XPS can also provide compositional information through in-depth analysis, using angle-resolved XPS or ion sputtering. The angle-resolved XPS is non-destructive, but ion sputtering is a destructive method. Both methods are possible to calculate the thickness and depth of thin films and obtain a depth profile. Angle-resolved XPS is able to gain valuable chemical state, composition and thickness information of ultra-thin films. Ion sputtering is easily accessible, and not limited to ultra-thin films. But it damages the sample, and cannot provide quantitative analysis and reliable chemical state analysis. Its depth resolution is hard to control and is affected by various parameters, such as sample surface roughness, sample composition and homogeneity, instrumental parameters, radiation-induced effects, etc [201].

2.3.6 X-ray absorption spectroscopy

X-ray absorption spectroscopy (XAS), also called X-ray absorption fine structure (XAFS), is a powerful and versatile technique for structural investigations of materials in chemistry, physics, biology, geology, and various other fields. It probes at atomic or molecular level, and determines the local geometry and electronic structure of elements of interest within a material. This technique doesn't require materials possessing long-range translational order, which is unlike XRD, and thus can be used on crystals, amorphous materials, disordered films, gases, liquids, and so on. The modern XAS is a synchrotron- based technique, benefiting from the high beam flux and tunable energy of synchrotron radiation. The synchrotron sources provide a wide range of X-ray energies that are applicable to most elements in the periodic table. The first useful synchrotron XAS was developed around 1970. Currently there are more than 60 synchrotron facilities in the world and mainly located in America, Europe, and Asia [10, 12, 202].

XAS is based on the X-ray photoelectric effect, and the wavenature of photoelectron determines the local structures around the selected atoms in material. When an incident X-ray is absorbed, a core electron with specific binding energy could be destroyed and emits a photoelectron. The energy overcomes different layers of binding energy is the absorption edge. It is named by the orbital shell the electron coming from. For example, a 1s electron is excited for a K-edge spectrum. The emitted photoelectron can be scattered from neighboring atoms, creating either constructive or destructive interference at the origin. The interference is controlled by interatomic distance, types and numbers of neighboring atoms, and the photoelectron wavelength. The interference leads to the oscillations in absorption probability of the XAS spectrum. The absorption probability is described with absorption coefficient $\mu(E)$ in XAFS. $\mu(E)$ generally decreases with energy as the energy increases and has an approximate inverse relationship with E^3. Of course the absorption edge is excluded here. At the absorption edge, the $\mu(E)$ suddenly increases and provides specific energy information corresponding to the targeting atoms in the material [10].

An XAS spectrum is usually divided into two regions, the X-ray absorption near edge structure (XANES) and the extended X-ray fine structure (EXAFS). XANES covers the region of pre-edge, absorption edge, and up to 100 eV above edge. In this region, the incident X-ray energy is sufficient to transfer the core electrons to higher unoccupied valence states. The wavelength of the photoelectron is larger than the interatomic distance between the absorbing atom and its nearest neighboring atoms. From XANES, we are able to obtain the geometry and symmetry information from the pre-edge peaks, and the valence state from the edge shift. XANES is experimentally simpler, and its signal is stronger. The EXAFS region is usually in a 500–1000 eV range, starting from the point where the photoelectron wavelength is equal to the atomic distance between central atom and its nearest neighboring atom to the point about 1000 eV past the inflection point of the absorption edge. In this region, the photoelectron wavelength is comparable to the distance between the absorbing atom and its neighboring atoms. The photoelectron wave is weakly backscattered by and among the neighboring atoms, and produces the EXAFS oscillations- the most important feature of EXAFS. The scattering in this region is mainly single scattering, while multiple scattering is more dominant in XANES. The analysis of EXAFS spectrum gives information of coordination number, kind, and distances of neighboring atoms from each other and the absorber [10, 11].

The setup of XAS in synchrotron facility is fairly complicated, including a wide variety of components such as X-ray monochromators, mirrors, slits, photon shutters, shielding, sample holder, detectors, data collection systems, etc [10]. For general users, the sample area, commonly called the hutch, is more important. In the hutch, the incident beam directed out from the storage ring interacts with samples, and the resulting signal is captured by detectors. There are two major ways in the hutch for data collection, the transmission mode and the fluorescence mode. For a sample with high concentration of the targeting element (typically > 10 %) or thin, transmission mode is more suitable. Otherwise, for low concentration (down to ppm level) or thick samples, the fluorescence mode is preferred. In the transmission mode, the relationship between the absorption coefficient and the incident and transmitted beam is simplified as $\mu(E) \propto \ln(I_0/I_t)$, while in the fluorescence mode the correlation is like this, $\mu(E) \propto I_f/I_0$ [203]. In the soft X-ray absorption spectroscopy, another mode called total electron yield mode is usually used, in which the sample is grounded and the neutralization current is measured.

It is important to keep in mind that XAS measures the "average" local structure of selected element in a sample, and it is not able to distinguish multiple sites structures within a sample. In most cases, X-rays are capable of penetrating through the sample, which makes XAFS a non-surface-sensitive technique. In addition, the data collection using XAS is relatively simple, however, the XAS data analysis and interpretation is rather challenging and requires expertise in dealing with analysis software and substantial understanding of tested samples. For more useful information about XAS, one can visit http://xafs.org/(now sponsored by IXAS and moved to http://www.ixasportal.net/wiki) for tutorials, workshops, software, and database [10, 203–205].

Wu and Zhou et al. reported a quasi *in situ* XAS study on the disproportionation of hydrogen peroxide promoted by ceria nanorods [19]. The experiment was carried out in a wet chemical environment, which was maintained by a Tris buffer solution to prevent from drying out and pH change. Ce L_3-edge was scanned at various reaction times up until 10 h. The XANES spectra in Figure 2.37(a) indicated an intensity

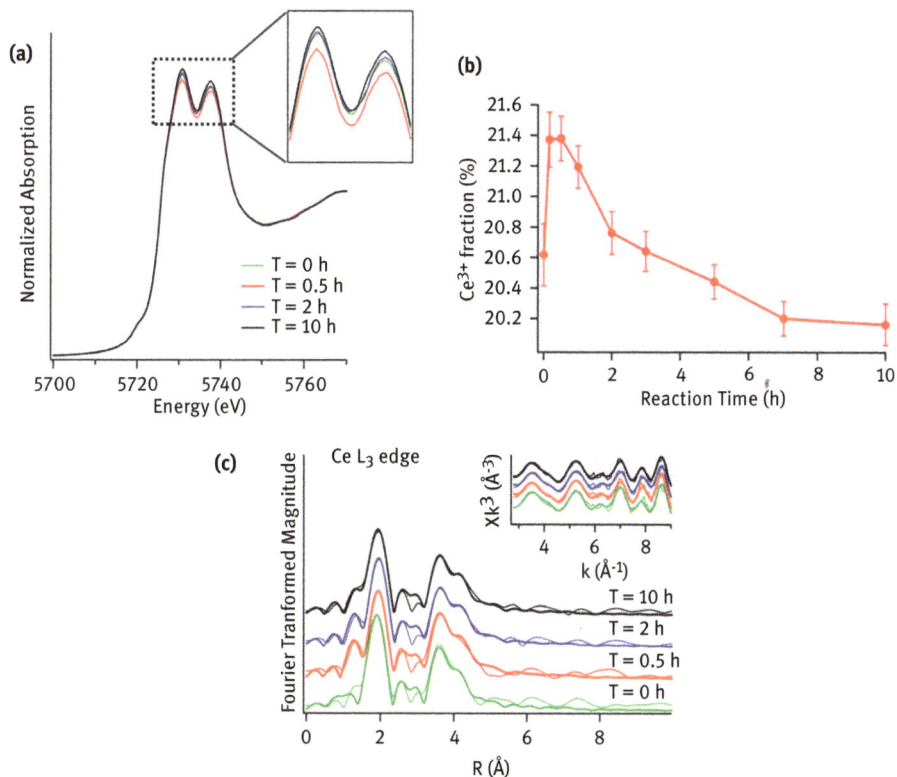

Figure 2.37: (a) Ce L_3-edge XANES spectra of ceria nanorods reacted with 10 mM H_2O_2 at different reaction times: T = 0 h, 0.5 h, 2 h and 10 h. (inset) Zoomed-in peak areas. (b) Ce^{3+} fraction in ceria nanorods as a function of reaction time in the 10 mM $H_2O_2/0.1$ M Tris buffer solution (c) Fourier transformed Ce L_3-edge EXAFS data of ceria nanorod sample with different reaction times in 10 mM $H_2O_2/0.1$ M Tris buffer solution. The inset shows the corresponding EXAFS spectra in k-space. Data: thick lines; fittings: thin lines. (Reproduced from Ref. [19] with permission from the Royal Society of Chemistry) [19].

change along reaction time. Its curve-fitting analysis provided a quantitative evalua-tion of Ce oxidation state evolution along with H_2O_2 disproportionation. According to Figure 2.37(b), Ce^{3+} fraction in ceria nanorods exhibited sharp increase in 0.5 h and decreased quickly to its original value ~ 20.6 % at T = 2 h. From 3 h to 10 h, the overall Ce^{3+} percentage varied insignificantly and stayed around the level of 20.2 %. This re-sult drew a conclusion that the surface Ce^{4+} was reduced to Ce^{3+} by H_2O_2 molecules in the beginning and subsequently slowly oxidized back to Ce^{4+} in 9 h. The slow oxi-dative process of ceria were attributed to the changes of H_2O_2 and surface Ce^{3+} con-centration, which caused the adjustment of reaction potential in the solution.

EXAFS data provides the details of atomic structures of selected element in a ma-terial, including identity of neighboring atoms, the atomic distances, and the coordi-nation numbers. Figure 2.37(c) shows the Fourier transformed Ce L_3-edge EXAFS

spectra of ceria nanorods with different reaction time. The data analysis was performed using IFEFFIT software package and its fitted parameters are shown in Table
2.5. The EXAFS data depicted the changes in local structure of Ce atoms with respect
to the reaction time. The coordination number of the first Ce-O shell at 2.30 Å of ceria
nanorods started from 5.8, then decreased to 4.7 and increased back to 5.8, corroborated the XANES results. The extra peak around 1.65 Å was possibly introduced by
H_2O_2, which caused the formation of Ce = O bond. The intensity change of this peak
was also consistent with the XANES data, confirmed the modification of local structures by H_2O_2. It was clear that, by combining the XANES and EXAFS data, the reversible changes in chemical state of cerium and its local structures could be fully
elucidated in the reaction of H_2O_2 disproportionation [19].

Table 2.5: Fitted structural parameters of the Ce L_3-edge EXAFS analysis for
ceria nanorod samples reacting with 10 mM H_2O_2/0.1 M Tris buffer solution.
N is the coordination number around the central atoms. R is the average bond
distance. σ^2 is the Debye–Waller factor. Italic marks indicate fixed parameters
in the fitting analysis. (Reproduced from Ref. [19] with permission from the
Royal Society of Chemistry) [19].

Reaction time	Atom	N	R (Å)	σ^2 $(10^{-3}Å^2)$
T = 0	O	5.8 ± 0.6	2.30 ± 0.01	4.7 ± 1.2
	Ce	5.6 ± 0.4	3.83	1.8 ± 0.6
T = 0.5 h	O	0.9 ± 0.3	1.65 ± 0.02	3.9 ± 0.8
	O	4.7 ± 0.4	2.32 ± 0.01	3.9 ± 0.8
	Ce	4.7 ± 0.4	3.83	1.2 ± 0.3
T = 1 h	O	0.7 ± 0.2	1.66 ± 0.02	3.6 ± 0.8
	O	4.7 ± 0.3	2.33 ± 0.01	3.6 ± 0.8
	Ce	4.5 ± 0.3	3.83	0.6 ± 0.4
T = 15 h	O	0.7 ± 0.3	1.68 ± 0.03	6.6 ± 0.8
	O	5.8 ± 0.4	2.33 ± 0.01	6.6 ± 0.8
	Ce	4.9 ± 0.3	3.83	1.1 ± 0.4

2.3.7 Raman spectroscopy

Raman spectroscopy has been used broadly to identify molecules and characterize
material structures in chemistry, physics biology, and medicine. It can observe the
vibrational, rotational, and other low-frequency modes in a system, similar to IR
spectroscopy. In contrast to the absorption effects in IR spectroscopy, Raman is dependent on inelastic scattering of light and hence its applicability is not hindered
by the presence of water in samples. This makes it very powerful in aqueous reaction systems in biology or medical research compared to IR. Nevertheless, in a

variety of materials or reaction systems, Raman and IR are considered as complementary techniques and can provide complementary information about the nature of materials [206, 207].

The Raman spectroscopy is based on the Raman effect or Raman scattering, which was firstly discovered in organic liquids by an Indian scientist V. C. Raman in 1928, and this discovery helped him win the Nobel Prize in physics in 1930. When a light reacts with a molecule or crystal, light can be absorbed, scattered, transmitted or reflected. The scattered light plays the important role in Raman spectroscopy. Among the scattered light, most is elastic, which has the same energy as the incident photons. This phenomenon is referred as Rayleigh scattering. A small fraction of light can be scattered inelastically, occurred with changes in molecular vibrations, rotational or electronic energy of a molecule. The energy transfer between the photons and molecules results in the energetic difference between the incident and scattered light. This process is called Raman scattering. There are two types of Raman scattering depending on the initial and final rovibronic state of molecules. One is Stokes Raman scattering, where the final state is higher in energy than the initial state, and the scattered photon will be shifted to a lower frequency or energy, called a Stokes shift. The other one is Anti-Stokes Raman scattering, where the final state is lower in energy and the scattered photon is shifted to a higher frequency, also called an anti- Stokes shift. Stokes-Raman scattering is more probable, and hence it is mainly measured in Raman spectroscopy [206, 207].

For a molecule to be visible under Raman, there must be a change in the molecular polarizability along the normal mode with a non-zero gradient at the equilibrium geometry. The Raman intensity corresponds to the polarizability- the larger the gradient of polarizability is, the higher the Raman scattering intensity will be. An increase in laser power or laser frequency also leads to stronger Raman scattering. Additionally, the advanced types of Raman spectroscopy such as resonance Raman and surface-enhanced Raman have been used to produce enhanced Raman scattering. In Raman spectra, the Raman scattering intensity is plotted against the Raman shift. The Raman shift is the difference between the measured frequency of scattered light and the incident beam. Therefore the Raman spectrum is roughly independent of the wavelength of light source. However, in cases that materials exhibit different physicochemical properties such as light absorption and fluorescence, Raman spectrum varies according to the light source [206–208].

Modern Raman instrumentation typically consists of a laser light source, a filter, a few lenses, a spectrograph, and a detector (CCD or ICCD). The excitation laser ranges from near UV, visible to near infrared light. The Raman scattering intensity is proportional to ν [4], where ν is the frequency of excitation laser. The shorter excitation wavelength also offers higher spatial resolution. However, for a lot of samples, the fluorescence is much stronger excited at near UV or blue region of light than the red or the near infrared light. The Raman signal is also positively correlated with laser power, which varies between a few microwatts to several hundred milliwatts. Strong laser power could cause absorption and lead to thermal decomposition of samples. Therefore, it is important to choose the right excitation laser with right laser power for individual sample. The filter, typically a notch or edge filter, filters out the much more intense Rayleigh scattering and anti-Stokes light, and only collects the Raman (Stokes) scattered light. The spectrographs make use of mirrors and gratings to bend the Raman shifted light according to the wavelength and disperse the signal onto the detector [209]. One of the most widely used spectrograph configurations for Raman spectroscopy is Czerny-Turner spectrograph. It applies collimators in an off-axis configuration and a planar reflective grating in the collimated space [210]. A CCD (charge-coupled device) detector, the most commonly equipped detector in Raman, records the signal and passes the signal to a readout system for analysis. It consists of an array of light sensitive Si-photodiodes, with each connected to a capacitor. The photodiodes capture the incoming photons and create electron-hole pairs, and the separated electrons are stored in the capacitors. There are a wide variety of CCD cameras with different sizes and cooling systems. The most important feature of a detector is the quantum efficiency. The multichannel two-dimensional CCD detectors have been reported with high quantum efficiency, and other advantages like extremely low level of the thermal noise, low read noise and the large spectral range. In Raman spectroscopy, a research grade optical microscope can be coupled to the excitation laser and the spectrometer to make a Raman microspectroscopy. The objective lenses of microscope focus the laser beam to a small sample area, usually several micrometers in diameter, and result in high photon flux than the conventional Raman setup. The Raman microspectroscopy is capable of obtaining both images and Raman spectra spontaneously. Additionally, it is advantageous in providing Raman maps or three-dimensional dataset through a motorized xyz microscope [209, 210].

Luo's group reported the investigation of surface properties of rare-earth (RE = Sm, Gd, Pr and Tb) doped ceria using UV (324 nm) and visible (514, 633 and 785 nm) Raman spectroscopy. It was found that the observed oxygen vacancy, calculated from A560/A460, was affected by the detecting laser wavelength, the doping ions, and the homogeneity of RE doped ceria samples. The intense peak at $460\,cm^{-1}$ was attributed to the Raman-active vibrational mode (F_{2g}) of fluorite-type structure. Herein it was due to the symmetrical stretching vibration of oxygen atoms around cerium ions. The peaks around $570\,cm^{-1}$ were assigned to oxygen vacancies of ceria, which was either introduced by dopants ($546\,cm^{-1}$) or the intrinsic oxygen vacancies in ceria nanopowder ($600\,cm^{-1}$). Figure 2.38 shows all samples exhibit similar Raman spectra, with a broad oxygen vacancy band at $570\,cm^{-1}$ excited by 325 nm laser. Sm- and Gd- doped ceria sample showed strong F_{2g} band and two much weaker oxygen vacancy bands under visible laser excitation. Pr- and Tb- doped samples, however, demonstrated very different spectra- a weaker F_{2g} band and a stronger band at 570 cm^{-1} when visible laser was applied. The results indicated that the UV Raman disclosed the surface information of samples. The visible laser lines penetrated into the

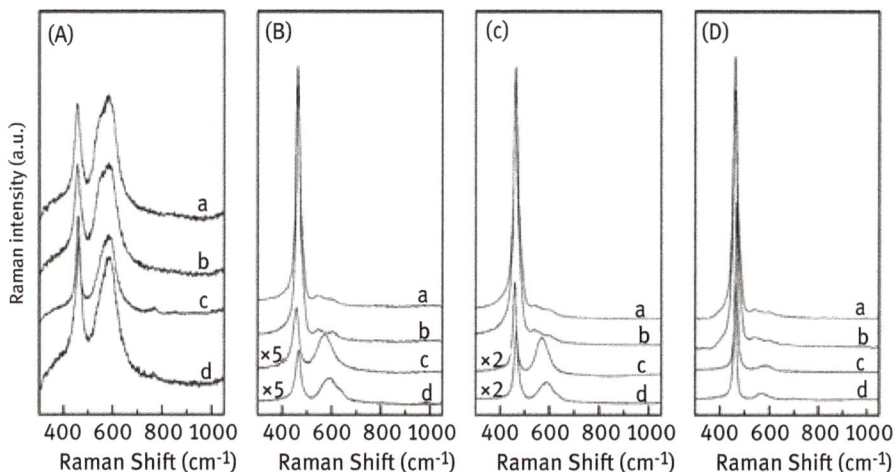

Figure 2.38: Raman spectra of (a) Sm- (b) Gd- (c) Pr- and (d) Tb-doped ceria nanopowder at (A) 325 nm (B) 514 nm, (C) 633 nm and (D) 785 nm excitation laser lines. (Reprinted with permission from Ref. [208]. Copyright (2011) American Chemical Society) [208].

inner layers of Sm- and Gd-doped ceria, while they could only investigate the outer surface structure of Pr- and Tb-doped samples, which was due to the strong absorption of visible light by these two samples [208].

Raman spectroscopy has also been combined with XRD to accurately determine the phase changes in materials. For instance, Yan's group showed that their mesoporous $Ce_{1-x}Zr_xO_2$ (x = 0.4–0.8) solid solutions all exhibited simple CeO_2 fluorite structures from the XRD patterns, and it was difficult to assert the structure by powder XRD diffraction. However, the Raman spectra revealed the presence of tetragonal phase and metastable phase as Ce/Zr ratio changed (see Figure 2.39). The six

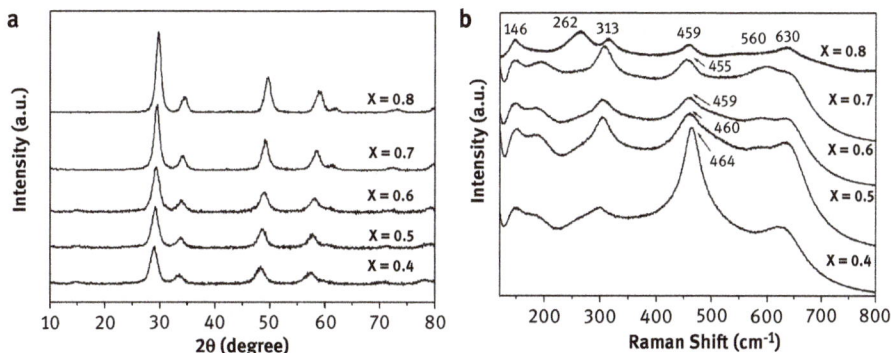

Figure 2.39: (a) PXRD patterns and (b) Raman spectra of mesostructured $Ce_{1-x}Zr_xO_2$ (x = 0.4– 0.8) calcined at 400 °C. (Reprinted with permission from Ref. [90]. Copyright (2007) American Chemical Society.) [90].

Raman-active modes of symmetry were observed for tetragonal ZrO_2, while for cubic CeO_2, only one F_{2g} mode centered at around $465\,cm^{-1}$ was Raman active. Combining the results of XRD and Raman data, it was concluded that the $Ce_{0.2}Zr_{0.8}O_2$ sample had a stable tetragonal structure, and for $x = 0.4$–0.7, the samples' structure were more corresponding to the metastable t'' phase [90].

2.3.8 Other characterization methods

The characterization methods in this chapter mainly focus on the extensively used microscopy and powerful spectroscopy. There are also a variety of other conventional and specific techniques employed in nanomaterials characterization, and their applications are largely dependent upon the physicochemical properties of materials, goals of experimentation, application of nanomaterials, reaction systems, ease of access, etc. Those techniques include traditional optical instruments like UV-Vis, IR, Fluorometer, DLS, etc., synchrotron-based techniques such as neutron scattering and X-ray fluorescence, thermal analysis methods such as thermal gravimetric analysis (TGA), physisorption and chemisorption, bioimaging techniques like magnetic resonance imaging (MRI) and fluorescence-lifetime imaging microscopy, electron paramagnetic resonance (EPR), and electrochemical methods of impedance spectroscopy, cyclic voltammetry and chronoamperometry, and many others.

Dynamic light scattering (DLS) is generally used to determine the size distribution of nanoparticles suspended in solution. It doesn't work for the nanopowder or nanoparticles without surfactant which are insoluble in solvents or cannot form a colloid or suspension. DLS has been exploited to study the interparticle interaction driven by surfactants as well. Yan's group showed a narrow size distributed Y_2O_3 nanoparticles of $14.0\,nm$ (S.D = 2.0 %) dispersed in cyclohexane, and a relatively broader size distribution of $17.7\,nm$ (S.D = 6.2 %) in cyclohexane/ethanol (v/v = 3:1). The polar solvent ethanol was revealed of compacting the oleic acid capping agent and resulting in self-aggregation of nanoparticles [43]. It has also been used to determine the degree of aggregation of nanoparticles by measuring the zeta potential. The higher the zeta potential, the lesser is the aggregation.

Synchrotron based X-ray fluorescence spectroscopy (SXRF) uses synchrotron radiation excitation source. XRF is used for determination of elemental composition of materials. An X-ray source is used to irradiate the sample and to cause the elements in the sample to emit (or fluoresce) their characteristic X-rays. A detector can measure the peaks of the emitted X-rays and determine the elements both qualitatively and quantitatively. SXRF permits multi-element determinations in the trace region with high precision compared with conventional XRF. It is becoming an advanced and essential technique in life and environmental sciences and nanomedicine applications. Micro-SXRF offers elemental imaging with high lateral spatial resolution and hence is the most attractive SXRF technique [211–213]. Hernandez-

Viezcas used *in situ* micro-SXRF mapping for detecting CeO_2 and ZnO nanoparticles in soybean plants. The experiments were performed at two beam lines. The micro-SXRF mapping conducted in ESRF using an incident beam with an energy of 5.8 KeV, a 0.60×1.1 μm^2 (V × H) focused beam, and a Si drift detector. The micro-SXRF experiment in SSRL were carried out with 0.5×0.5 μm^2 step size and 250 ms dwell time using a Si (111) monochromator for energy selective and Vortex (SII) detector for X-ray fluorescence detection. The results showed that CeO_2 nanoparticles in soil could be taken up by food crops, yet ZnO nanoparticles were not accumulated in the grains [15].

Thermal gravimetric analysis (TGA) is a thermal analysis method, and provides information about phase transitions, physical and chemical absorption and desorption, thermal decomposition, and solid-gas reactions [7, 82, 123]. The setup of this instrument is quite simple, typically consists of a precision balance with a sample pan located inside a furnace. The sample mass is monitored during heating. The thermal reaction can be conducted under various atmospheres, such as air, CO_2, H_2, H_2S, etc. TGA is usually coupled with differential thermal analysis (DTA) and/or differential scanning calorimetry (DSC) to yield more accurate information about thermal reaction of samples. Singh and co-workers conducted TGA and TGA/DTA experiments to study the catalytic activity of CeO_2, Pr_2O_3, and Nd_2O_3 nanoparticles on thermal decomposition of ammonium perchlorate and composite solid propellants. It was found that rare earth oxide catalysts increased mass loss of ammonium perchlorate but lowered the high-temperature decomposition. DTA/TGA data indicated a three-stage decomposition of ammonium perchlorate, and proved CeO_2 as the most effective oxide for this reaction [7].

Physisorption is a well-established technique for evaluating surface area, porosity, and pore structure of nanomaterials. It is specifically very important for the characterization of micro- and mesoporous nanomaterials [214]. N_2 physisorption analysis is the most commonly used one, a method which relies on physical adsorption of N_2 gas at liquid nitrogen temperature (77 K). By measuring the adsorbed amounts of N_2 as a function of a pressure ratio between absolute pressure and saturation pressure, it is able to elucidate the mechanism of pore filling and provides information of surface area, pore volume, and pore size. Physisorption is a reversible process, but the desorption mechanism might be different from adsorption. Physical models are used for the interpretation of isotherms of porosity and pore structure. Li et al. synthesized a series of high-surface-area mesoporous ceria microspheres and characterized them with N_2 adsorption–desorption isotherms. All samples were outgassed in vacuum at either 100 °C or 300 °C before measurement. The surface areas of samples were estimated by Brunauer–Emmett–Teller (BET) method, and Barrett–Joyner–Halanda (BJH) method was employed to analyze pore size distributions derived from the desorption branches of the isotherms. The total pore volume was estimated at a relative pressure of 0.985 [62].

In contrast with physisorption is chemisorption, which involves with chemical bond formation between adsorbate and sample surface. Chemisorption is largely used in heterogeneous catalysts analysis, and allows the investigation of surface area, active sites, degree of dispersion, and acid-base sites. The temperature programmed desorption (TPD), and reduction/oxidation (TPR/TPO) are more commonly used in metal oxide catalysis systems for the determinations of surface/bulk active sites, activation energy determination, reduction/oxidation degree of active sites. In the thermal analysis methods, the thermal change at a given temperature depends on many factors like the nature of the catalyst system, types of gas used, flow rates, and pressure, hence it is very important to pay attention to all these experimental aspects. The TPD/R/O experiments are carried out by flowing a suitable reactive gas or gas mixture through the samples placed in a tubular sample holder. Samples need to be cleaned and pre-treated before real measurement. In TPD, samples are previously saturated with the chosen adsorbates by flowing the reactive gas or executing a pulse chemisorptions analysis. In TPR and TPO, dilute gas of H_2 or O_2 (~5 %) in inert gas, with N_2 or Ar in TPR, and He for TPO, are generally used [47, 215].

2.4 Critical safety considerations

There is no doubt that rare earth oxide nanoparticles have a huge potential for new and emerging applications. Rare earth nanoparticles exhibit high availability and great applications as catalysts, superconductors, imaging agents, fertilizer additives, and feed additives. Notably, cerium oxide nanoparticles have been among the most highly used nanoparticles in industry. Their high oxygen storage capacities give rise to the catalysis excellency for internal combustion and oil-cracking processes [216]. Moreover, their applications in biomedical research for antioxidants and drug delivery have been exploited dramatically in the past decade [217, 218].

It is well known that nanoparticles are mostly toxic, and are classified from harmful to extremely toxic. Some of them are even more toxic than the notorious biocide pentachlorophenol which is banned in most countries [219]. Rare earth oxide nanoparticles are also believed to exhibit impact on health and environment safety, both short term and long term. Some groups have devoted efforts to investigate their toxicity on plants, animals and human health. Hernanzed-Viezas and co-workers performed studies of the effect of CeO_2 nanoparticles on soybean seeds from germination to full maturity. The organic farm soil was amended with CeO_2 NPs at 1000 mg/kg. CeO_2 NPs remained with the plant during the entire growth process with a small percentage of Ce (IV) biotransformed to Ce (III) [15]. Gao et al. studied the cytotoxicity of La_2O_3, Eu_2O_3, Dy_2O_3 and Yb_2O_3 nanoparticles from 2.5 to 80 µg/ml in macrophages. The high atomic number oxide Yb_2O_3 was non-toxic while the low atomic number oxides, La_2O_3, Eu_2O_3 and Dy_2O_3, induced 75.1 %,

53.6 % and 20.7 % dead cells, respectively. The rare earth nano-oxides induced cellular membrane permeabilization through phosphate sequestration and membrane damage through accumulating calcium by ~2.4 fold [220]. Zhang's group conducted root elongation experiments to research on the phytotoxicity of four rare earth oxide nanoparticles, CeO_2, La_2O_3, Gd_2O_3 and Yb_2O_3, on seven higher plant species including radish, rape, tomatoes, lettuce, wheat, cabbage, and cucumber. They concluded that the effects on root growth were strongly correlated with the nanoparticles, the plant species and plant growth process. The studied CeO_2 nanoparticles had negligible effect on the root elongation of all plants except lettuce. The 2000 mg/L suspensions of nano-La_2O_3, Gd_2O_3 and Yb_2O_3 drastically inhibited the root elongation of all investigated plants, but the extent of inhibition varied with plants and nanoparticles. In the studied nanoparticle concentrations from 20 to 2000 mg/L, the released RE^{3+} ion from the nanoparticles displayed insignificant effects on plants' root elongation [221].

Despite all the existing toxicity studies of rare earth oxide nanoparticles, due to its intrinsic complexity and strategy deficiency, the evaluation of impact on the environment and human health from rare earth nanoparticles is still limited, and some of the fields remain almost untouched. It is worth noting that due to the inconsistency in preparation, particle size, shape, and surface characteristic, nanoparticles used in different investigations usually perform distinctly and exhibit dissimilar safety impacts. Currently, most of the studies are performed after end-user applications or simplified conditions. For example, the ceria nanoparticles used as antioxidants are typically conducted in simple buffer solutions, and the investigations in biological media, cell, tissues, or even animals are still rare and immature. The *in situ* investigations on environment and human health impact are absent. Although it is necessary and critical to have such studies, numerous questions are extremely difficult to answer under real application conditions, especially for their biomedical applications. Questions of these include the evolution of nanoparticles along time, agglomeration, species change, and surface group modification, influences induced by concentration, temperature, and light exposure factors, and reactions occurring in different organs of biosystems. Moreover, valid models for risk assessment of nanoparticles used in different systems like water, soil, animal, and human are still in their infancy and need further improvement and development.

2.5 Conclusions and future perspectives

In the last decade, reports of synthesis approach and various applications of rare earth oxide nanoparticles have dramatically increased. These reports provide numerous synthetic routes including physical, chemical, biological, and hybrid methods to design nanoparticles with a variety of properties and complex functionalities. The rare earth nanoparticles displayed a wide range of monodisperse or well-defined crystalline size, sophisticated crystallite shape, compositions, and crystal structures. The well-controlled nanoparticles usually demonstrated good stability and high performance when employed in catalysis, optics, sensor, and biomedical applications.

In all reported methods, solution-based chemical methods generally gave access to better controlled nanoparticles, and were less costly and easier to manipulate than other routes. The aqueous methods such as hydrothermal and co-precipitation method are more straightforward, simpler, and usually used for mass-production. The resulting nanoparticles are more robust. The concerns of organic solvents disposal and their safety are absent. The non-aqueous solution-based methods like microemulsion synthesis are more advantageous in yielding highly crystalline nanostructures at moderate temperatures. The organic species are crucial in determining crystallization pathway and formation mechanism, and thus control the particle size, shape, dimension, and property of nanoparticles. Current synthesis is mostly dependent upon trial-and-error approaches due to the lack of understandings in fundamental mechanisms. Most studies focused on the size and shape influences induced by reaction parameters, i.e. metal precursors, temperature, reaction time, and solvents. A few of them investigated the reaction mechanism and proposed rational synthesis strategies. However, the rational systematic design is still challenging and yet it is strongly demanded in the near future. The challenge is even more severe in the synthesis of metal oxides with both a well-defined pore-solid architecture and highly crystalline walls. The design strategies of these porous oxides were more complicated and were strongly correlated with rare earth elements, precursors, and solvents. Rational systematic design of these nanostructures is largely limited by their complex structure and more complicated formation mechanism.

Although new synthetic methodologies are important for materials design, and most of the research efforts in past years are dedicated to the development of new synthesis routes, there is no doubt that efforts of in-depth investigation of nanoparticles' physical and chemical properties should be made and intensified. It is well known that nanoparticles exhibit size-dependent properties, thus obtaining knowledge regarding their properties would in turn provide clues for synthetic strategies. Furthermore, characterization of nanomaterials is a key step involved in implementing designed materials into technological applications. In the last decade, the characterization techniques were not limited to laboratory instruments. A larger number of structural characterizations have extended to synchrotron-based techniques. X-ray absorption, XPS, XRD, and neutron scattering become more and more popular in nanomaterials characterization. Some of the techniques are now used as routine techniques for studying the structural-property relationship and other fundamental principles. Synchrotron-based techniques provide higher resolution, greater precision, and more details about nanomaterials in much shorter experimental time than lab instruments. However, due to the extremely high energy flux and focused beam of synchrotron radiation, it is important to be aware that the synchrotron techniques damage samples frequently, especially for those with organic functional groups, and thus one has to choose the synchrotron techniques wisely, and be careful of the data interpretation.

To further gain insights into nanomaterials in reaction systems, it is essential to utilize *in situ* methods under real experimental conditions. *In situ* methods enable monitoring the changes of chemical species and evolution of particles during a reaction, and hence provide clues for crystallization process, reaction mechanism, and rational synthetic strategies. Another way for understanding the guiding principles of nanocrystals growth is the computational modeling. *In situ* methods are even more useful for nanomaterial application systems and become central for some fields, for instance, the catalysis field. The *in situ* methods can offer answers of active sites or species, structure–activity relationship, reaction kinetics, activation and deactivation mechanism.

Last but not least, it is significant to search for "greener" synthesis procedures and to work on the industrial up-scale of the laboratory methodologies. Despite a variety of unsolved issues and challenges, the environmental friendly and low energy routes gain much more attention and have been pursued due to the severe environment and energy crisis in recent years. Not only the synthesis, but also the disposal and treatment of nanomaterials after use should also be "greener". Their influences on water, soil, animal, and human health need better evaluation. For the large-scale synthesis and application of nanomaterials, it is far less developed compared with the laboratory scale. Issues regarding the scalability, reproducibility, and cost-effective production are detrimental and require significant improvement.

References

[1] Arole VM, Munde SV. Fabrication of nanomaterials by top-down and bottom-up approaches-an overview. J Adv Appl Sci Technol. 2014;1:89–93.

[2] Niederberger M, Pinna N. Metal oxide nanoparticles in organic solvents: synthesis, formation, assembly and application. London: Springer, 2009:1–5.

[3] Kumar KY, Muralidhara HB, Nayaka YA, Balasubramanyam J, Hanumanthappa H. Low-cost synthesis of metal oxide nanoparticles and their application in adsorption of commercial dye and heavy metal ion in aqueous solution. Powder Technol. 2013;246:125–36.

[4] Yan Z-G, Yan C-H. Controlled synthesis of rare earth nanostructures. J Mater Chem. 2008;18:5046–59.

[5] Gedanken A, Mastai Y. The chemistry of nanomaterials: synthesis, properties and applications. Weinheim, Germany: Wiley- Interscience, 2005:113–69.

[6] Niederberger M, Pinna N. Metal oxide nanoparticles in organic solvents: synthesis, formation, assembly and application. London: Springer; 2009:7–18.

[7] Singh S, Srivastava P, Kapoor IP, Singh G. Preparation, characterization, and catalytic activity of rare earth metal oxide nanoparticles. J Therm Anal Calorim. 2013;111:1073–82.

[8] Patzke GR, Ying Z, Roman K, Franziska C. Oxide nanomaterials: synthetic developments, mechanistic studies, and technological innovations. Angew Chem Int Ed. 2011;50:826–59.

[9] Williams DB, Carter CB. Transmission electron microscopy: a textbook for materials science. Boston, MA: Springer US; 2009:3–22.

[10] Bunker G. Introduction to XAFS: A practical guide to X-ray absorption fine structure spectroscopy. Cambridge: Cambridge University Press, 2010.

[11] Lytle FW. X-ray absorption spectroscopy. Berichte der Bunsengesellschaft für physikalische Chemie. 1987;91:1251–7.

[12] Yano J, Yachandra VK. X-ray absorption spectroscopy. Photosynth Res. 2009;102:241.

[13] Zhou Y, Lawrence NJ, Wang L, Kong L, Wu T-S, Liu J, et al. Resonant photoemission observations and DFT study of s–d hybridization in catalytically active gold clusters on ceria nanorods. Angew Chem Int Ed. 2013;52:6936–9.

[14] Crozier PA, Wang R, Sharma R. In situ environmental TEM studies of dynamic changes in cerium-based oxides nanoparticles during redox processes. Ultramicroscopy. 2008;108:1432–40.

[15] Hernandez-Viezcas JA, Castillo-Michel H, Andrews JC, Cotte M, Rico C, Peralta-Videa JR, et al. In situ synchrotron x-ray fluorescence mapping and speciation of CeO_2 and ZnO nanoparticles in soil cultivated soybean (glycine max). ACS Nano. 2013;7:1415–23.

[16] Li F-B, Newman RC, Thompson GE. In situ atomic force microscopy studies of electrodeposition mechanism of cerium oxide films: nucleation and growth out of a gel mass precursor. Electrochim Acta. 1997;42:2455–64.

[17] Liu Z, Duchoň T, Wang H, Grinter DC, Waluyo I, Zhou J, et al. Ambient pressure XPS and IRRAS investigation of ethanol steam reforming on Ni–ceO$_2$(111) catalysts: an in situ study of C–C and O–H bond scission. Phys Chem Chem Phys. 2016;18:16621–8.

[18] Sharma R. An environmental transmission electron microscope for in situ synthesis and characterization of nanomaterials. J Mater Res. 2005;20:1695–707.

[19] Wu T-S, Zhou Y, Sabirianov RF, Mei W-N, Soo Y-L, Cheung CL. X-ray absorption study of ceria nanorods promoting the disproportionation of hydrogen peroxide. Chem Commun. 2016;52:5003–6.

[20] Sato S, Takahashi R, Kobune M, Gotoh H. Basic properties of rare earth oxides. Appl Catal A: Gen. 2009;356:57–63.

[21] Gai S, Li C, Yang P, Lin J. Recent progress in rare earth micro/nanocrystals: soft chemical synthesis, luminescent properties, and biomedical applications. Chem Rev. 2014;114:2343–89.

[22] Trovarelli A, Fornasiero P. Catalysis by ceria and related materials. Vol. 12. 2nd ed. London: Imperial Colledge Press, 2013.

[23] Apostolov AT, Apostolova IN, Wesselinowa JM. Magnetic properties of rare earth doped SnO_2, TiO_2 and CeO_2 nanoparticles. Phys Status Solidi (B). 2018;255:1800179.

[24] Kumar V, Ntwaeaborwa OM, Soga T, Dutta V, Swart HC. Rare earth doped zinc oxide nanophosphor powder: a future material for solid state lighting and solar cells. ACS Photonics. 2017;4:2613–37.

[25] Alam U, Khan A, Ali D, Bahnemann D, Muneer M. Comparative photocatalytic activity of sol–gel derived rare earth metal (La, Nd, Sm and Dy)-doped ZnO photocatalysts for degradation of dyes. RSC Adv. 2018;8:17582–94.

[26] Liu G, Chen X. Handbook on the physics and chemistry of rare earths. vol. 37. Amsterdam: Elsevier, 2007:99–169.

[27] Esch F, Fabris S, Zhou L, Montini T, Africh C, Fornasiero P, et al. Electron localization determines defect formation on ceria substrates. Science. 2005;309:752–5.

[28] Lawrence NJ, Brewer JR, Wang L, Wu T-S, Wells-Kingsbury J, Ihrig MM, et al. Defect engineering in cubic cerium oxide nanostructures for catalytic oxidation. Nano Lett. 2011;11:2666–71.

[29] Yuan Q, Duan -H-H, Li -L-L, Sun L-D, Zhang Y-W, Yan C-H. Controlled synthesis and assembly of ceria-based nanomaterials. J Colloid Interface Sci. 2009;335:151–67.

[30]   Zhang D, Du X, Shi L, Gao R. Shape-controlled synthesis and catalytic application of ceria nanomaterials. Dalton Tran. 2012;41:14455–75.

[31] Bazzi R, Flores-Gonzalez MA, Louis C, Lebbou K, Dujardin C, Brenier A, et al. Synthesis and luminescent properties of sub-5-nm lanthanide oxides nanoparticles. J Lumin. 2003;102-103:445–50.

[32] Abdelaal HM. Facile hydrothermal fabrication of nano-oxide hollow spheres using monosaccharides as sacrificial templates. ChemistryOpen. 2015;4:72–5.

[33] Gao Y, Zhao Q, Fang Q, Xu Z. Facile fabrication and photoluminescence properties of rare-earth-doped Gd_2O_3 hollow spheres via a sacrificial template method. Dalton Tran. 2013;42:11082–91.

[34] Liu R, Wu K, Li L-D, Sun L-D, Yan C-H. Self-sacrificed two-dimensional $REO(CH_3COO)$ template-assisted synthesis of ultrathin rare earth oxide nanoplates. Inorg Chem Front. 2017;4:1182–6.

[35] Yada M, Mihara M, Mouri S, Kuroki M, Kijima T. Rare earth (Er, Tm, Yb, Lu) oxide nanotubes templated by dodecylsulfate assemblies. Adv Mater. 2002;14:309–13.

[36] Wu GS, Xie T, Yuan XY, Cheng BC, Zhang LD. An improved sol–gel template synthetic route to large-scale CeO_2 nanowires. Mater Res Bull. 2004;39:1023–8.

[37] Si R, Zhang Y-W, You L-P, Yan C-H. Rare-earth oxide nanopolyhedra, nanoplates, and nanodisks. Angew Chem Int Ed. 2005;44:3256–60.

[38] Panda AB, Glaspell G, El-Shall MS. Microwave synthesis and optical properties of uniform nanorods and nanoplates of rare earth oxides. J Phys Chem C. 2007;111:1861–1864.

[39] Wang D, Kang Y, Ye X, Murray CB. Mineralizer-assisted shape-control of rare earth oxide nanoplates. Chem Mater. 2014;26:6328–32.

[40] Bierman MJ, Van Heuvelen KM, Schmeißer D, Brunold TC, Jin S. Ferromagnetic semiconducting euo nanorods. Adv Mater. 2007;19:2677–81.

[41] Hamm CM, Alff L, Albert B. Synthesis of microcrystalline Ce_2O_3 and formation of solid solutions between cerium and lanthanum oxides. Zeitschrift für anorganische und allgemeine Chemie. 2014;640:1050–3.

[42] Goto A, Ohta Y, Kitayama M. Solid-state synthesis of metastable ytterbium (ii) oxide. J Mater Sci Cheml Eng. 2018;06:15.

[43] Si R, Zhang Y-W, Zhou H-P, Sun L-D, Yan C-H. Controlled-synthesis, self-assembly behavior, and surface-dependent optical properties of high-quality rare-earth oxide nanocrystals. Chem Mater. 2007;19:18–27.

[44] Reddy BM, Kumar TV, Durgasri N. Catalysis by ceria and related materials. London: Imperial College Press, 2013:397–464.

[45] Cao C-Y, Cui Z-M, Chen C-Q, Song W-G, Cai W. Ceria hollow nanospheres produced by a template-free microwave-assisted hydrothermal method for heavy metal ion removal and catalysis. J Phys Chem C. 2010;114:9865–70.

[46] Wang X, Li Y. Synthesis and characterization of lanthanide hydroxide single-crystal nanowires. Angew Chem. 2002;114:4984–7.

[47] Zhou Y, Lawrence NJ, Wu T-S, Liu J, Kent P, Soo Y-L, et al. Pd/CeO_{2-x} nanorod catalysts for CO oxidation: insights into the origin of their regenerative ability at room temperature. ChemCatChem. 2014;6:2937–46.

[48] Yang J, Quan Z, Kong D, Liu X, Lin J. Y_2O_3:eu^{3+} Microspheres: solvothermal synthesis and luminescence properties. Cryst Growth Des. 2007;7:730–5.

[49] Devaraju MK, Yin S, Sato T. A rapid hydrothermal synthesis of rare earth oxide activated Y $(OH)_3$ and Y_2O_3 nanotubes. Nanotechnology. 2009;20:305302.

[50] Xun W, Yadong L. Synthesis and characterization of lanthanide hydroxide single-crystal nanowires. Angew Chem Int Ed. 2002;41:4790–3.

[51] Xu A-W, Fang Y-P, You L-P, Liu H-Q. A simple method to synthesize $Dy(OH)_3$ and Dy_2O_3 nanotubes. J Am Chem Soc. 2003;125:1494–5.

[52] David C, Seon F United States Patent; Grant, Rhone-Poulenc CHimie: France 1996; vol. US5496528A.

[53] Mai H-X, Sun L-D, Zhang Y-W, Si R, Feng W, Zhang H-P, et al. Shape-selective synthesis and oxygen storage behavior of ceria nanopolyhedra, nanorods, and nanocubes. J Phys Chem B. 2005;109:24380–5.

[54] Tang CC, Bando Y, Liu BD, Golberg D. Cerium oxide nanotubes prepared from cerium hydroxide nanotubes. Adv Mater. 2005;17:3005–9.

[55] Yan L, Yu R, Chen J, Xing X. Template-free hydrothermal synthesis of CeO_2 nano-octahedrons and nanorods: investigation of the morphology evolution. Cryst Growth Des. 2008;8:1474–7.

[56] Wang X, Li L, Zhang YG, Wang S, Zhang Z, Fei L, et al. High-yield synthesis of NiO nanoplatelets and their excellent electrochemical performance. Cryst Growth Des. 2006;6:2163–5.

[57] Wu Q, Zhang F, Xiao P, Tao H, Wang X, Hu Z, et al. Great influence of anions for controllable synthesis of CeO_2 nanostructures: from nanorods to nanocubes. J Phys Chem C. 2008;112:17076–80.

[58] Zhang Y-W, Liu J-H, Si R, Yan Z-G, Yan C-H. Phase evolution, texture behavior, and surface chemistry of hydrothermally derived scandium (hydrous) oxide nanostructures. J Phys Chem B. 2005;109:18324–31.

[59] Aruna ST, Mukasyan AS. Combustion synthesis and nanomaterials. Curr Opin Solid State Mater Sci. 2008;12:44–50.

[60] Taekyung Y, Byungkwon L, Younan X. Aqueous-phase synthesis of single-crystal ceria nanosheets. Angew Chem Int Ed. 2010;49:4484–7.

[61] Zhong L-S, Hu J-S, Cao A-M, Liu Q, Song W-G, Wan L-J. 3D Flowerlike ceria micro/nanocomposite structure and its application for water treatment and CO removal. Chem Mater. 2007;19:1648–55.

[62] Li H, Lu G, Dai Q, Wang Y, Guo Y, Guo Y. Hierarchical organization and catalytic activity of high-surface-area mesoporous ceria microspheres prepared via hydrothermal routes. ACS Appl Mater Interfaces. 2010;2:838–46.

[63] Zhang Y, Zhang L, Deng J, Dai H, He H. Controlled synthesis, characterization, and morphology-dependent reducibility of ceria–zirconia–yttria solid solutions with nanorod-like, microspherical, microbowknot-like, and micro-octahedral shapes. Inorg Chem. 2009;48:2181–92.

[64] Yang Z, Han D, Ma D, Liang H, Liu L, Yang Y. Fabrication of monodisperse CeO_2 hollow spheres assembled by nano-octahedra. Cryst Growth Des. 2010;10:291–5.

[65] Wang X, Zhuang J, Peng Q, Li Y. Hydrothermal synthesis of rare-earth fluoride nanocrystals. Inorg Chem. 2006;45:6661–5.

[66] Chen G, Chen F, Liu X, Ma W, Luo H, Li J, et al. Hollow spherical rare-earth-doped yttrium oxysulfate: a novel structure for upconversion. Nano Res. 2014;7:1093–102.

[67] Ren X, Zhang P, Han Y, Yang X, Yang H. The studies of Gd_2O_3: Eu^{3+} hollow nanospheres with magnetic and luminescent properties. Mater Res Bull. 2015;72:280–5.

[68] Zhang D, Fu H, Shi L, Pan C, Li Q, Chu Y, et al. Synthesis of CeO_2 nanorods via ultrasonication assisted by polyethylene glycol. Inorg Chem. 2007;46:2446–51.

[69] Du Y, Zhang S, Wang J, Wu J, Dai H. Nb_2O_5 nanowires in-situ grown on carbon fiber: a high-efficiency material for the photocatalytic reduction of Cr(VI). J Environ Sci. 2018;66:358–67.

[70] Fu L, Liu ZM, Liu YQ, Han BX, Wang JQ, Hu PA, et al. Coating carbon nanotubes with rare earth oxide multiwalled nanotubes. Adv Mater. 2004;16:350–2.

[71] Strandwitz NC, Stucky GD. Hollow microporous cerium oxide spheres templated by colloidal silica. Chem Mater. 2009;21:4577–82.

[72] Guo Z, Jian F, Du F. A simple method to controlled synthesis of CeO_2 hollow microspheres. Scr Mater. 2009;61:48–51.

[73] Titirici -M-M, Antonietti M, Thomas A. A generalized synthesis of metal oxide hollow spheres using a hydrothermal approach. Chem Mater. 2006;18:3808–12.

[74] Yang S-C, Su W-N, Lin SD, Rick J, Hwang B-J. Preparation of highly dispersed catalytic Cu from rod-like $CuO–CeO_2$ mixed metal oxides: suitable for applications in high performance methanol steam reforming. Catal Sci Technol. 2012;2:807–12.

[75] Sun C, Li H, Chen L. Nanostructured ceria-based materials: synthesis, properties, and applications. Energy Environ Sci. 2012;5:8475–505.

[76] Demazeau G. Solvothermal reactions: an original route for the synthesis of novel materials. J Mater Sci. 2008;43:2104–14.

[77] Chen G, Xu C, Song X, Xu S, Ding Y, Sun S. Template-free synthesis of single-crystalline-like CeO_2 hollow nanocubes. Cryst Growth Des. 2008;8:4449–53.

[78] Chen G, Zhu F, Sun X, Sun S, Chen R. Benign synthesis of ceria hollow nanocrystals by a template-free method. CrystEngComm. 2011;13:2904–8.

[79] Chengyun W, Yitai Q, Xie Y, Changsui W, Yang L, Guiwen Z. A novel method to prepare nanocrystalline (7 nm) ceria. Mater Sci Eng: B. 1996;39:160–2.

[80] Tianshu Z, Hing P, Huang H, Kilner J. Ionic conductivity in the CeO_2-Gd_2O_3 system (0.05≤Gd/Ce≤0.4) prepared by oxalate coprecipitation. Solid State Ionics. 2002;148:567–73.

[81] Higashi K, Sonoda K, Ono H, Sameshima S, Hirata Y. Synthesis and sintering of rare-earth-doped ceria powder by the oxalate coprecipitation method. J Mater Res. 2011;14:957–67.

[82] Li J-G, Ikegami T, Mori T, Wada T. Reactive $Ce_{0.8}RE_{0.2}O_{1.9}$ (RE = La, Nd, Sm, Gd, Dy, Y, Ho, Er, and Yb) powders via carbonate coprecipitation. 1. Synthesis and characterization. Chem Mater. 2001;13:2913–20.

[83] Collins E, Voit SL, Vedder R. Evaluation of Coprecipitation Processes for the Synthesis of Mixed-Oxide Fuel Feedstock Materials, 2011.

[84] Danks AE, Hall SR, Schnepp Z. The evolution of 'sol–gel' chemistry as a technique for materials synthesis. Mater Horiz. 2016;3:91–112.

[85] Boettcher SW, Fan J, Tsung C-K, Shi Q, Stucky GD. Harnessing the sol–gel process for the assembly of non-silicate mesostructured oxide materials. Acc Chem Res. 2007;40:784–92.

[86] Hu J, Deng W, Chen D. Ceria hollow spheres as an adsorbent for efficient removal of acid dye. ACS Sustainable Chem Eng. 2017;5:3570–82.

[87] Aerogel market size worth $3.29 Billion by 2025|CAGR 22.6 %; grand view research: press room, 2018. https://www.grandviewresearch.com/press-release/global-aerogel-market

[88] Yang J, Lukashuk L, Li H, Föttinger K, Rupprechter G, Schubert U. High surface area ceria for CO oxidation prepared from cerium t-butoxide by combined sol–gel and solvothermal processing. Catal Lett. 2014;144:403–12.

[89] Hajizadeh-Oghaz M, Razavi RS, Barekat M, Naderi M, Malekzadeh S, Rezazadeh M. Synthesis and characterization of Y_2O_3 nanoparticles by sol–gel process for transparent ceramics applications. J Sol-gel Sci Technol. 2016;78:682–91.

[90] Yuan Q, Liu Q, Song W-G, Feng W, Pu W-L, Sun L-D, et al. Ordered mesoporous $Ce_{1-x}Zr_xO_2$ solid solutions with crystalline walls. J Am Chem Soc. 2007;129:6698–9.

[91] Patra A, Friend CS, Kapoor R, Prasad PN. Upconversion in Er^{3+}: ZrO_2 nanocrystals. J Phys Chem B. 2002;106:1909–12.

[92]   Patra A, Friend CS, Kapoor R, Prasad PN. Fluorescence upconversion properties of Er^{3+}-doped TiO_2 and $BaTiO_3$ nanocrystallites. Chem Mater. 2003;15:3650–5.

[93]   Saha S, Chowdhury PS, Patra A. Luminescence of Ce^{3+} in Y_2SiO_5 nanocrystals: role of crystal structure and crystal size. J Phys Chem B. 2005;109:2699–702.

[94]   Hussein GA. Rare earth metal oxides : formation, characterization and catalytic activity Thermoanalytical and applied pyrolysis review. J Anal Appl Pyrolysis. 1996;37:111–49.

[95]   Cao YC. Synthesis of square gadolinium-oxide nanoplates. J Am Chem Soc. 2004;126:7456–7.

[96]   Xiao X, Zhang DE, Zhang F, Gong JY, Zhang XB, Wang YH, et al. Synthesis of feather-like CeO_2 microstructures and enzymatic electrochemical catalysis for trichloroacetic acid. Funct Mater Lett. 2018;11:1850036.

[97]   Imagawa H, Sun S. Controlled synthesis of monodisperse CeO_2 nanoplates developed from assembled nanoparticles. J Phys Chem C. 2012;116:2761–5.

[98]   Mai H-X, Zhang Y-W, Si R, Yan Z-G, Sun L-D, You L-P, et al. High-quality sodium rare-earth fluoride nanocrystals: controlled synthesis and optical properties. J Am Chem Soc. 2006;128:6426–36.

[99]   Zhou J, Liu Z, Li F. Upconversion nanophosphors for small-animal imaging. Chem Soc Rev. 2012;41:1323–49.

[100] Hua M, Zhang S, Pan B, Zhang W, Lv L, Zhang Q. Heavy metal removal from water/wastewater by nanosized metal oxides: a review. J Hazard Mater. 2012;211-212:317–31.

[101] Li X, Zhang F, Zhao D. Lab on upconversion nanoparticles: optical properties and applications engineering via designed nanostructure. Chem Soc Rev. 2015;44:1346–78.

[102] Jeong J, Kim N, Kim M-G, Kim W. Generic synthetic route to monodisperse sub-10 nm lanthanide oxide nanodisks: a modified digestive ripening process. Chem Mater. 2016;28:172–9.

[103] Zhou Z, Hu R, Wang L, Sun C, Fu G, Gao J. Water bridge coordination on the metal-rich facets of Gd_2O_3 nanoplates confers high T1 relaxivity. Nanoscale. 2016;8:17887–94.

[104] Wang D, Kang Y, Doan-Nguyen V, Chen J, Küngas R, Wieder NL, et al. Synthesis and oxygen storage capacity of two-dimensional Ceria nanocrystals. Angew Chem Int Ed. 2011;50:4378–81.

[105] Jadhav KR, Shaikh IM, Ambade KW, Kadam VJ. Applications of microemulsion based drug delivery system. Curr Drug Deliv. 2006;3:267–73.

[106] Jha SK, Dey S, Karki R. Microemulsions- potential carrier for improved drug delivery. Asian J Biomedl Pharm Sci. 2011;1:5.

[107] Winsor PA. Hydrotropy, solubilisation and related emulsification processes. Trans Faraday Soc. 1948;44:376–98.

[108] Capek I. Preparation of metal nanoparticles in water-in-oil (w/o) microemulsions. Adv Colloid Interface Sci. 2004;110:49–74.

[109] Burda C, Chen X, Narayanan R, El-Sayed MA. Chemistry and properties of nanocrystals of different shapes. Chem Rev. 2005;105:1025–102.

[110] Zarur AJ, Ying JY. Reverse microemulsion synthesis of nanostructured complex oxides for catalytic combustion. Nature. 2000;403:65.

[111] Gröger H, Kind C, Leidinger P, Roming M, Feldmann C. Nanoscale sollow spheres: microemulsion-based synthesis, structural characterization and container-type functionality. Materials. 2010;3:4355–86.

[112] Solans C, García-Celma MJ. Surfactants for microemulsions. Curr Opin Colloid Interface Sci. 1997;2:464–71.

[113] Bumajdad A, Zaki MI, Eastoe J, Pasupulety L. Microemulsion-based synthesis of CeO_2 powders with high surface area and high-temperature stabilities. Langmuir. 2004;20:11223–33.

[114] Kockrick E, Schrage C, Grigas A, Geiger D, Kaskel S. Synthesis and catalytic properties of microemulsion-derived cerium oxide nanoparticles. J Solid State Chem. 2008;181:1614–1620.

[115] Li F-T, Ran J, Jaroniec M, Qiao SZ. Solution combustion synthesis of metal oxide nanomaterials for energy storage and conversion. Nanoscale. 2015;7:17590–610.

[116] Varma A, Mukasyan AS, Rogachev AS, Manukyan KV. Solution combustion synthesis of nanoscale materials. Chem Rev. 2016;116:14493–586.

[117] Bianchetti MF, Juárez RE, Lamas DG, de Reca NE, Pérez L, Cabanillas E. Synthesis of nanocrystalline CeO$_2$–Y$_2$O$_3$ powders by a nitrate–glycine gel-combustion process. J Mater Res. 2011;17:2185–8.

[118] Kang W, Ozgur DO, Varma A. Solution combustion synthesis of high surface area CeO$_2$ nanopowders for catalytic applications: reaction mechanism and properties. ACS Appl Nano Mater. 2018;1:675–85.

[119] Liu Q, Dong X, Xiao G, Zhao F, Chen F. A novel electrode material for symmetrical SOFCs. Adv Mater. 2010;22:5478–82.

[120] Mello PA, Barin JS, Guarnieri RA. Microwave-assisted sample preparation for trace element analysis. Amsterdam: Elsevier, 2014:59–75.

[121] Adam D. Out of the kitchen. Nature. 2003;421:571.

[122] Gabriel C, Gabriel S, Grant EH, Grant EH, Halstead BS, Michael P. Mingos D. Dielectric parameters relevant to microwave dielectric heating. Chem Soc Rev. 1998;27:213–24.

[123] Kumar E, Selvarajan P, Muthuraj D. Synthesis and characterization of CeO$_2$ nanocrystals by solvothermal route. Mater Res. 2013;16:269–76.

[124] Khachatourian AM, Golestani-Fard F, Sarpoolaky H, Vogt C, Vasileva E, Mensi M, et al. Microwave synthesis of Y$_2$O$_3$: Eu^{3+}nanophosphors: A study on the influence of dopant concentration and calcination temperature on structural and photoluminescence properties. J Lumin. 2016;169:1–8.

[125] Feldman D. Sonochemistry, theory, applications and uses of ultrasound in chemistry, by Timothy J. Mason and J. Phillip Lorimer, Wiley-Interscience, New York, 1989, 252, Journal of Polymer Science Part C: Polymer Letters 1989, 27, 537-537.

[126] Xu H, Zeiger BW, Suslick KS. Sonochemical synthesis of nanomaterials. Chem Soc Rev. 2013;42:2555–67.

[127] Sáez V, Mason T. Sonoelectrochemical synthesis of nanoparticles. Molecules. 2009;14:4284.

[128] Ohl CD, Kurz T, Geisler R, Lindau O, Lauterborn W. Bubble dynamics, shock waves and sonoluminescence. Philos Trans R Soc London. Ser A: Math Phys Eng Sci. 1999;357:269–94.

[129] Zhong H-X, Ma Y-L, Cao X-F, Chen X-T, Xue Z-L. Preparation and characterization of flowerlike Y$_2$(OH)$_5$NO$_3$•1.5H$_2$O and Y$_2$O$_3$ and their efficient removal of Cr(VI) from aqueous solution. J Phys Chem C. 2009;113:3461–6.

[130] Therese GH, Kamath PV. Electrochemical synthesis of metal oxides and hydroxides. Chem Mater. 2000;12:1195–204.

[131] Golden T, Shang Y, Qang Q, Zhou T. Electrochemical synthesis of rare earth ceramic oxide coatings. Eds. London: IntechOpen, 2015.

[132] Switzer J. Electrochemical synthesis of ceramic films and powders. American Ceramic Society Bulletin, American Ceramic Society, 1987;66.

[133] Zhou Y, Phillips RJ, Switzer JA. Electrochemical synthesis and sintering of nanocrystalline cerium(IV) oxide powders. J Am Ceram Soc. 1995;78:981–5.

[134] Aldykiewicz AJ, Davenport AJ, Isaacs HS. Studies of the formation of cerium-rich protective films using X-ray absorption near-edge spectroscopy and rotating disk electrode methods. J Electrochem Soc. 1996;143:147–54.

[135] Lu X-H, Huang X, Xie S-L, Zheng D-Z, Liu Z-Q, Liang C-L, et al. Facile electrochemical synthesis of single crystalline CeO$_2$ octahedrons and their optical properties. Langmuir. 2010;26:7569–73.

[136] Lu X, Zhai T, Cui H, Shi J, Xie S, Huang Y, et al. Redox cycles promoting photocatalytic hydrogen evolution of CeO$_2$ nanorods. J Mater Chem. 2011;21:5569–72.
[137] Lei C, Zhong-Hai L, Lei S, Jun B, Wen-Han L, Chen G. Ultrafine nano suspensions of rare earth oxides prepared by high-energy ball milling in pure water. Acta Phys Chim Sin. 2004;20:722–6.
[138] Salah N, Habib SS, Khan ZH, Memic A, Azam A, Alarfaj E, et al. High-energy ball milling technique for ZnO nanoparticles as antibacterial material. Int J Nanomedicine. 2011;6:863–9.
[139] Mooney JB, Radding SB. Spray pyrolysis processing. Annu Rev Mater Sci. 1982;12:81–101.
[140] Perednis D, Gauckler LJ. Thin film deposition using spray pyrolysis. J Electroceram. 2005;14:103–11.
[141] Hao J, Cocivera M. Optical and luminescent properties of undoped and rare-earth-doped Ga$_2$O$_3$ thin films deposited by spray pyrolysis. J Phys D: Appl Phys. 2002;35:433.
[142] Elidrissi B, Addou M, Regragui M, Monty C, Bougrine A, Kachouane A. Structural and optical properties of CeO$_2$ thin films prepared by spray pyrolysis. Thin Solid Films. 2000;379:23–7.
[143] Xu Y, Yan X-T. Chemical vapour deposition: an integrated engineering design for advanced materials. London: Springer, 2010:1–28.
[144] Creighton JR, Ho P. Chemical vapor deposition. Materials Park: ASM International, 2001.
[145] Jiang Y, Song H, Ma Q, Meng G. Deposition of Sm$_2$O$_3$ doped CeO$_2$ thin films from Ce(DPM)$_4$ and Sm(DPM)$_3$ (DPM=2,2,6,6-tetramethyl-3,5-heptanedionato) by aerosol-assisted metal–organic chemical vapor deposition. Thin Solid Films. 2006;510:88–94.
[146] Todokoro H, Ezumi M United States Patent; HItachi, Ltd.: US, 1996; vol. US005872358A:26.
[147] Goldstein JI, Newbury DE, Echlin P, Joy DC, Lyman CE, Lifshin E, et al. Scanning electron microscopy and X-ray microanalysis. 3rd ed. Boston, MA: Springer US, 2003:21–60.
[148] Takaya M, Shinohara Y, Serita F, Ono-Ogasawara M, Otaki N, Toya T, et al. Dissolution of functional materials and rare earth oxides into pseudo alveolar fluid. Ind Health. 2006;44:639–44.
[149] Menéndez CL, Zhou Y, Marin CM, Lawrence NJ, Coughlin EB, Cheung CL, et al. Preparation and characterization of Pt/Pt: CeO$_{2-x}$nanorod catalysts for short chain alcohol electrooxidation in alkaline media. RSC Adv. 2014;4:33489–96.
[150] Park E-J, Choi J, Park Y-K, Park K. Oxidative stress induced by cerium oxide nanoparticles in cultured BEAS-2B cells. Toxicology. 2008;245:90–100.
[151] Goldstein JI, Newbury DE, Echlin P, Joy DC, Lyman CE, Lifshin E, et al. Scanning electron microscopy and X-ray microanalysis. 3rd ed. Boston, MA: Springer US, 2003:61–98.
[152] Zanfoni N, Avril L, Imhoff L, Domenichini B, Bourgeois S. Direct liquid injection chemical vapor deposition of platinum doped cerium oxide thin films. Thin Solid Films. 2015;589:246–51.
[153] Goldstein JI, Newbury DE, Echlin P, Joy DC, Lyman CE, Lifshin E, et al. Scanning electron microscopy and X-ray microanalysis. 3rd ed. Boston, MA: Springer US, 2003:271–96.
[154] Goldstein JI, Newbury DE, Echlin P, Joy DC, Lyman CE, Lifshin E, et al. Scanning electron microscopy and X-ray microanalysis. 3rd ed. Boston, MA: Springer US, 2003:297–353.
[155] Silicon drift detector energy dispersive spectroscopy (SDD EDS/EDX) setup. 2018 http://sam.zeloof.xyz/eds/
[156] Men Y, Gnaser H, Zapf R, Hessel V, Ziegler C, Kolb G. Steam reforming of methanol over Cu/CeO$_2$/γ-Al$_2$O$_3$ catalysts in a microchannel reactor. Appl Catal A: Gen. 2004;277:83–90.
[157] Williams DB, Carter CB. Transmission electron microscopy: a textbook for materials science. Boston, MA: Springer US, 2009:371–88.
[158] Williams DB, Carter CB. Transmission electron microscopy: a textbook for materials science. Boston, MA: Springer US, 2009:389–405.

[159] Haigh SJ, Young NP, Sawada H, Takayanagi K, Kirkland AI. Imaging the active surfaces of cerium dioxide nanoparticles. ChemPhysChem. 2011;12:2397–9.

[160] Lin Y, Wu Z, Wen J, Poeppelmeier KR, Marks LD. Imaging the atomic surface structures of CeO$_2$ nanoparticles. Nano Lett. 2014;14:191–6.

[161] Liu X, Zhang C, Li Y, Niemantsverdriet JW, Wagner JB, Hansen TW. Environmental transmission electron microscopy (ETEM) studies of single iron nanoparticle carburization in synthesis gas. ACS Catal. 2017;7:4867–75.

[162] Wagner JB, Cavalca F, Damsgaard CD, Duchstein LDL, Hansen TW. Exploring the environmental transmission electron microscope. Micron. 2012;43:1169–75.

[163] Gao P, Kang Z, Fu W, Wang W, Bai X, Wang E. Electrically driven redox process in cerium oxides. J Am Chem Soc. 2010;132:4197–201.

[164] Hatsujiro H, Toshio N, Terukazu E, Kishio F. High temperature gas reaction specimen chamber for an electron microscope. Jpn J Appl Phys. 1968;7:946.

[165] Boyes ED, Gai PL. Environmental high resolution electron microscopy and applications to chemical science. Ultramicroscopy. 1997;67:219–32.

[166] Williams DB, Carter CB. Transmission electron microscopy: a textbook for materials science. Boston, MA: Springer US, 2009:511–32.

[167] Williams DB, Carter CB. Transmission electron microscopy: a textbook for materials science. Boston, MA: Springer US, 2009:141–71.

[168] Pennycook SJ, Jesson DE. High-resolution Z-contrast imaging of crystals. Ultramicroscopy. 1991;37:14–38.

[169] Howie A. Image contrast and localized signal selection techniques. J Microsc. 1979;117:11–23.

[170] Zhou Y, Menéndez CL, Guinel MJ, Needels EC, González-González I, Jackson DL, et al. Influence of nanostructured ceria support on platinum nanoparticles for methanol electrooxidation in alkaline media. RSC Adv. 2014;4:1270–5.

[171] Krumeich F, Müller E, Wepf RA, Nesper R. Characterization of catalysts in an aberration-corrected scanning transmission electron microscope. J Phys Chem C. 2011;115:1080–3.

[172] Egerton RF. Electron energy-loss spectroscopy in the electron microscope. Boston, MA: Springer US, 2011:1–28.

[173] Egerton RF. Electron energy-loss spectroscopy in the electron microscope. Boston, MA: Springer US, 2011:293–397.

[174] Williams DB, Carter CB. Transmission electron microscopy: a textbook for materials science. Boston, MA: Springer US, 2009:679–98.

[175] Mullins DR, Overbury SH, Huntley DR. Electron spectroscopy of single crystal and polycrystalline cerium oxide surfaces. Surf Sci. 1998;409:307–19.

[176] Hansma PK, Tersoff J. Scanning tunneling microscopy. J Appl Phys. 1987;61:R1–R24.

[177] Binnig G, Rohrer H. Scanning tunneling microscopy. Surf Sci. 1983;126:236–44.

[178] Chen CJ. Introduction to scanning tunneling microscopy. Oxford: Oxford University Press, 2007.

[179] Meyer E, Hug HJ, Bennewitz R. Scanning probe microscopy: the lab on a tip. Berlin, Heidelberg: Springer Berlin Heidelberg, 2004:15–44.

[180] Binnig G, Quate CF, Gerber C. Atomic force microscope. Phys Rev Lett. 1986;56:930–3.

[181] Meyer E, Hug HJ, Bennewitz R. Scanning probe microscopy: the lab on a tip. Berlin, Heidelberg: Springer Berlin Heidelberg, 2004:45–95.

[182] Fabris S, Vicario G, Balducci G, de Gironcoli S, Baroni S. Electronic and atomistic structures of clean and reduced ceria surfaces. J Phys Chem B. 2005;109:22860–7.

[183] Torbrügge S, Reichling M, Ishiyama A, Morita S, Custance Ó. Evidence of subsurface oxygen vacancy ordering on reduced CeO$_2$(111). Phys Rev Lett. 2007;99:056101.

[184] Epp J. Materials characterization using nondestructive evaluation (NDE) methods. Sawston: Woodhead Publishing, 2016:81–124.

[185] Bragg WH, Bragg WL. The reflection of X-rays by crystals. Proc R Soc London Ser A. 1913;88:428–38.

[186] Granqvist G. Award Memory Speech. 1915 https://www.nobelprize.org/prizes/physics/1915/ceremony-speech/

[187] He BB. Introduction to two-dimensional X-ray diffraction. Powder Diffr. 2003;18:71–85.

[188] International Centre for Diffraction Data. 2016 https://en.wikipedia.org/wiki/International_Centre_for_Diffraction_Data

[189] Inorganic Crystal Structure Database. https://cds.dl.ac.uk/cds/datasets/crys/icsd/llicsd.html

[190] Samsonov GV. The oxide handbook. Boston, MA: Springer US, 1973:9–35.

[191] Scherrer P. Bestimmung der Größe und der inneren Struktur von Kolloidteilchen mittels Röntgenstrahlen. Nachrichten von der Gesellschaft der Wissenschaften zu Göttingen, Mathematisch-Physikalische Klasse. 1918;1918:98–100.

[192] Langford JI, Wilson AJ. Scherrer after sixty years: a survey and some new results in the determination of crystallite size. J Appl Crystallogr. 1978;11:102–13.

[193] Bourja L, Bakiz B, Benlhachemi A, Ezahri M, Villain S, Gavarri JR. Synthesis and characterization of nanosized $Ce_{1-x}Bi_xO_{2-\delta}$ solid solutions for catalytic applications. J Taibah Univ Sci. 2010;4:1–8.

[194] Lussier JA, Souza DH, Whitfield PS, Bieringer M. Structural competition and reactivity of rare-earth oxide phases in $Y_xPr_{2-x}O_3$ (0.05 ≤ x ≤ 0.80). Inorg Chem. 2018;57:14106.

[195] Swartz WE. X-ray photoelectron spectroscopy. Anal Chem. 1973;45:788A-800a.

[196] Watts F. Surface science techniques. vol. 46. Oxford, U.K.: Elsevier, 1994:5–23.

[197] Alford TL, Feldman LC, Mayer JW. Fundamentals of nanoscale film analysis. Boston, MA: Springer US, 2007:199–213.

[198] Karslıoğlu O, Bluhm H. Operando research in heterogeneous catalysis. Frenken J, Groot I, editors. Cham: Springer International Publishing, 2017:31–57.

[199] Hollander JM, Jolly WL. X-ray photoelectron spectroscopy. Acc Chem Res. 1970;3:193–200.

[200] Haasch RT. Practical materials characterization. New York, NY: Springer, 2014:93–132.

[201] Hubin A, Terryn H. Comprehensive analytical chemistry. vol. 42. Amsterdam: Elsevier, 2004:277–312.

[202] Light Sources of the World. 2015 http://www.light2015.org/Home/LearnAboutLight/Lightsources-of-the-world.html

[203] Newville M. Fundamentals of XAFS. Chicago, 2004:V1.7. http://xafs.org/Tutorials?action=AttachFile&do= get&target=Newville_xas_fundamentals.pdf.

[204] IXAS wiki. http://www.ixasportal.net/wiki/TopPage

[205] XAFS. http://xafs.org/

[206] Dietzek B, Cialla D, Schmitt M, Popp J. Confocal Raman microscopy. Dieing T, Hollricher O, Toporski J, editors. Berlin, Heidelberg: Springer Berlin Heidelberg, 2011:21–42.

[207] Wachs IE. Raman and IR studies of surface metal oxide species on oxide supports: supported metal oxide catalysts. Catal Today. 1996;27:437–55.

[208] Guo M, Lu J, Wu Y, Wang Y, Luo M. UV and visible raman studies of oxygen vacancies in rare-earth-doped ceria. Langmuir. 2011;27:3872–7.

[209] Hollricher O. Confocal Raman microscopy. Berlin, Heidelberg: Springer, 2011:43–60.

[210] Raman spectroscopy- overview and product solutions for Raman spectroscopy. 2010 https://www.azooptics.com/Article.aspx?ArticleID=309

[211] Iida, A. Synchrotron Radiation X-Ray Fluorescence Spectrometry. In Meyers RA, editor. Encyclopedia of Analytical Chemistry, 2013. doi:10.1002/9780470027318.a9329.

[212] Knöchel A, Petersen W, Tolkiehn G. X-ray fluorescence spectrometry with synchrotron radiation. Anal Chim Acta. 1985;173:105–16.

[213] Zhu Y, Cai X, Li J, Zhong Z, Huang Q, Fan C. Synchrotron-based X-ray microscopic studies for bioeffects of nanomaterials. Nanomed: Nanotechnol Biol Med. 2014;10:515–24.

[214] Thommes M, Kaneko K, Neimark Alexander V, Olivier James P, Rodriguez-Reinoso F, Rouquerol J, et al. Physisorption of gases, with special reference to the evaluation of surface area and pore size distribution (IUPAC Technical Report). Pure Appl Chem. 2015;87:1051.

[215] Fadonia M, Lucarelli L. Temperature programmed desorption, reduction, oxidation and flow chemisorption for the characterisation of heterogeneous catalysts. Theoretical aspects, instrumentation and applications. Stud Surf Sci Catal. 1999;120A:177–225.

[216] Bozek F, Mares J, Bozek M, Huzlik J Proceedings of the 5th WSEAS International Conference on Waste Management, Water Pollution, Air Pollution, Indoor Climate Iasi, Romania, 2011:170–5.

[217] Xu C, Qu X. Cerium oxide nanoparticle: a remarkably versatile rare earth nanomaterial for biological applications. Npg Asia Mater. 2014;6:e90.

[218] Caputo F, Mameli M, Sienkiewicz A, Licoccia S, Stellacci F, Ghibelli L, et al. A novel synthetic approach of cerium oxide nanoparticles with improved biomedical activity. Sci Rep. 2017;7:4636.

[219] Kahru A, Dubourguier H-C. From ecotoxicology to nanoecotoxicology. Toxicology. 2010;269:105–19.

[220] Gao J, Li R, Wang F, Liu X, Zhang J, Hu L, et al. Determining the cytotoxicity of rare earth element nanoparticles in macrophages and the involvement of membrane damage. Environ Sci Technol. 2017;51:13938–48.

[221] Ma Y, Kuang L, He X, Bai W, Ding Y, Zhang Z, et al. Effects of rare earth oxide nanoparticles on root elongation of plants. Chemosphere. 2010;78:273–9.

Supplementary Material

The online version of this article offers supplementary material (DOI:https://doi.org/10.1515/psr-2018-0084).

Bionote

Yunyun Zhou was born in Qingdao, China in 1986. She received her bachelor degree in chemical engineering from the Ocean University of China in 2008, and obtained her Ph.D. in chemistry at the University of Nebraska-Lincoln under the supervision of Dr. Chin Li (Barry) Cheung in 2015. Her Ph.D. research focused on the investigation of rare earth oxides-based catalytic systems using spectroscopic, electron microscopic and electrochemical techniques. Zhou became an ORISE postdoctoral fellow after graduation and started to work at National Energy Technology Lab (NETL) since 2015 under the supervision of Dr. Christopher Matranga. Her postdoctoral work was studying iron-based catalyst for Fischer-Tropsch synthesis. Later, Zhou became a research scientist in 2016 and continued her Fischer-Tropsch research at NETL. She has also been involved in the DOE-sponsored oxygen separation and carbon material fabrication projects.

Jingfang Zhang, Yifu Yu and Bin Zhang

3 Synthesis and characterization of size controlled alloy nanoparticles

Abstract: Bimetallic and multimetallic alloy nanoparticles are emerging as a class of critical nanomaterials in electronic, optical and magnetic fields due to their unique physic-chemical properties. In particular, precise control of the nanoparticle size can endow them with broad versatility and high selectivity. This chapter reviews some tremendous achievements in the development of size controlled bimetallic and multimetallic alloy nanoparticles, with special emphasis on general preparation methods, characterization methodologies and instrumentation techniques. Some key factors and future perspectives on the development of size-controlled bimetallic and multimetallic alloy nanoparticles are also discussed.

Keywords: Bimetallic and multimetallic alloy nanoparticles, Controlled size, Chemical reduction, Thermal decomposition, Electrochemical deposition

3.1 Introduction

Over the past decades, metal nanoparticles hold a significant position in various application fields, and they are naturally attracting considerable interest of researchers. The earlier preparation of monometal nanoparticles can be traced back to 1857 when Faraday successfully prepared "finely divided" gold (Au), though the Au particle size was confirmed to be nanometer level (6 ± 2 nm) nearly a century later [1, 2]. Faraday's observation serves as a landmark for nanoscience and nanotechnology of Au nanoparticles. Nowadays, Au nanoparticles have been used widely in catalysis [3–5], sensors [6], detection [7] and so on. With the development of synthetic methodology and characterization technique, more and more bimetallic and multimetallic alloy nanoparticles have come out. Atomic mixing pattern of alloy nanoparticles plays an important role in their physical and chemical properties, which mostly depends on (1) relative strength of the bond, (2) surface energies of two different metals, (3) relative atomic sizes, (4) charge transfer, (5) strength of binding to stabilizing ligands, (6) specific electronic/magnetic effects [8]. According to mixing pattern between different metal elements, bimetallic alloy nanoparticles can be classified into four types (Figure 3.1) [8]: (1) core-shell segregated alloys. A kind of metal atom as a core is surrounded by a shell of another metal atom to form a core-shell segregated structure. This type exists in a great amount of alloy

This article has previously been published in the journal *Physical Sciences Reviews*. Please cite as: Zhang, J., Yu, Y., Zhang, B. Synthesis and characterization of size controlled alloy nanoparticles *Physical Sciences Reviews* [Online] **2020**, 5. DOI: 10.1515/psr-2018-0046.

https://doi.org/10.1515/9783110345001-003

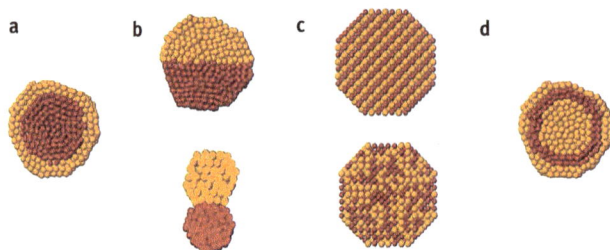

Figure 3.1: Schematic representation of four possible mixing patterns: core-shell (a), subcluster segregated (b), homogeneously mixed (c), multi-shell (d) alloys. The diagrams show cross sections of the clusters. Reproduced with permission from ref. 8. Copyright 2008, American Chemical Society.

nanoparticles. (2) subcluster segregated nanoalloys. The two metals share a pseudo-planar interface or are connected through a very small amount of metal bonds. (3) homogeneously mixed alloys. The mixing forms of two metals can be either in an atomically ordered or a statistically random manner. They are commonly referred to "ordered nanoalloys"and "random nanoalloys", respectively. (4) multi-shell nanoalloys. This type of structure is also vividly called "onion-like structure", consisting of more than one metal layer as shells surrounding by a core of metal atoms or mixed alloys. A number of multi-shell nanostructures have been reported, such as multishelled Pd-Pt [9], triple-layered core-shell Au-Pd-Pt [10, 11], Ag-Co-Ni [12] and Au-Co-Fe [13]. Generally, bimetallic and multimetallic alloy nanoparticles take an advantage over momometallic ones in achieving high-performance applications because of synergistic effect between/among different metals. For instance, our group fabricated Pt-Ni and Pt-Co alloy nanoparticle networks (NNs) with controlled composition, which exhibited markedly improved formic acid oxidation performance than monometallic Pt NNs and commercial Pt/C catalyst [14, 15]. Ternary FePtCu alloy nanorods show higher catalytic activity and durability for the oxygen reduction reaction (ORR) compared to binary FePt counterparts and commercial Pt catalyst [16]. The electronic structure of Pt can be modified by alloying other metals with Pt because of the electron transfer from introduced metals to Pt, leading to enhanced catalytic performance.

Size control of bimetallic and multimetallic alloy nanoparticles is a critical issue for synthetic chemistry, also necessary to understand the effect of nanoparticle size on their performance in heterogeneous catalysis. The precise control of nanoparticle size will provide practical possibility for modifying their properties and improving application performance. Taking electrocatalytic ORR as an example, Shao et al. reported a control synthesis of Pt nanoparticles in the range of 1 ~ 5 nm and investigated the size-activity relationship [17]. When the size increases from 1.3 to 2.2 nm, the specific activity improves greatly by 4-fold. The specific activity improves slowly as particle size increases from 2.2 to 5 nm. Further study proves that different sizes of nanoparticles have different oxygen binding energies on Pt sites, leading to different electrocatalytic activities.

Fabricating bimetallic and multimetallic alloy nanoparticles with controlled size has become a prolific area of research for a number of applications. However, the synthesis of size controlled alloy nanoparticles faces great challenges, and the main reason has the following points: (1) the redox potential difference between/among different metals results in inconsistent reduction rate. (2) the heat of mixing of different metals is another factor to determine whether alloy will be formed. (3) nanoscale metals tend to be low surface areas in order to minimize the surface free energy, making it difficult to regulate their sizes. In this chapter, we seek to overview the common preparation methods of size controlled alloy nanoparticles, including chemical reduction, thermal decomposition of precursor metal complexes as well as electrochemical deposition. We focus on discussing the diverse and vital experimental parameters and conditions (such as temperature, reaction time, capping and reducing agents) as well as different growth control mechanism in the synthesis of nanoscale sized bimetallic and multimetallic alloys. And we will demonstrate the available characterization methodologies and instrumentation techniques to study the morphology and structure information. Furthermore, we will discuss the prospective and challenges on the development of these size controlled bimetallic and multimetallic alloy nanoparticles. We hope this chapter can attract considerable interest from other intersectional fields and inject new vitality into nanoscience and nanotechnology.

3.2 Preparation methods

The preparation of bimetallic and multimetallic alloy nanoparticles has a variety of ways that relies on chemical reductive reagent, or molecular beams (such as laser, pulsed arc cluster ion source, ion/magnetron sputtering), or γ-ray irradiation and so on. Due to the increased synthesis complexity by the bimetallic and multimetallic elements as well as the dependence of final structure and properties of nanomaterials on their preparation methods, it is very necessary for researchers to select proper methods to synthesize the required nanostructures. In this section, we will in detail describe the preparation procedures, especially chemical synthesis, for different size controlled bimetallic and multimetallic alloy nanostructures.

3.2.1 Chemical reduction

Chemical reduction is one of the most popular methods for preparing various alloys, because this method is very simple and reproducible. This method usually employs chemical reducing agent (e. g. $NaBH_4$, N_2H_4, H_2 or ascorbic acid) to reduce the metal precursors (metal ions or metal complexes) in proper solvent, generating colloidal metallic nanoparticles with the presence of surfactant (e. g. citrate, alkylthiols, or thioethers) or polymeric ligands (e. g. polyvinyl pyrrolidone (PVP), polyvinyl alcohol (PVA), dendrimers) as the stabilizer of the nanoparticle surface [18, 19]. The size, structure and morphology of alloys can be controlled delicately by altering the reaction conditions, such as the amount and variety of stabilizer and reduction agent, concentration of metal ions, reaction temperature and time, and so forth. However, the use of some additives, such as stabilizer, can cause problems for subsequent removal of the ligands absorbed on the surface of alloy nanoparticles.

Turkevich proposed that the formation of colloidal nanoclusters need to experience nucleation, growth, and subsequent agglomeration mechanism [1]. The nucleation process plays an important role, which can not only influence the crystallographic growth of the produced nanoparticle, but also change the reaction kinetics of nanocrystal growth [20]. The nucleation process can be divided into homogeneous and heterogeneous nucleation. In homogeneous nucleation, different metal precursors are firstly mixed together and then the followed nucleation and growth proceeds by a same chemical process in a one-pot reaction. In heterogeneous nucleation, small particles (<1 nm in diameter) are considered as relatively stable in solution and can act as "seed" nucleus for further growth. The "seed" metal particle is prepared in advance to be added into the reaction mixture, thus the nucleation and growth are carried out separately. The two different nucleation types make co-reduction and successive reduction distinguishable (introduced in Sections 3.2.1.1 and 3.2.1.2). In addition, metal complexes can also be precursors to prepare metallic nanoparticles (introduced in Section 3.2.1.3). In some cases, the reduction of metal precursors needs other additional metal to produce the specific nanostructure (introduced in Section 3.2.1.4).

In the reduction process, nucleation rate is an important factor for size control of the resulted nanoparticles [21]. As shown in Figure 3.2, if the nucleation rate is fast, more particle seeds serve as nucleation centers, resulting in nanoparticles of smaller sizes. In contrast, if the nucleation rate is slow, less nucleation seeds are generated, resulting in the increased sizes of produced nanoparticles.

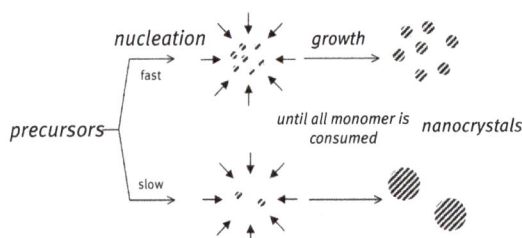

nucleation   growth

precursors   fast

                until all monomer is   nanocrystals
                consumed

               slow

Figure 3.2: Schematic representation of nanocrystal synthesis. Reproduced with permission from ref. 21. Copyright 2003, American Chemical Society.

3.2.1.1 Co-reduction

The co-reduction technique utilizes a one-pot reaction to reduce the mixture of metal salts in an appropriate solution. During the reduction process, there are usually two possible cases:

(1) If a mild reductant is employed, the metal salts with the highest redox potential are firstly reduced to small particle as a core, and the other metals deposited on the surface of core as a shell. In other words, the formation of a core-shell nanostructure benefits from the slow reduction kinetics of metal salts. For example, Han group reported a one-step synthesis of Au@Pd core-shell nanoparticle using a co-reduction method [22]. In a typical synthesis of Au@Pd core-shell bimetallic nanoparticles, 1 mL of a 5 mM aqueous solution of $HAuCl_4$/K_2PdCl_4 mixtures in molar ratios of 1:1 was added to 5 mL of 50 mM cetyltrimethylammonium chloride (CTAC). The whole system was sealed, heated, and maintained at 90 °C in a conventional forced-convection drying oven for 48 h. The resulting hydrosol was subjected to centrifugation (9000 rpm for 5 min) to remove excess CTAC. Figure 3.3(a) shows the scanning electron microscopy (SEM) image of the as-synthesized Au@Pd octahedral nanoparticles with an average edge length of 41.1 ± 3.5 nm. The transmission electron microscopy (TEM) image (inset in Figure 3.3a), high-angle annular dark-field scanning TEM (HAADF-STEM) image and STEM-EDS (energy dispersive spectroscopy) line spectra (Figure 3.3b), and elemental mapping images (Figure 3.3c) confirms the core-shell nanostructure of Au@Pd nanooctahedron. Because of the larger redox potential of Au(III) ($AuCl_4^-$/Au, + 1.002 V vs SHE (standard hydrogen electrode) than Pd(II) ($PdCl_4^{2-}$/Pd, + 0.591 V vs SHE), the Au core can be formed first and served as the a nucleic center for the growth of outer Pd layer, where CTAC with weak reducing power was used as the reducing agent in this experiment. Note that a suitable reducing agent is essential for the successful formation of Au@Pd core-shell nanostructure. If a stronger reductant ascorbic acid was used at 90 °C, homogeneously mixed Au-Pd alloy nanoparticles were produced instead of core-shell nanoparticles. In addition, the size of Au@Pd core-shell nanoparticles can be tuned by controlling the reaction times (Figure 3.3d). More recently, co-reduction of multimetal precursors has been used to produce multi-shell nanoparticles. Yamauchi group successfully prepared Au@Pd@Pt triple-layered core-shell structured nanoparticles via a very simple, one-step, and efficient route [10]. The preparation began by mixing 20 mM $HAuCl_4$ (2.5 mL), 20 mM Na_2PdCl_4 (2.5 mL), 20 mM K_2PtCl_4 (2.5 mL), and PVP (0.1 g) to prepare an aqueous precursor solution (9.0 mL). Then, ascorbic acid solution (0.4 M, 1 mL) was quickly added under sonication, giving the final $HAuCl_4$, Na_2PdCl_4, and K_2PtCl_4 precursor amounts of 0.05, 0.05, and 0.05 mmol, respectively. The mixture solution was placed for 6 h at room temperature after being sonicated for 5 min. After the reaction was finished, the obtained product was collected by centrifugation at 10,000 rpm for 20 min and then residual PVP was removed by consecutive washing/centrifugation cycles. As shown in Figure 3.4, the resulting nanoparticles were triple-layered nanostructures consisting of a Au core, a Pd inner layer, and a Pt outer shell.

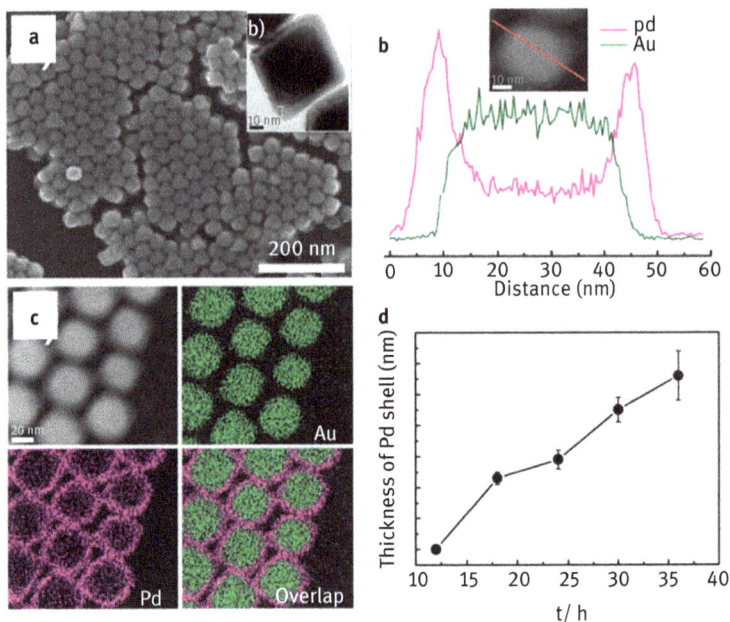

Figure 3.3: (a) SEM and TEM (inset) images of the Au@Pd nanooctahedra. (b) HAADF-STEM image and cross-sectional compositional line profiles of a Au@Pd nanooctahedron. (c) HAADF-STEM-EDS mapping images of the Au@Pd nanooctahedra. (d) Change in the Pd shell thickness during the reaction estimated from TEM analysis. Reproduced with permission from ref. 22. Copyright 2009, American Chemical Society.

(2) If a strong reductant is employed, the fast reduction rate of metal salts makes their nucleation and growth unseparated, allowing the production of alloy nanostructures. Xu group demonstrated the successful synthesis of Ni-Fe alloy nanoparticles using a co-reduction process [23]. Typically, 0.024 g of $NiCl_2 \cdot 6H_2O$ and 0.028 g of $FeSO_4 \cdot 7H_2O$ were dissolved in 2.5 mL of distilled water containing 0.100 g of hexadecyltrimethylammonium bromide by vigorous shaking, to which an aqueous solution of sodium borohydride (1.5 mL, 0.010 g) was added. The content of the flask is vigorously shaken to obtain the NiFe nanocatalyst as a black suspension. The obtained nanocatalysts were washed with water and ethanol and dried in vacuum. Sodium borohydride was used as strong reducing agent in an aqueous solution to reduce simultaneously Ni^{2+} and Fe^{2+} ions. Taking Ni-Ru system as another example, there is a large standard reduction potentials difference between the two elements, separate nucleation of Ni and Ru is more easily to occur unless a fast reduction process is conducted. Chen et al. presented the synthesis of Ni-Ru alloy nanoparticles in oleylamine at high temperature [24]. In a typical synthesis process,

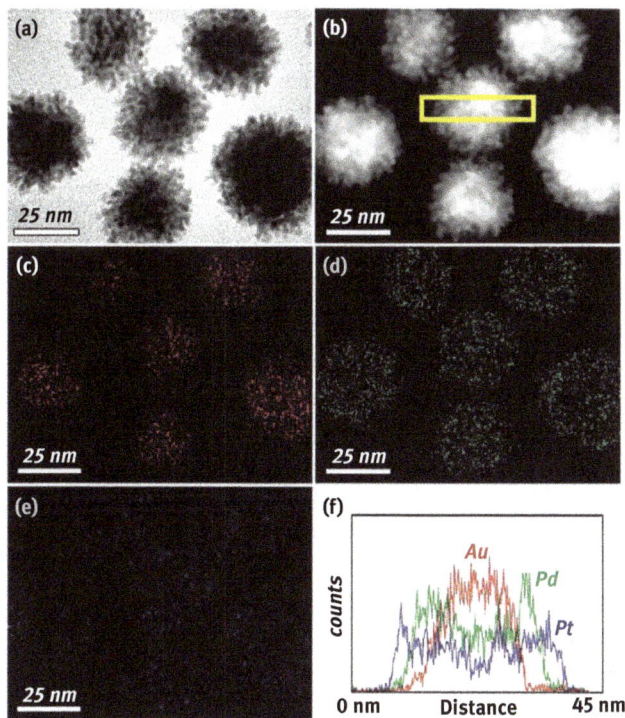

Figure 3.4: (a) Bright-field and (b) dark-field TEM images of trimetallic Au@Pd@Pt core-shell nanoparticles. Elemental mappings of (c) Au, (d) Pd, and (e) Pt, respectively. (f) Elemental distribution along a single nanoparticle indicated by the square in part b. Reproduced with permission from ref. 10. Copyright 2011, American Chemical Society.

nickel(II) acetylacetonate (64 mg) was dissolved in diphenyl ether (5 mL) containing oleic acid (0.75 mL). The solution was heated to about 120 °C and kept at this temperature for 20 min to remove humidity and oxygen. Next, the solution was cooled to about 90 °C, and ruthenium(III) acetylacetonate (33.2 mg) and Super-Hydride solution (1 mL, 1.0 m lithium triethylborohydride in tetrahydrofuran) was quickly injected into the solution. After 1 min, the as-prepared mixture was transferred into a flask containing oleylamine (15 mL), which had been preheated to about 300 °C. The mixture was kept at 250 °C for 15 min and then cooled to room temperature. Ethanol (60 mL) was added to precipitate the nanoparticles, and the product was collected by centrifugation at 9000 rpm for 10 min. The strong reducing agent superhydride, a preheated precursor solution and a high-boiling solvent in this study ensured the rapid reduction of Ni^{2+} and Ru^{3+} precursors and avoided the separate nucleation, resulting in the formation of Ni-Ru alloy nanoparticles.

In addition to the use of strong reducing agent, an effective strategy to prevent the separate nucleation of each metal is to modify the reactivity of the metal precursors [20]. One possible approach is to add correct stabilizer ligands in a proper reaction medium to form metal-surfactant or metal-polymer complexes and thus decrease the reduction rate of metal ions even in a favorable case [25, 26]. Han and co-workers reported one-pot aqueous synthetic method for preparing Au-Pd alloy nanocrystals by co-reduction of Au and Pd precursors with ascorbic acid in the presence of cetyltrimethylammonium chloride (CTAC) [27]. 1 mL of a 5 mM aqueous solution of $HAuCl_4/K_2PdCl_4$ (molar ratio 1/1) was added to an aqueous solution of CTAC (5 mL, 50 mM), and then an aqueous solution of ascorbic acid (100 mM, 50 mL) was added to this mixture. The resultant reaction solution was kept under ambient conditions for about 1 h. The resultant hydrosol was subjected to centrifugation (6000 rpm for 5 min, twice) to remove excess reagent. Au-Pd nanocrystals with an average edge length of 48 ± 7 nm were finally prepared. The alloy nanostructure was proved by HAADF-STEM-EDS images. Interestingly, the size of nanocrystals can be not only controlled by different reaction time, but also by varying the amount of CTAC. When increased twofold in the amount of CTAC with other conditions unchanged, Au-Pd nanocrystals with an edge length of 75 nm were obtained. This is mainly because that the larger amount of CTAC capped more Au and Pd atoms, leading to fewer nucleation seeds and thus larger nanocrystals.

Lou and co-workers reported the synthesis of intermetallic $PtCu_3$ nanocages by controlling the reduction rates of the two metals in the presence of the coordinating ligand cetyltrimethylammonium bromide (CTAB) [28]. In a typical synthesis, 131 mg of H_2PtCl_6 solution (8 wt%) and 20 mg of $Cu(acac)_2$ were added into a solution containing 8 mL of oleylamine and 50 mg of CTAB. After about 30 min of ultra-sonication, the solution was transferred into a 15 mL Teflon-lined autoclave. The autoclave was maintained at 170 °C for 24 h, and then cooled down to room temperature. The black precipitate was washed by centrifugation and re-dispersion into ethanol and toluene for several times before drying at 80 °C overnight. $PtCu_3$ nanocages with an edge size of 20 nm were finally prepared, as shown in Figure 3.5(a-b). The elemental mapping analysis (Figure 3.5(c-d)) shows the uniform distribution of both Pt and Cu elements. In addition to the reducing action, CTAB may change the reduction rates of Pt and Cu species, and Cu species was first reduced prior to Pt species despite more positive standard reduction potential for Pt^{II}/Pt pair (1.18 V) than that of Cu^{II}/Cu (0.34 V). Then, Pt ions reacted with reduced Cu via galvanic replacement reaction, leading to the formation of intermetallic $PtCu_3$ nanocages (Figure 3.5).

3.2.1.2 Successive reduction

Successive reduction method utilizes a pre-formed nanoparticle of one metal as preferential seed site, and other metal atoms can be deposited on the seed surface

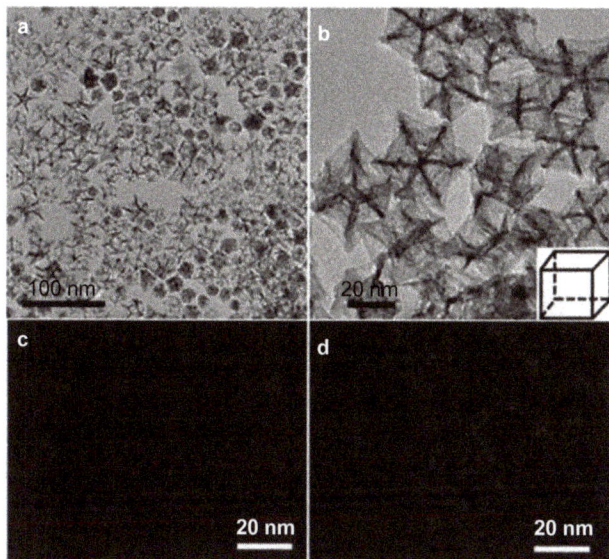

Figure 3.5: (a) Low magnification and (b) high magnification TEM images of PtCu$_3$ nanocages. Elemental mapping images for (c) Cu and (d) Pt. Reproduced with permission from ref. 28. Copyright 2012, American Chemical Society.

to grow into larger nanostructures. Therefore, this reaction process is performed in a separate synthetic step, detaching nanocrystal nucleation from growth. In Murphy's work, they proposed a two-step mechanism for seed-mediated nucleation in the chemical reduction of gold salts to gold nanoparticles: an slow nucleation at the beginning of the reaction, followed by a burst of nucleation mediated by seeds and growth (Figure 3.6) [29]. As the reaction progresses, the nucleation has experienced a course of fast to slow, and eventually replaced by the growth process. This method is usually used to prepare bimetallic and multimetallic nanocrystals with specific nanostructures, which are very difficult to prepare through traditional synthetic methods. For example, this method has been used to synthesize icosahedral Au-Pd core-shell nanocrystals [30]. Initially, monodisperse icosahedral gold nanocrystal seeds (7.1 ± 0.4 nm in size) were synthesized. The gold seeds were added to the reaction mixture containing hexadecylamine (0.12 g) and palladium acetylacetonate (15.2 mg) in toluene, and a total solvent volume of 1 mL. The bottle was then filled with 1 bar of hydrogen gas and placed in an oven at 60 °C for 2 h. The solution was then removed and the product washed twice by centrifugation at 14,000 rpm using an equal amount of methanol. The resulted Au-Pd nanocrystals are 11.4 ± 0.4 nm in size and maintain the icosahedral morphology of the seed. The small lattice mismatch (<5 %) between Pd and Au allows for epitaxial growth of Pd onto the Au seeds and the formation of Au-Pd core-shell nanostructures. Most importantly, this

1. Slow nucleation

2. Seed-medicated nucleation and growth

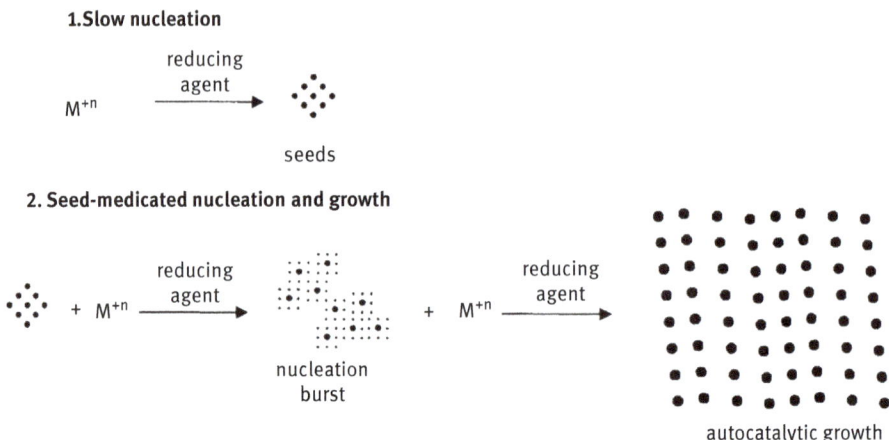

Figure 3.6: Generalized two-step mechanism for solution-phase M nanoparticle synthesis. Reproduced with permission from ref. 29. Copyright 2001, American Chemical Society.

method makes it simple to control the growth at different reaction times to obtain desired Pd shell thicknesses of 0.4, 0.8, 1.5, 2.2, and 3.2 nm after 30, 60, 90, 120, and 180 min, respectively.

In the successive reduction method, the activation energy of metal deposition onto a pre-formed particle is remarkably lower than that of homogeneous nucleation in solution [31]. Hence, the successive reduction method is able to generate highly precise and structural/compositional complex nanostructures (such as nanocrystals with high-index facets, or hollow structures) even in the condition of milder reducing agent, lower temperature, or aqueous media [32–35]. Skrabalak and coworkers represented this method to successfully synthesize various architecturally controlled Au-Pd bimetallic nanostructures [36]. The reaction used ascorbic acid as reducing agent at a lower reaction temperature (25 °C) in aqueous solution. 2 mL of either water or HCl acid solution was added to the entire Au seed solution. This procedure was followed by the simultaneous addition via separate pipettes of 2 mL of H_2PdCl_4 and 0.1 mL of $HAuCl_4$ solution. The vials were gently mixed by inversion followed promptly by the addition of 0.5 mL of ascorbic acid (0.1 M) solution. The reaction vial was capped and allowed to sit undisturbed in a 25 °C oil bath for 24 h. By careful manipulation of growth kinetics, concave, cuboctahedral, octopodal, and octahedral Au-Pd alloy nanocrystals were prepared (Figure 3.7).

3.2.1.3 Reduction of metal complexes
Complex containing all the metal species is another kind of metal precursor to prepare metallic nanoparticles. For example, silver(I) bis(oxalato)palladate(II) and

Figure 3.7: SEM images of various Au-Pd nanocrystals. Reproduced with permission from ref. 36. Copyright 2012, American Chemical Society.

silver(I) bis(oxalato)platinate(I) as precursors for Ag-Pd and Ag-Pt bimetallic nano-particles was pioneered by Esumi and co-workers [37, 38]. More recently, Tang and Chen demonstrated a kind of coordinate inorganic polymer, cyanogel, as metal precursors and synthesized Pt-Co bimetallic alloys by cyanogel-reduction method [39, 40]. In a typical synthesis, 2 mL of 50 mM K_2PtCl_4 and 1 mL of 50 mM $K_3Co(CN)_6$ aqueous solutions were added into a 10 mL Teflon-lined stainless-steel autoclave, and was then heated at 95 °C for 36 hours to generate the yellow $K_2PtCl_4/K_3Co(CN)_6$ cyanogel. After being cooled to room temperature, 5 mL of 0.5 M $NaBH_4$ solution was added into the yellow $K_2PtCl_4/K_3Co(CN)_6$ cyanogel and the resulting mixed solution left to stand for 1 h. After reaction, the black Pt-Co alloy were separated by centrifugation at 15,000 rpm for 10 min, washed three times with water and then

dried at 40 °C in a vacuum dryer for 12 hours. This method is simple to obtain bimetallic nanostructures with a high alloying degree, and can be scaled up readily in
the absence of surfactants and templates.

3.2.1.4 Specific metal assisted reduction

(a) Noble metal induced reduction. For some weak reducing agents, such as octade-
cylamine [41], noble metal ions (Pd^{2+}, Pt^{4+}, Au^{3+}, Ru^{3+}, Ir^{3+}, and Rh^{3+}) can be easily
reduced to metal nanoparticles. However, some transition metal oxides instead of
metal nanoparticles were obtained when transition metal ions were reduced by these
weak reducing agents [42]. In principle, reduction of transition metal ions by octadecylamine is thermodynamically unviable due to their lower reduction potentials.
Surprisingly, Li group found that the coexistence of noble metal (Au) and octadecylamine allows for the reduction of transition metal ions (Co^{2+} and Ni^{2+}) [43]. They proposed the noble metal induced reduction method to prepare bimetallic nanoalloys
comprising a noble and a transition metal, and a schematic representation of this
method is given in Figure 3.8(a). Firstly, Au^{3+} ion obtains electrons derived from octadecylamine to form Au atom. At the same time, the formed Au surface is surrounded
by electron cloud generated from free electrons of other Au atoms. The electron cloud
draws transition metal ions (M^{2+}) nearby and a part of electron cloud transfer to the
empty orbital of M^{2+}, leading to an elliptical electron cloud distribution. The contin-
ual electron supply from octadecylamine to Au makes the electron cloud shift from
Au to M continuously until M^{2+} is completely reduced. In a typical synthesis, 0.05 g
of $HAuCl_4$ and 0.25 g of $Co(NO_3)_2 \cdot 6H_2O$ (or 0.25 g of $Ni(NO_3)_2 \cdot 6H_2O$) were added into
10 mL of octadecylamine at 120 °C. Then, the system was heated to 200 °C. After 10
min of magnetically stirring, the products were collected and washed with ethanol
several times. Au-Co core-shell nanocrystals and Au-Ni spindly nanostructures were
obtained (Figure 3.8(b-c)). Moreover, this strategy can generally applied to prepare

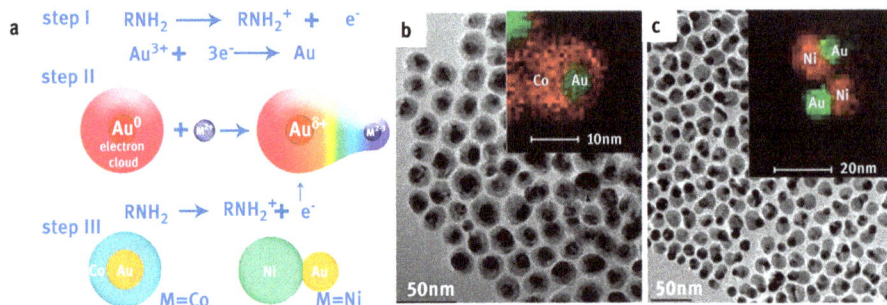

Figure 3.8: (a) Schematic illustration of the noble metal induced reduction method synthetic
strategy. TEM images of (b) Au-Co core-shell nanocrystals and (c) Au-Ni spindly nanostructures.
Reproduced with permission from ref. 43. Copyright 2010, American Chemical Society.

various bimetallic nanocrystals consisting of other noble (Pt, Pd, Rh, Ru, etc.) and transition (Fe, Co, Ni, Cu, Zn, etc.) metals.

(b) Underpotential deposition (UPD). UPD refers to the surface coverage of metal sub- and monolayer on a foreign metal substrate at potentials that can be more positive than that for deposition on the same metal surface [44–46]. The UPD phenomenon can occur on various metal systems, such as Cu, Ag, Pb, Hg on Au substrate, Cu, Ag on Pt substrate, Pb, Tl on Ag substrate and so on [44]. UPD can act as a bridge to realize simultaneous reduction of different metallic precursors. Zhang et al. reported a Cu^{2+}-assisted synthesis of Au-Pd alloy nanocrystals [47]. A typical synthesis method is shown as follows: an H_2PdCl_4 aqueous solution (1.0 mL, 1.0 mmol/L), a octadecyl trimethyl ammonium chloride glycol solution (3.0 mL, 0.10 mol/L) and an $Cu(CH_3COO)_2$ aqueous solution (0.05 mL, 1.0 mmol/L) were added into an aqueous $HAuCl_4$ solution (3.0 mL, 1.0 mmol/L) in order. After homogeneous mixing, a freshly prepared L-ascorbic acid aqueous solution (0.20 mL, 0.10 mol/L) was quickly added with a gentle shaking to this solution and left undisturbed for 12 hours at 15 °C. A representative TEM image indicated the successful synthesis of hexoctahedral Au-Pd nanocrystals with an average size of 55 nm (Figure 3.9a). Elemental mapping images and the crosssectional compositional line-scanning profile demonstrated the alloy structure of Au-Pd nanocrystals (Figure 3.9b). The control experiment confirmed that phase separation occurred in the absence of Cu^{2+} ions. In consequence, the UPD behavior greatly improves the alloying of Pd into a Au lattice.

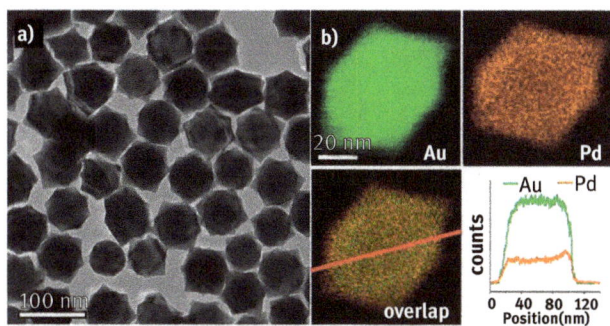

Figure 3.9: (a) TEM image of the as-prepared Au-Pd nanocrystals. (b) Elemental scanning images and the crosssectional compositional line-scanning profile of the AuPd alloy nanocrystals. Reproduced with permission from ref. 47. Copyright 2011, American Chemical Society.

Similar examples employing UPD for preparing bimetallic and multimetallic nanostructures can be found in other systems (such as Cu UPD for Pt-Cu alloy [48], Pt@Cu core-shell [49], decorated Pt@Au [50], Pt monolayer@Co@Pd core-shell [51], and Au@Pt core-shell [52, 53]).

(c) Galvanic Replacement. Galvanic replacement reaction is a redox process, in which a metal as a sacrificial template reacts with ions of a second metal and the second metal will be deposited onto the surface of the template. As such, the size of resulted nanostructures can be controlled by selecting template metal with proper size. A galvanic replacement reaction involves two half reactions: the anodic oxidation/dissolution of a metal, and the cathodic reduction/deposition of the ions of a second metal. The driving force for the galvanic replacement reaction is difference of the reduction potentials between the two metals, with the potential of the second metal being higher than that of the first metal [54, 55]. Galvanic replacement provides a versatile and effective method for generating metal nanostructures with controlled size, composition and shape. For example, Xia and co-workers demonstrated the morphological and structural evolutions at various stages of the replacement reaction between Ag nanocubes and $HAuCl_4$ (Figure 3.10a) [56]. In a typical synthesis, Ag nanocubes were firstly prepared. The SEM image in Figure 3.10(b) shows that Ag nanocubes have smooth surfaces. The as-obtained diluted dispersion of Ag nanocubes was refluxed for 10 min before a specific volume of 1 mM $HAuCl_4$ aqueous solution was added dropwise. The mixture was refluxed for another 20 min until its color became stable. Vigorous magnetic stirring was maintained in the entire process. The AgCl precipitate at the bottom of the container was removed by adding a saturated solution of NaCl. Then, the solid was rinsed with water and centrifuged six more times and finally dispersed with water. At the initial stage, Ag nanocubes react with $HAuCl_4$ to form Au atoms that are deposited on the surface of

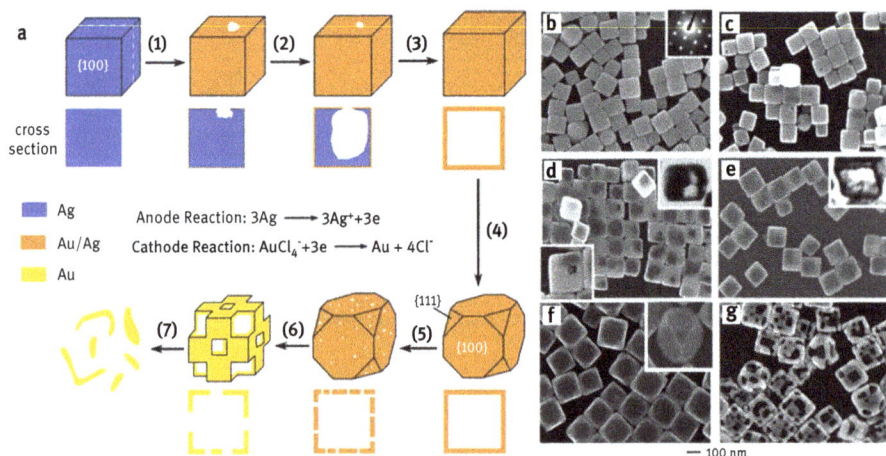

Figure 3.10: (a) Schematic illustration summarizing all morphological and structural changes involved in the galvanic replacement reaction between a Ag nanocube and $HAuCl_4$. (b) SEM image of Ag nanocubes. (c-g) SEM images at steps 1, 2, 3, 4 and 6 in galvanic replacement reaction of (a), respectively. Insets at the upper right-hand corners in (d) and (e) are corresponding TEM images. Reproduced with permission from ref. 56. Copyright 2004, American Chemical Society.

each cube. Meanwhile, Ag atoms also diffuse into the Au to generate a very thin shell of Au-Ag alloys. Because of the Kirkendall effect caused by the different diffusion rate of Ag and Au atoms, a small hole is observed on the surface of each cube (Figure 3.10c). As the reaction time goes on, larger and larger holes appear on the surface and inside (Figure 3.10d). As a result, cubic Au-Ag alloy nanoboxes with uniform walls are generated (Figure 3.10e). Afterwards, dealloying process occur on cubic Au-Ag alloy nanoboxes. Selective removal of Ag atoms from the Au-Ag alloy wall lead to the morphological reconstruction from cubic Au-Ag nanoboxes to truncated Au-Ag nanoboxes (Figure 3.10f), to Au nanoboxes with large pores in the walls (Figure 3.10g). More importantly, this replacement reaction between Ag and $HAuCl_4$ can be extended to prepare nanotubes and nanocages with larger pores from Ag nanowires and Ag nanospheres. Based on the above mechanism, the composition and shape of resulted nanostructures can be controlled by adjusting the reaction time or the amount of ions of a second metal.

3.2.2 Thermal decomposition of metal complex precursors

Thermal decomposition has been considered as a common method to prepare very small metallic alloy nanoparticles from target metal complexes (usually organometallic compounds and clusters) [57–62]. But this synthesis process is tedious, which usually needs the protection of inert gas, higher temperature and longer reaction time.

A pioneering work by Thomas and co-workers reported the synthesis of a series of Ru-based bimetallic (Ru_6Pd_6, Ru_6Sn, $Ru_{10}Pt_2$, Ru_5Pt, $Ru_{12}Cu_4$, and $Ru_{12}Ag_4$) nanoparticles through thermolysis of their organometallic clusters with carbonyl and phosphine ligands at approximately 200 °C in vacuum for 2 h [59]. Sun et al. demonstrated the synthesis of monodisperse FePt nanoparticles by reduction of platinum acetylacetonate and decomposition of iron pentacarbonyl, and a schematic diagram for preparing FePt nanoalloys is shown in Figure 3.11 [63, 64]. The detailed synthetic procedure is as follows: under airless condition, platinum acetylacetonate (197 mg, 0.5 mmol), 1,2-hexadecanediol (390 mg, 1.5 mmol), and dioctylether (20 mL) were mixed and heated to 100 °C. Oleic acid (0.16 mL, 0.5 mmol), oleylamine (0.17 mL, 0.5 mmol), and $Fe(CO)_5$ (0.13 mL, 1 mmol) were added, and the mixture was heated to reflux (297 °C). The refluxing was continued for 30 min. The heat source was then removed, and the reaction mixture was allowed to cool to room temperature. The inert gas protected system could be opened to ambient environment at this point. The black product was precipitated by adding ethanol (about 40 mL) and separated by centrifugation. The yellow-brown supernatant is discarded. The black precipitate is dispersed in hexane (25 mL) in the presence of oleic acid (0.05 mL) and oleylamine (0.05 mL) and precipitated out by adding ethanol (20 mL) and centrifuging. The product is dispersed in hexane (20 mL), centrifuged to

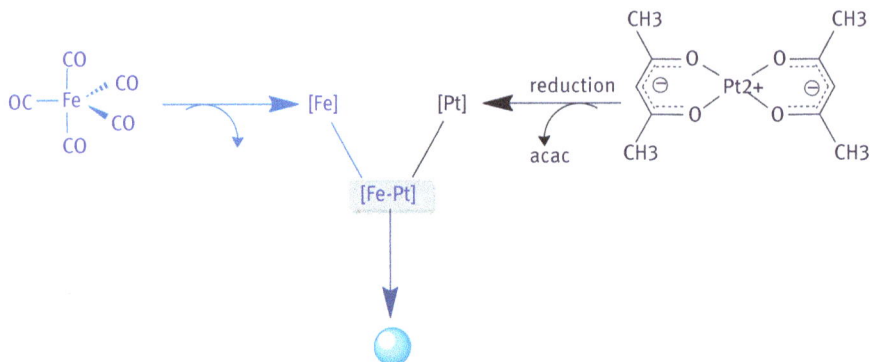

Figure 3.11: Schematic illustration of the formation of FePt nanoparticle from the decomposition of Fe(CO)$_5$ and reduction of Pt(acac)$_2$.

remove any insoluble precipitate (almost no precipitation was found at this stage), and precipitated out by adding ethanol (15 mL) and centrifuging. The FePt nanoparticles were redispersed in hexane and stored under N$_2$. After ligand exchange with hexanoic acid/hexylamine, a cubic packed multilayer of 6-nm Fe$_{50}$Pt$_{50}$ particles were obtained. In addition, the size of FePt particle can be tunable from 3 to 10 nm by first in situ growing 3-nm monodisperse seed particles and then adding more reagents to enlarge the existing seeds to the desired size.

The key to obtain alloy structure via this method is to ensure the simultaneous decomposition of the two metal complex precursors, otherwise irregular nanostructures would be generated through "heterogeneous" decomposition. For example, during the synthesis of Fe-Co nanoparticles, due to the different decomposition temperature of Fe(CO)$_5$ (200 °C) and Co$_2$(CO)$_8$ (150 °C), separate Fe and Co monometallic clusters formed. When aluminium trialkyl was added as a catalyst, the temperature of Fe(CO)$_5$ can reduced to around 150 °C, which allows for the formation of Fe-Co alloys [65].

3.2.3 Electrochemical deposition

Electrodeposition is an elelctrochemical precipitation method through reducing metal precursor ions from an electrolyte solution at low temperature [66]. With the simple, fast, high yield and binder-free advantages, electrodeposition has been considered as a powerful and versatile method to generate bimetallic and multimetallic nanoparticles in solution based on a two- or three-electrode electrochemical system [8, 67]. The electrodeposition behavior can be controlled by deposition conditions (the applied potential, current density and time duration on the electrode) and bath chemistry (precursor solutions and electrode materials), realizing the regulation of morphology, particle size, and composition of the resulted nanomaterials [68–72]. Nanomaterials with complex nanostructures can also be obtained when suitable templates or additional reagents were used [73–75]. Of note, because the deposition process relies on the electric drive, a fluctuation in the voltage and current may lead to the inhomogeneity of nanoparticle film.

A variety of electrochemical techniques have been employed to deposit nano-materials onto electrode, mainly including:

(a) Chronoamperometry. Xu et al. reported the electrodeposition of AuPdCu alloy nanoparticles on a multiwalled carbon nanotube (MWCNT) film coated glassy carbon electrode (GCE) [76]. Prior to electrodeposition, MWCNT was modified on GCE. AuPdCu nanoparticles were electrodeposited on the MWCNT/GCE from the aqueous solution containing 1.5 mM $HAuCl_4$, 0.5 mM $PdCl_2$, 1.0 mM $CuSO_4$ and 0.2 M Na_2SO_4. The electrodeposition potential was set at -0.2 V (vs SCE) and the electrodeposition time was 200 s. The resulting AuPdCu nanoparticles were washed carefully with redistilled water and then dried at room temperature. The AuPdCu alloy nanoparticles with diameters of about 80 nm were well dispersed on the surface of modified GCE.

(b) Cyclic voltammetry. Quan et al. deposited Pt-Pd alloys onto carbon papers using cyclic voltammetry electrodeposition method [77]. The electrolyte was prepared by dissolving chloroplatinic acid ($H_2PtCl_6 \bullet 6H_2O$, 0.5 mM) and disodium tetrachloropalladate (Na_2PdCl_4, 0.5 mM) in 0.5 M H_2SO_4 solution, and then added by sodium dodecylbenzenesulfonate ($C_{18}H_{29}NaO_3S$, 1 mM). A carbon paper, a titanium sheet, and Ag/AgCl electrode were used as working electrode, counter electrode, and reference electrode, respectively. Cyclic voltammetry was performed between −0.4 V and 0.4 V at a scan rate of 50 mV/s under N_2 atmosphere in the electrolyte solution. The obtained electrodes were washed with deionized water and dried at room temperature. The SEM results showed that multiwall carbon nanotubes (MWCNTs) modified electrode can reduce the average Pt-Pd size from 210 nm to 90 nm and improve the dispersity of particle.

(c) Potential step experiments. Quan et al. deposited Pt-Pb alloy nanoparticles on MWCNTs [78]. The electrodeposition was carried out from a 0.1 M KCl solution containing 0.67 mM K_2PtCl_6 and 0.33 mM $Pb(NO_3)_2$ by applying 60–120 of potential steps (stepped from + 0.5 to -0.4 V, step width: 10 s). After electrodeposition of Pt-Pb alloy nanoparticles, the MWCNT electrode was gently washed in water and air-dried at room temperature. Pt-Pb alloy nanoparticles with a diameter of 10–40 nm are sparsely dispersed on the sidewalls of MWCNTs.

(d) Potential pulse experiments. Xiao et al. fabricated PtRuNi ternary alloy nanoparticles on MWNTs using a potential pulse technique [79]. The pulsed electrodeposition was performed in 0.2 M H_2SO_4 aqueous solution containing 1.0 mM H_2PtCl_6 + 1.0 mM $RuCl_3$ + 1.0 mM $NiSO_4$. Prior to electrochemical deposition, the solution was deoxygenated with nitrogen gas. The potential range was 0.9 to 0 V, the pulse width was 0.25 s and the number of steps was 300. The obtained PtRuNi electrode was washed carefully with redistilled water and then dried at room temperature. The PtRuNi nanoclusters on the surface of MWNTs have an average diameter of 80 nm. And the particle diameter can become much smaller in the presence of ionic liquid.

(e) Chronopotentiometry. Ueda et al. deposited Co-Cu alloys on Cu substrate using the chronopotentiometry technique [66]. Prior to electrodeposition, the electrolyte was prepared by mixing 28–39 g of $CoSO_4 \bullet 7H_2O$, 15–25 g of $CuSO_4 \bullet 5H_2O$, 76 g of $Na_3C_6H_5O_7$ and 2 g of NaCl in 1 L H_2O. Electrodeposition was performed with a current density of 2 mA/cm^2 in electrolyte solution at a pH value of 6.

(f) Current pulse experiments. Muntean et al. deposited Pt-Co alloy on carbon nanofibers using a current pulse technique [80]. The Pt-Co electrodes are obtained applying a current density of 100 mA cm^{-2}, 20 ms for t_{on}, 100 ms for t_{off}, during 5000 cycles from the deposition solution containing the K_2PtCl_4 (5 mM) and a proper amount of $CoCl_2$. The geometric surface of the working electrode was 1 cm^2. The particle size and chemical composition of Pt-Co alloy can be easily controlled by adjusting the deposition parameters as t_{on}, t_{off}, as well as the current density i.

3.3 Characterization methodologies and instrumentation techniques

3.3.1 Microscopy

For the as-prepared bimetallic and multimetallic nanoparticles, it is very important to study their morphology, size, composition, and crystalline degree. Microscopy provides many channels for researchers to access to these information. For the fine structure analysis, the use of single particle spectroscopy is necessary [81].

3.3.1.1 Electron microscopy

Electron microscope is a useful tool to investigate nanomaterials by using a beam of accelerated electrons as a source of illumination.

Scanning electron microscopy (SEM) is a technique by scanning the surface of sample with the electron beam to produce various signals that contain the surface information about morphology and size. Samples are generally required to be dispersed on a specimen holder or a conductive support and are observed in high vacuum in conventional SEM. The resolution of an SEM image is limited by the instrument's electron-optical performance, the contrast produced by the specimen/detector system, and the sampling volume of the signal within the specimen [82]. Although the resolution of the SEM is not high enough to image individual atoms because of the relative large electron spot and sampling volume compared to the distance between atoms, SEM has a greater depth of view and can provide a comparatively large area image of the specimen.

Transmission electron microscopy (TEM) is a technique in which a beam of electrons pass through a sample to form an image associated with morphological and structural information as well as size of the sample. Samples are needed to be dispersed on a specimen grid by the deposition of a dilute sample solution and then placed into the specimen holder to get TEM images under vacuum. The resolution of TEM can be accurate to Ångstrom level and enables the acquisition of structural information of sample by high-resolution TEM (HRTEM) [8]. From the HRTEM image, we will get a basic overview of crystalline degree of samples.

Scanning transmission electron microscopy (STEM) is a mode of TEM in which the electron beam is scanned over the sample. This technique is always combined with high-angle annular dark field (HAADF) or "Z-contrast" imaging technique to reveal the internal structure at atomic resolution. Because the atomic contrast is directly related to the atomic number (Z-contrast image), HAADF-STEM image can provide the images of crystalline materials with strong compositional sensitivity [83].

3.3.1.2 Scanning probe microscopy
Scanning probe microscopy (SPM) is a branch of microscopy in which a physical probe scans the specimen to form an image of surface.

Atomic force microscopy (AFM) is a very-high-resolution type of SPM, by getting close to the surface of sample using a mechanical probe. The topographic map of the sample surface is produced by raster scanning the position of the sample and recording the height of the probe that corresponds to interactions between the probe and sample [84].

Scanning tunneling microscopy (STM) is used for imaging surfaces at the atomic level. A voltage bias is applied when the conducting tip is brought close to the sample, and a tunneling current is generated [85]. Current vs voltage (I-V) curves are recorded by measuring tunneling current as a function of applied voltage. In the constant current and height mode, we can obtain the information about the electronic structure and topographic images.

3.3.2 X-ray spectroscopy

3.3.2.1 X-ray photoelectron spectroscopy
X-ray photoelectron spectroscopy (XPS) is a surface-sensitive quantitative spectroscopic technique by irradiating a material with a beam of X-rays under high vacuum, which is concerned with the measurement of core-electron binding energies. The atom/molecule on the surface of sample absorbs X-rays with sufficient energy

to cause the ejection of electrons, and the kinetic energies of these photoelectrons are measured by an electron analyzer [86]. A typical XPS spectrum is a plot of the number of electrons versus the binding energy of the electrons. The electron binding energy values are corresponding to characteristic XPS peaks of each element. XPS can be used to analyze the surface chemistry of a material, such as elemental composition (top 0–10 nm usually) and chemical or electronic state as well as local bonding information [87–89].

3.3.2.2 X-ray absorption spectroscopy

X-ray absorption spectroscopy (XAS) is widely used to determine the local geometric/electronic structure of a material by synchrotron radiation facilities. XAS is a type of absorption spectroscopy that electronic transitions from a core states of the metal to the excited states and the continuum, named as X-ray absorption near-edge structure (XANES) and extended X-ray absorption fine structure (EXAFS), respectively [90]. The XANES uses radiation up to 40 eV from the X-ray absorption edge of an element in the material, and can give the information about the oxidation state and coordination environment of the metal atoms as well as the symmetry of the metal site. The EXAFS spectra are typically produced in the range of 500–1000 eV before an absorption edge of an element being examined and provide the atomic numbers, types, and distances between the neighboring atom from the absorbing atom [91].

3.3.3 Other characterization techniques

3.3.3.1 Energy-dispersive X-ray spectroscopy

Energy-dispersive X-ray spectroscopy (EDS) is an element analysis technique for a sample, which is usually used in combination with SEM. It relies on the bombardment of high-energy beam of charged particles (electron beam or X-ray beam) on the sample, and X-rays were emitted from a specimen. The number and energy of the X-rays can be measured by an energy-dispersive spectrometer. Due to the characteristic X-ray energy of each element, EDS can provide the information about chemical composition of samples [92].

3.3.3.2 Diffraction

A diffraction pattern is obtained by illuminating the crystal with a beam of incident X-rays. X-ray diffraction (XRD) can be used for determining the crystalline structure, lattice spacing, and particle size of samples [88]. Electron diffraction (ED) is usually performed in a TEM to deduce the crystal structure of samples [86].

3.3.3.3 Ultraviolet-visible (UV-vis) spectroscopy

UV-vis refers to electronic transitions of atoms and molecules in the ultraviolet-visible spectral region. It can be used for studying the size, concentration and the chemical composition of nanoparticles during their formation process [93, 94].

3.3.3.4 Infrared (IR) spectroscopy

During the synthesis process of nanoparticles, some small molecules as capping agents are often employed to control their size and shape. The surface of as-prepared nanoparticles is always absorbed small molecules, and needed to be cleaned several times with proper agents. IR spectroscopy is used to probe the surface composition and structure of samples [95–97].

3.3.3.5 Nuclear magnetic resonance

Nuclear magnetic resonance (NMR) spectroscopy is one of the most powerful characterization methods for obtaining local electronic environment of metal atoms by virtue of the chemical shift due to the electronic molecular orbital coupling to the external magnetic field [18]. According to the line shape and the numbers of peaks, the information about symmetry of the atomic environment can be obtained by NMR of quadrupolar nuclei.

^{13}C NMR and ^{1}H NMR have been used to determine the structure of adsorbed organic molecules onto the surface of metal nanoparticles [98–100].

3.4 Critical safety considerations

The extraordinary chemical and physical properties of alloy nanoparticles have driven intensive research on the large scale synthesis and related commercial applications. The use of alloy nanoparticles should be considered not only for the contribution to the multifunctional applications but also in the potential adverse effect on environment and human body during their construction and post-treatment process. The nanomaterials can be incidentally or accidentally released to the environment at various stages in their lifecycle (Figure 3.12). To use these nanomaterials environmentally, healthily and safely, it is necessary and important to consider the lifecycle exposure pathways including occupational exposure, community exposure and environmental release [101]. In addition, the stability of nanoparticles should be considered to fully estimate their risk in their lifecycle. The behaviors of nanoparticles in the environment usually include two aspects: (1) the aggregation/disaggregation of nanoparticles and (2) the dissolution/release of nanoparticles [102]. The aggregation/disaggregation and dissolution/release behaviors are related with the transport of nanoparticles, which is affected by various environmental conditions like pH, ionic strength and natural organic matter. Based on the Derjaguin-

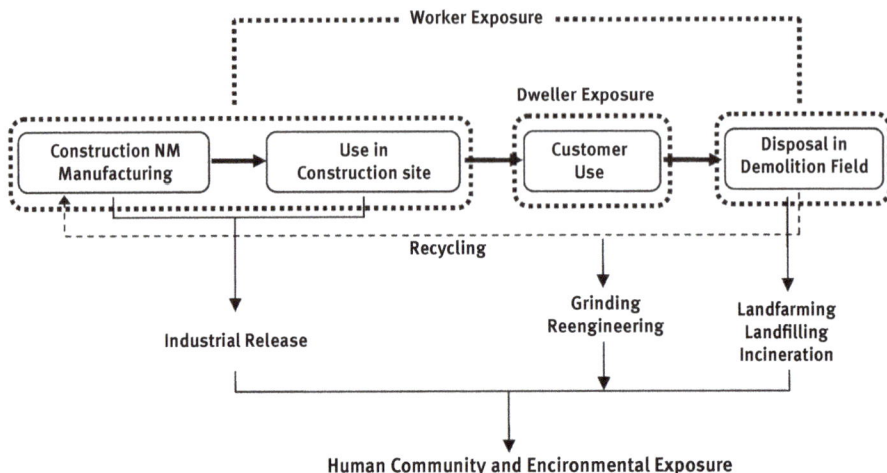

Figure 3.12: Possible exposure scenarios during the lifecycle of manufactured nanomaterials used in construction. Reproduced with permission from ref. 101. Copyright 2010, American Chemical Society.

Landau-Verwey-Overbeek (DLVO) theory, electrostatic repulsion, steric, and van der Waals forces between particles are main factors to influence the stability of nanoparticles [103].

During the construction, renovation and demolition process, workers pose themselves at great tremendous risk in inhalation, skin contact and accidental ingestion [101, 104, 105]. Results of biokinetic studies on nano-sized particles are shown in Figure 3.13. The various routes of nanoparticle uptake are illustrated, especially in the aspects of skin exposure and organs effects. Therefore, a couple of issues are important to note here. (1) Operation staff should know clearly the safety standards of original and other possible forms during the lifecycles of nanoparticles based on the Material Safety Data Sheets (MSDS) before work, especially for the toxic nanoparticles. (2) Air monitoring is needed to periodically perform over entire operation areas because of the dust contamination and damage to human body. Personal protective equipment (e. g., masks, coveralls, and gloves) need to be provided for users. If there is an emergency, medical reserve should be supported immediately to reduce injuries. (3) For the manufacturers of nanotechnology products, regulations on the dose toxicity before access to markets should be thoroughly discussed and verified experimentally.

3.5 Conclusions and future perspective

In this chapter, we provide recent advances on the preparation methods and characterization techniques for bimetallic and multimetallic alloy nanoparticles with controllable size. Each preparation method has its feature to enrich the toolbox of preparing

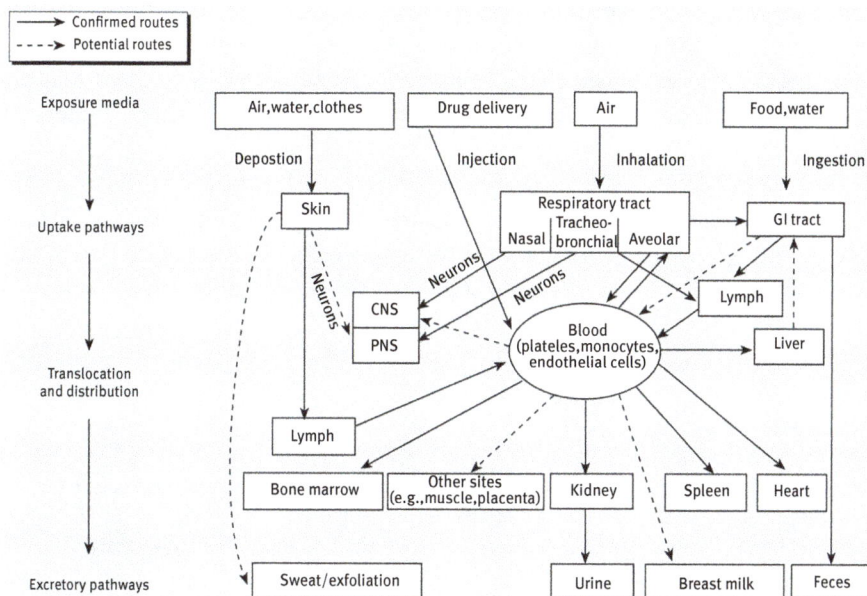

Figure 3.13: Biokinetics of nano-sized particles [105]. Reproduced with permission from Environmental Health Perspectives.

different types of alloy nanostructures. The improved high-resolution of characteriza-tion techniques enable access to detailed information including electronic and geo-metric structures as well as chemical ordering of an individual nanoparticle. The surprising results may provide potential opportunities in multidisciplinary research fields to promote the rapid developments of nanoscience and nanotechnology. Beyond the astonishing progress made to date, there are still many challenges ahead in the development of novel alloy nanoparticles. (1) The ultimate goal for the synthesis of size-controlled alloys is to gain high-performance nanomaterials for various appli-cations. It is highly desirable to develop general and scalable production methods to prepare high-quality alloy nanoparticles with defined size and surface properties. Moreover, the precise control of the uniform dispersion and effective interaction of nanoparticles on the support is also critical. (2) The influence of surface absorbed cap-ping agents on the properties of nanoparticles remains to be explored. It is necessary to develop a strategy without changing the size and structure of nanoparticles to re-move the capping agents, obtaining a "clean" surface and getting the real information on its property. (3) The better understanding of growth mechanisms of alloy nanopar-ticles may provide potential opportunities for developing new synthetic methods and creating novel architectures. The various advanced *in situ* characterization techniques should be introduced to further probe the evolution process of nanoparticles [106]. (4) More comprehensive theoretical methods should be established to figure out which factor is most important for controlling the size and alloy structure, such as surface strain engineering [107] and ligand passivation. With the development of novel preparation methods and advanced characterization techniques, it is anticipated that the exploration of alloy nanoparticles and their promising applications will go deeper and wider.

Funding: This work was financially supported by the National Natural Science Foundation of China (No. 21805069), the Natural Science Foundation of Tianjin City (No. 17JCJQJC44700 and No. 16JCZDJC30600) and the Scientific Research Foundation of Hebei Agricultural University (No. ZD201716).

References

[1] Turkevich J, Stevenson PC, Hillier J. A study of the nucleation and growth processes in the synthesis of colloidal gold. Discuss Faraday Soc. 1951;11:55–75.

[2] Faraday MX. The Bakerian Lecture. —experimental relations of gold (and other metals) to light. Philos Trans R Soc London. 1857;147:145–81.

[3] Daniel M-C, Astruc D. Gold nanoparticles: assembly, supramolecular chemistry, quantum-size-related properties, and applications toward biology, catalysis, and nanotechnology. Chem Rev. 2004;104:293–346.

[4] Corma A, Garcia H. Supported gold nanoparticles as catalysts for organic reactions. Chem Soc Rev. 2008;37:2096–126.

[5] Chen Y, Qiu J, Wang X, Xiu J. Preparation and application of highly dispersed gold nanoparticles supported on silica for catalytic hydrogenation of aromatic nitro compounds. J Catal. 2006;242:227–30.

[6] Liu J, Lu Y. Adenosine-dependent assembly of aptazyme-functionalized gold nanoparticles and its application as a colorimetric biosensor. Anal Chem. 2004;76:1627–32.

[7] Jv Y, Li B, Cao R. Positively-charged gold nanoparticles as peroxidiase mimic and their application in hydrogen peroxide and glucose detection. Chem Commun. 2010;46:8017–9.

[8] Ferrando R, Jellinek J, Johnston RL. Nanoalloys: from theory to applications of alloy clusters and nanoparticles. Chem Rev. 2008;108:845–910.

[9] Zhang H, Jin M, Wang J, Kim MJ, Yang D, Xia Y. Nanocrystals composed of alternating shells of Pd and Pt can be obtained by sequentially adding different precursors. J Am Chem Soc. 2011;133:10422–5.

[10] Wang L, Yamauchi Y. Strategic synthesis of trimetallic Au@Pd@Pt core-shell nanoparticles from poly(vinylpyrrolidone)-based aqueous solution toward highly active electrocatalysts. Chem Mater. 2011;23:2457–65.

[11] Wang L, Yamauchi Y. Autoprogrammed synthesis of triple-layered Au@Pd@Pt core-shell nanoparticles consisting of a Au@Pd bimetallic core and nanoporous pt shell. J Am Chem Soc. 2010;132:13636–8.

[12] Qiu F, Liu G, Li L, Wang Y, Xu C, An C, et al. Synthesis of triple-layered Ag@Co@Ni core-shell nanoparticles for the catalytic dehydrogenation of ammonia borane. Chem A Eur J. 2014;20:505–9.

[13] Aranishi K, Jiang H-L, Akita T, Haruta M, Xu Q. One-step synthesis of magnetically recyclable Au/Co/Fe triple-layered core-shell nanoparticles as highly efficient catalysts for the hydrolytic dehydrogenation of ammonia borane. Nano Res. 2011;4:1233–41.

[14]   Xu Y, Yuan Y, Ma A, Wu X, Liu Y, Zhang B. Composition-tunable Pt-Co alloy nanoparticle
       networks: facile room-temperature synthesis and supportless electrocatalytic applications.
       Chemphyschem. 2012;13:2601–9.
[15]   Xu Y, Hou S, Liu Y, Zhang Y, Wang H, Zhang B. Facile one-step room-temperature synthesis of
       Pt_3Ni nanoparticle networks with improved electro-catalytic properties. Chem Commun.
       2012;48:2665–7.
[16]   Zhu H, Zhang S, Guo S, Su D, Sun S. Synthetic control of FePtM nanorods (M = Cu, Ni) to
       enhance the oxygen reduction reaction. J Am Chem Soc. 2013;135:7130–3.
[17]   Shao M, Peles A, Shoemaker K. Electrocatalysis on platinum nanoparticles: particle size
       effect on oxygen reduction reaction activity. Nano Lett. 2011;11:3714–9.
[18]   Toshima N, Yonezawa T. Bimetallic nanoparticles-novel materials for chemical and physical
       applications. New J Chem. 1998;22:1179–201.
[19]   Bönnemann H, Richards RM. Nanoscopic metal particles-synthetic methods and potential
       applications. Eur J Inorg Chem. 2001;2001:2455–80.
[20]   Tao AR, Habas S, Yang P. Shape control of colloidal metal nanocrystals. Small.
       2008;4:310–25.
[21]   Shevchenko EV, Talapin DV, Schnablegger H, Kornowski A, Festin Ö, Svedlindh P, et al. Study
       of nucleation and growth in the organometallic synthesis of magnetic alloy nanocrystals: the
       role of nucleation rate in size control of $CoPt_3$ nanocrystals. J Am Chem Soc.
       2003;125:9090–101.
[22]   Lee YW, Kim M, Kim ZH, Han SW. One-step synthesis of Au@Pd core-shell nanooctahedron.
       J Am Chem Soc. 2009;131:17036–7.
[23]   Singh SK, Singh AK, Aranishi K, Xu Q. Noble-metal-free bimetallic nanoparticle-catalyzed
       selective hydrogen generation from hydrous hydrazine for chemical hydrogen storage. J Am
       Chem Soc. 2011;133:19638–41.
[24]   Chen G, Desinan S, Rosei R, Rosei F, Ma D. Synthesis of Ni-Ru alloy nanoparticles and their
       high catalytic activity in dehydrogenation of ammonia borane. Chem A Eur J.
       2012;18:7925–30.
[25]   Cui X, Li W, Ryabchuk P, Junge K, Beller M. Bridging homogeneous and heterogeneous
       catalysis by heterogeneous single-metal-site catalysts. Nat Catal. 2018;1:385–97.
[26]   Huang W, Kang X, Xu C, Zhou J, Deng J, Li Y, et al. 2D PdAg alloy nanodendrites for enhanced
       ethanol electroxidation. Adv Mater. 2018;30:1706962.
[27]   Lee YW, Kim M, Kang SW, Han SW. Polyhedral bimetallic alloy nanocrystals exclusively
       bound by {110} facets: au-Pd rhombic dodecahedra. Angew Chem Int Ed.
       2011;50:3466–70.
[28]   Wang A, Li J, Zhang T. Heterogeneous single-atom catalysis. Nat Rev Chem. 2018;2:65–81.
[29]   Jana NR, Gearheart L, Murphy CJ. Evidence for seed-mediated nucleation in the chemical
       reduction of gold salts to gold nanoparticles. Chem Mater. 2001;13:2313–22.
[30]   Henning AM, Watt J, Miedziak PJ, Cheong S, Santonastaso M, Song M, et al. Gold-palladium
       core-shell nanocrystals with size and shape control optimized for catalytic performance.
       Angew Chem Int Ed. 2013;52:1477–80.
[31]   Markov IV. Crystal growth for beginners: fundamentals of nucleation. Singapore: World
       Scientific, 2003.
[32]   Lu C-L, Prasad KS, Wu H-L, Ho J-AA, Huang MH. Au nanocube-directed fabrication of Au-Pd
       core-shell nanocrystals with tetrahexahedral, concave octahedral, and octahedral structures
       and their electrocatalytic activity. J Am Chem Soc. 2010;132:14546–53.
[33]   Yang C-W, Chanda K, Lin P-H, Wang Y-N, Liao C-W, Huang MH. Fabrication of Au-Pd core-shell
       heterostructures with systematic shape evolution using octahedral nanocrystal cores and
       their catalytic activity. J Am Chem Soc. 2011;133:19993–20000.

[34] Wang F, Li C, Sun L-D, Wu H, Ming T, Wang J, et al. Heteroepitaxial growth of high-index-faceted palladium nanoshells and their catalytic performance. J Am Chem Soc. 2011;133:1106–11.

[35] Liu H, Qu J, Chen Y, Li J, Ye F, Lee JY, et al. Hollow and cage-bell structured nanomaterials of noble metals. J Am Chem Soc. 2012;134:11602–10.

[36] DeSantis CJ, Sue AC, Bower MM, Skrabalak SE. Seed-mediated co-reduction: A versatile route to architecturally controlled bimetallic nanostructures. ACS Nano. 2012;6:2617–28.

[37] Torigoe K, Esumi K. Preparation of bimetallic silver-palladium colloids from silver(I) bis (oxalato)palladate(II). Langmuir. 1993;9:1664–7.

[38] Torigoe K, Nakajima Y, Esumi K. Preparation and characterization of colloidal silver-platinum alloys. J Phys Chem. 1993;97:8304–9.

[39] Xu J, Liu X, Chen Y, Zhou Y, Lu T, Tang Y. Platinum-Cobalt alloy networks for methanol oxidation electrocatalysis. J Mater Chem. 2012;22:23659–67.

[40] Zhang L, Wan L, Ma Y, Chen Y, Zhou Y, Tang Y, et al. Crystalline palladium-cobalt alloy nanoassemblies with enhanced activity and stability for the formic acid oxidation reaction. Appl Catal B Environ. 2013;138:229–35.

[41] Zheng H, Smith RK, Jun Y-W, Kisielowski C, Dahmen U, Alivisatos AP. Observation of single colloidal platinum nanocrystal growth trajectories. Science. 2009;324:1309–12.

[42] Wang DS, Xie T, Peng Q, Zhang SY, Chen J, Li YD. Direct thermal decomposition of metal nitrates in octadecylamine to metal oxide nanocrystals. Chem A Eur J. 2008;14:2507–13.

[43] Wang D, Li Y. One-Pot protocol for Au-based hybrid magnetic nanostructures via a noble-metal-induced reduction process. J Am Chem Soc. 2010;132:6280–1.

[44] Herrero E, Buller LJ, Abruña HD. Underpotential deposition at single crystal surfaces of Au, Pt, Ag and other materials. Chem Rev. 2001;101:1897–930.

[45] Kokkinidis G. Underpotential deposition and electrocatalysis. J Electroanal Chem Interfacial. 1986;201:217–36.

[46] Szabó S. Underpotential deposition of metals on foreign metal substrates. Int Rev Phys Chem. 1991;10:207–48.

[47] Zhang L, Zhang J, Kuang Q, Xie S, Jiang Z, Xie Z, et al. Cu^{2+}-assisted synthesis of hexoctahedral Au-Pd alloy nanocrystals with high-index facets. J Am Chem Soc. 2011;133:17114–7.

[48] Jiang Y, Jia Y, Zhang J, Zhang L, Huang H, Xie Z, et al. Underpotential deposition-induced synthesis of composition-tunable Pt-Cu Nanocrystals and their catalytic properties. Chem A Eur J. 2013;19:3119–24.

[49] Carino EV, Crooks RM. Characterization of Pt@Cu core@shell dendrimer-encapsulated nanoparticles synthesized by Cu underpotential deposition. Langmuir. 2011;27:4227–35.

[50] Yu Y, Hu Y, Liu X, Deng W, Wang X. The study of Pt@Au electrocatalyst based on Cu underpotential deposition and Pt redox replacement. Electrochim Acta. 2009;54:3092–7.

[51] Shao M, Sasaki K, Marinkovic N, Zhang L, Adzic R. Synthesis and characterization of platinum monolayer oxygen-reduction electrocatalysts with Co-Pd core-shell nanoparticle supports. Electrochem Commun. 2007;9:2848–53.

[52] Yancey DF, Carino EV, Crooks RM. Electrochemical synthesis and electrocatalytic properties of Au@Pt dendrimer-encapsulated nanoparticles. J Am Chem Soc. 2010;132:10988–9.

[53] Zhai J, Huang M, Dong S. Electrochemical designing of Au/Pt core shell nanoparticles as nanostructured catalyst with tunable activity for oxygen reduction. Electroanal. 2007;19:506–9.

[54] Xia X, Wang Y, Ruditskiy A, Xia Y. 25th anniversary article: galvanic replacement: a simple and versatile route to hollow nanostructures with tunable and well-controlled propertiese. Adv Mater. 2013;25:6313–33.

[55] Thota S, Wang Y, Zhao J. Colloidal Au-Cu alloy nanoparticles: synthesis, optical properties and applications. Mater Chem Front. 2018;2:1074–89.

[56] Sun Y, Xia Y. Mechanistic study on the replacement reaction between silver nanostructures and chloroauric acid in aqueous medium. J Am Chem Soc. 2004;126:3892–901.

[57] Esumi K, Tano T, Torigoe K, Meguro K. Preparation and characterization of bimetallic palladium-copper colloids by thermal decomposition of their acetate compounds in organic solvents. Chem Mater. 1990;2:564–7.

[58] Hermans S, Raja R, Thomas JM, Johnson BFG, Sankar G, Gleeson D. Solvent-free, low-temperature, selective hydrogenation of polyenes using a bimetallic nanoparticle Ru-Sn catalyst. Angew Chem Int Ed. 2001;40:1211–5.

[59] Thomas JM, Johnson BFG, Raja R, Sankar G, Midgley PA. High-performance nanocatalysts for single-step hydrogenations. Acc Chem Res. 2003;36:20–30.

[60] Thomas JM. Bimetallic catalysts and their relevance to the hydrogen economy. Ind Eng Chem Res. 2003;42:1563–70.

[61] Rutledge RD, Morris WH, Wellons MS, Gai Z, Shen J, Bentley J, et al. Formation of FePt nanoparticles having high coercivity. J Am Chem Soc. 2006;128:14210–1.

[62] Robinson I, Zacchini S, Tung LD, Maenosono S, Thanh NTK. Synthesis and characterization of magnetic nanoalloys from bimetallic carbonyl clusters. Chem Mater. 2009;21:3021–6.

[63] Sun S, Murray CB, Weller D, Folks L, Moser A. Monodisperse FePt nanoparticles and ferromagnetic FePt nanocrystal superlattices. Science. 2000;287:1989–92.

[64] Sun S. Recent advances in chemical synthesis, self-assembly, and applications of FePt nanoparticles. Adv Mater. 2006;18:393–403.

[65] Bönnemann H, Brand RA, Brijoux W, Hofstadt HW, Frerichs M, Kempter V, et al. Air stable Fe and Fe-Co magnetic fluids-synthesis and characterization. Appl Organomet Chem. 2005;19:790–6.

[66] Ueda Y, Ito M. Magnetoresistance in Co-Cu alloy films formed by electrodeposition method. Jpn J Appl Phys. 1994;30:L1403.

[67] Mohanty US. Electrodeposition: a versatile and inexpensive tool for the synthesis of nanoparticles, nanorods, nanowires, and nanoclusters of metals. J Appl Electrochem. 2010;41:257–70.

[68] Lu D-L, Domen K, Tanaka K-I. Electrodeposited Au-Fe, Au-Ni, and Au-Co Alloy nanoparticles from aqueous electrolytes. Langmuir. 2002;18:3226–32.

[69] Chen A, Holt-Hindle P. Platinum-based nanostructured materials: synthesis, properties, and applications. Chem Rev. 2010;110:3767–804.

[70] Lu D-L, Tanaka K-I. Gold particles deposited on electrodes in salt solutions under different potentials. J Phys Chem. 1996;100:1833–7.

[71] Huang H, Yang X. One-step, shape control synthesis of gold nanoparticles stabilized by 3-thiopheneacetic acid. Colloids Surf A. 2005;255:11–7.

[72] OFinot M, Braybrook GD, TMcDermott M. Characterization of electrochemically deposited gold nanocrystals on glassy carbon electrodes. J Electroan Chem. 1999;466:234–41.

[73] Baber S, Zhou M, Lin QL, Naalla M, Jia QX, Lu Y, et al. Nanoconfined surfactant templated electrodeposition to porous hierarchical nanowires and nanotubes. Nanotechnol. 2010;21:165603.

[74] Saedi A, Ghorbani M. Electrodeposition of Ni-Fe-Co alloy nanowire in modified AAO template. Mater Chem Phys. 2005;91:417–23.

[75] Valizadeh S, Hultman L, George JM, Leisner P. Template synthesis of Au/Co multilayered nanowires by electrochemical deposition. Adv Funct Mater. 2002;12:766–72.

[76]   Xu F, Zhao L, Zhao F, Deng L, Hu L, Zeng B. Electrodeposition of AuPdCu alloy nanoparticles on a multiwalled carbon nanotube coated glassy carbon electrode for the electrocatalytic oxidation and determination of hydrazine. Int J Electrochem Sci. 2014;9:2832–47.

[77]   Quan X, Mei Y, Xu H, Sun B, Zhang X. Optimization of Pt-Pd alloy catalyst and supporting materials for oxygen reduction in air-cathode microbial fuel cells. Electrochim Acta. 2015;165:72–7.

[78]   Cui H-F, Ye J-S, Liu X, Zhang W-D, Sheu F-S. Pt-Pb alloy nanoparticle/carbon nanotube nanocomposite: a strong electrocatalyst for glucose oxidation. Nanotechnol. 2006;17:2334–9.

[79]   Xiao F, Zhao F, Zeng J, Zeng B. Novel alcohol sensor based on PtRuNi ternary alloy nanoparticles-multi-walled carbon nanotube-ionic liquid composite coated electrode. Electrochem Commun. 2009;11:1550–3.

[80]   Muntean R. Carbon Nanofibers Decorated with Pt-Co alloy nanoparticles as catalysts for electrochemical cell applications. i. synthesis and structural characterization. Int J Electrochem Sci. 2017;12:4597–609.

[81]   Thota S, Chen S, Zhou Y, Zhang Y, Zou S, Zhao J. Structural defect induced peak splitting in gold-copper bimetallic nanorods during growth by single particle spectroscopy. Nanoscale. 2015;7:14652–8.

[82]   Goldstein JI, Newbury DE, Echlin P, Joy DC, Fiori C, Lifshin E. Scanning electron microscopy and x-ray microanalysis || image formation in the scanning electron microscope. Boston, MA, USA: Springer, 1981.

[83]   Pennycook SJ, Jesson DE. High-resolution Z-contrast imaging of crystals. Ultramicroscopy. 1991;37:14–38.

[84]   Allen S, Davies MC, Roberts CJ, Tendler SJB, Williams PM. Atomic force microscopy in analytical biotechnology. Trends Biotechnol. 1997;15:101–5.

[85]   Chen CJ. Introduction to scanning tunneling microscopy. Oxford, UK: Oxford University Press, 1993.

[86]   Swartz WE, Jr X-ray photoelectron spectroscopy. Anal Chem. 1973;45:788A-800a.

[87]   Koyasu K, Mitsui M, Nakajima A, Kaya K. Photoelectron spectroscopy of palladium-doped gold cluster anions; Au_nPd^- (n= 1-4). Chem Phys Lett. 2002;358:224–30.

[88]   Zhang J, Xu Y, Zhang B. Facile synthesis of 3D Pd-P nanoparticle networks with enhanced electrocatalytic performance towards formic acid electrooxidation. Chem Commun. 2014;50:13451–3.

[89]   Zhang J, Li K, Zhang B. Synthesis of dendritic Pt-Ni-P alloy nanoparticles with enhanced electrocatalytic properties. Chem Commun. 2015;51:12012–5.

[90]   Koningsberger DC, Prins R. X-ray absorption: principles, applications, techniques of EXAFS, SEXAFS, and XANES. Hobeken, NJ, USA: Wiley, 1988.

[91]   Yano J, Yachandra VK. X-ray absorption spectroscopy. Photosyn Res. 2009;102:241–54.

[92]   Goldstein JI, Newbury DE, Echlin P, Joy DC, Lyman CE, Lifshin E, et al. Scanning electron microscopy and X-ray microanalysis || special topics in electron beam X-Ray microanalysis. Boston, MA, USA: Springer, 2003.

[93]   Haiss W, Thanh NTK, Aveyard J, Fernig DG. Determination of size and concentration of gold nanoparticles from UV– Vis spectra. Anal Chem. 2007;79:4215–21.

[94]   Han SW, Kim Y, Kim K. Dodecanethiol-derivatized Au/Ag bimetallic nanoparticles: TEM, UV/VIS,XPS, and FTIR analysis. J Colloid Interface Sci. 1998;208:272–8.

[95]   Ma M, Zhang Y, Yu W, Shen H-Y, Zhang H-Q, Gu N. Preparation and characterization of magnetite nanoparticles coated by amino silane. Colloids Surf A. 2003;212:219–26.

[96]   Badia A, Singh S, Demers L, Cuccia L, Brown GR, Lenno RB. Self-assembled monolayers on gold nanoparticles. Chem A Eur J. 1996;2:359–63.

[97]   Liu X, Wang A, Li L, Zhang T, Mou C-Y, Lee J-F. Structural changes of Au-Cu bimetallic catalysts in CO oxidation: in situ XRD, EPR, XANES, and FT-IR characterizations. J Catal. 2011;278:288–96.

[98]   Bradley JS, Millar JM, Hill EW. Surface chemistry on colloidal metals: a high-resolution NMR study of carbon monoxide adsorbed on metallic palladium crystallites in colloidal suspension. J Am Chem Soc. 1991;113:4016–7.

[99]   Zhou H, Du F, Li X, Zhang B, Li W, Yan B. Characterization of organic molecules attached to gold nanoparticle surface using high resolution magic angle spinning 1H NMR. J Phyl Chem C. 2008;112:19360–6.

[100] Liu X, Yu M, Kim H, Mameli M, Stellacci F. Determination of monolayer-protected gold nanoparticle ligand-shell morphology using NMR. Nat Commun. 2012;3:1182.

[101] Lee J, Mahendra S, Alvarez PJJ. Nanomaterials in the construction industry: a review of their applications and environmental health and safety considerations. ACS Nano. 2010;4:3580–90.

[102] Kim H-A, Choi YJ, Kim K-W, Lee B-T, Ranville JF. Nanoparticles in the environment: stability and toxicity. Rev Environ Health. 2012;27:175–9.

[103] Badawy AME, Luxton TP, Silva RG, Scheckel KG, Suidan MT, Tolaymat TM. Impact of environmental conditions (ph, ionic strength, and electrolyte type) on the surface charge and aggregation of silver nanoparticles suspensions. Environ Sci Technol. 2010;44:1260–6.

[104] Robertson TA, Sanchez WY, Roberts MS Are commercially available nanoparticles safe when applied to the skin? J Biomed Nanotechnol. 2010;6:452–68.

[105] Oberdörster G, Oberdörster E, Oberdörster J. Nanotoxicology: an emerging discipline evolving from studies of ultrafine particles. Environ Health Perspect. 2005;113:823–39.

[106] Wu S, Sun Y. In situ techniques for probing kinetics and mechanism of hollowing nanostructures through direct chemical transformations. Small Methods. 2018;2:1800165.

[107] Luo M, Guo S. Strain-controlled electrocatalysis on multimetallic nanomaterials. Nat Rev Mater. 2017;2:17059.

Bionotes

Jingfang Zhang received her Ph.D. degree in Chemistry from Tianjin University in 2017 (with Prof. Bin Zhang). Currently, she is an associate professor in the chemistry department at Hebei Agricultural University. Her research focuses on the development of metal-based nanomaterials for electrocatalytic applications.

Yifu Yu received his B.E. and Ph.D. degrees in Chemical Engineering from Tianjin University. He carried out postdoctoral research in Nanyang Technological University (2014.7–2017.7). Currently, He is an associate professor in the chemistry department at Tianjin University. His research interest includes the controlled transformation synthesis of advanced nanomaterials for catalytic applications.

Bin Zhang received his Ph.D. degree from University of Science and Technology of China in 2007. He carried out postdoctoral research in University of Pennsylvania (July 2007 to July 2008) and worked as an Alexander von Humboldt fellow in Max Planck Institute of Colloids and Interfaces (August 2008 to July 2009). Currently, he is a professor in the chemistry department at Tianjin University and Collaborative Innovation Center of Chemical Science and Engineering (Tianjin). He mainly focuses on the controlled synthesis of advanced nanomaterials for catalytic applications.

C. Tojo, D. Buceta and M. A. López-Quintela

4 On the minimum reactant concentration required to prepare Au/M core-shell nanoparticles by the one-pot microemulsion route

Abstract: The minimum reactant concentration required to synthesize Au/M (M = Ag, Pt, Pd, Ru …) core-shell nanoparticles by the one-pot microemulsion route was calculated by a simulation model under different synthesis conditions. This minimum concentration was proved to depend on the reduction potential of the slower metal M and on the rigidity of the surfactant film composing the microemulsion. Model results were tested by comparing with Au/M nanoparticles taken from literature. In all cases, experimental data obey model predictions. From this agreement, one can conclude that the smaller the standard potential of the slower reduction metal, the lower the minimum concentration needed to obtain core-shell nanoparticles. In addition, the higher the surfactant flexibility, the higher the minimum concentration to synthesize metal segregated nanoparticles. Model prediction allows to quantify which is the best value of concentration to prepare different pairs of core-shell Au/M nanoparticles in terms of nature of M metal in the couple and microemulsion composition. This outlook may become an advanced tool for fine-tuning Au/M nanostructures.

Graphical Abstract:

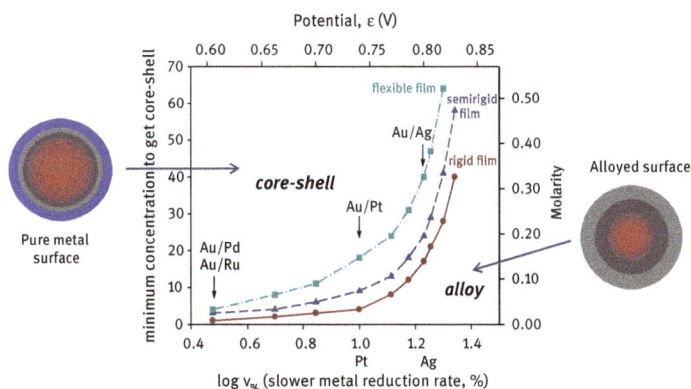

This article has previously been published in the journal *Physical Sciences Reviews*. Please cite as: Tojo, C., Buceta, D., López-Quintela, M. A. On the minimum reactant concentration required to prepare Au/M core-shell nanoparticles by the one-pot microemulsion route. *Physical Sciences Reviews* [Online] **2020**, 5. DOI: 10.1515/psr-2018-0045.

https://doi.org/10.1515/9783110345001-004

Keywords bimetallic nanoparticles, nanocatalysts, microemulsion, simulation, one-pot method

4.1 Introduction

The idea of combining two different metals in the same nanocatalyst was first proposed by Sinfelt [1, 2]. Since then and up to today, bimetallic nanoparticles are of particular interest because of, among others, their potentially high catalytic activity as compared to monometallic ones [3–10] even at lower temperatures [11, 12]. Specifically, the effectiveness of chemical reaction and catalytic selectivity depend on the arrangement of the atoms located at the surface of the nanoparticle, that is, within the first few covering layers. The second metal modifies the interatomic interactions and produces changes on the nanostructure including the surface. As a result, controlling the distribution of metal atoms in a nanoparticle composing by two metals not only provides a fundamental knowledge, but also is interesting for applied nanotechnologies, ranging from catalysis to plasmonics [13–15].

The preparation of bimetallic nanoparticles with specific metal distribution requires a special control of the experimental conditions, because the intraparticle structure strongly determines the final nanoparticle properties. Water in oil microemulsions route is one of the preferred methods to synthesize nanoparticles due to their ability to govern over composition, size and shape of the particles. The development in this field of study has been impressive in the last few years, so different pairs of bimetallic particles have been obtained by this strategy [16–19]. However, an optimized control of the metal distribution in the bimetallic nanostructure is still a challenging task. In fact, theoretical studies are scarce and an immense trial and error approach is necessary to tune the particle intrastructure. With the aim of predicting the atomic arrangement of final nanoparticle, a simulation model was developed to study the synthesis of bimetallic nanoparticles by the microemulsion method. The motivation arises from the conviction that the kinetics of nanoparticle synthesis inside reverse micelles determines the final nanoarrangement. To check the model, Au/Pt nanoparticles were prepared under similar experimental conditions than those explored by simulation [20] and characterized using High-Resolution Scanning Transmission Electron Microscopy (HR-STEM). Likewise, STEM profiles were computed from the simulated structures. The remarkable agreement between experimental and simulated STEM profiles supports the model, and shows that the metal segregation in Au/Pt nanoparticles can be simply modified by using different reactant concentrations. Au/Pt bimetallic nanoparticles were chosen due to the high catalytic activity for different reactions [21–25]. Au/Pt pair has been obtained by different preparation routes [21–23, 25–33] including the microemulsion method [11, 34]. The catalytic behaviour of Au/Pt nanoparticles depends on the nanoparticle architecture. As an illustration of this point, the better nanoarrangement for formic acid electrooxidation [35] and oxygen reduction reaction [36, 37] is an Au core covered by a Pt shell. By contrast, electro-oxidation of methanol [38] is enhanced by using a Pt-Au alloyed shell. If the nanoarrangement could be previously optimized by choosing the most convenient synthetic conditions, it would imply an important breakthrough. Thereby, a bimetallic nanocatalyst would be not only characterized at the end of the synthesis, but also pre-designed and obtained with a particular nanoarchitecture for the reaction of interest. In the paper at hand, our aim is to define the best experimental conditions to obtain different pairs of core-shell Au/M nanoparticles in terms of nature of M metal in the couple and microemulsion composition. For this purpose, the simulation model can be used to establish concrete and practical guidelines for chemical synthesis of nanomaterials.

This work is restricted to bimetallic nanoparticles composed by Au combined with another metal whose reduction was slower. This selection is done according to kinetic results which suggest that the ability of the microemulsion to alter final metal distribution is restricted to couples of metals including a fast reduction metal. The presence of Au, whose chemical reduction is instantaneous, gives greater relevance to the reaction media. That is, Au chemical reduction takes place as fast as the intermicellar exchange of matter manages the reactants encounter. As a result, the limiting step in Au reduction is the material exchange between micelles (see reference [39] for a deeper kinetic study). This is not the case in slower metals, whose reduction rate depends on both reduction rate and material exchange rate between micelles. Furthermore, the reduction rate can be so slow that the compartmentalization of reaction media could be irrelevant due to chemical control [40]. Because the relative rates of the two metals dictate the final metal segregation in the nanostructure, the microemulsion loses its potential impact on intrastructure when the two chemical reductions are slow [40]. Therefore, this study is restricted to the synthesis of Au/M nanoparticles, in which final bimetallic structures can be manipulated by a change in the synthesis conditions (microemulsion composition and concentration).

In the following, after a short explanation of the experimental procedure and prediction model (Section 4.2), we first focus on depicting the main rules to obtain core-shell structures. Next, we describe and discuss results on the intrastructure of different Au/M nanoparticles obtained by a one pot-method using the microemulsion route (Section 4.3). By this way, rather than carrying out a full analysis of all available researches, only those studies that fulfil the requirements (Au/M nanoparticles obtained from microemulsions by a one-pot method) have been considered. Excellent review articles on bimetallic nanoparticles, including preparation methods, metal distribution and their role as catalysts, are already available [10, 15, 41–45]. In this contribution, we also explore which is the minimum value of concentration to prepare well-defined core-shell nanoparticles depending on the standard potential of the slow metal M and the microemulsion composition. This unique outcome may become an advanced tool for fine-tuning Au/M nanostructures.

4.2 Preparation methods and computational model

4.2.1 One-pot method to prepare bimetallic nanoparticles in microemulsions

A water in oil microemulsion consists of water domains of nanometre-sized dimensions dispersed in an oil medium. The interface between water and oil is covered by a monolayer of surfactant. The water nanodroplets (also called reverse micelles) play as nanoreactors within which nanomaterials can be prepared. The nanodroplets provide a confined reaction media that restricts particle growth and prevents agglomeration.

This synthetic method consists of, first, preparing the nonpolar phase of a water dispersed in oil microemulsion by mixing the appropriate quantities of surfactant and oil. Then two different aqueous solutions, one containing the metal precursors and another containing the reducing agent, are prepared. Next the water solution containing the precursors is added to the solution of surfactant/oil under stirring, giving rise to a microemulsion whose water droplets contain the precursors. Another similar microemulsion is obtained by using the reducing agent aqueous solution as water phase. In this way, reactants are solved within the micelles composing the microemulsion. Then the two microemulsions, each one carrying one of the reactants, are mixed. The droplets of a microemulsion are not static, but they move and repeatedly collide with one another. Only if an interdroplet collision is energetic enough, the surfactant film breaks up, and a channel communicating colliding droplets can be stablished. Reactants can cross the channel allowing the mixture of the droplets content. Then chemical reactions, nucleation and growth of the particles can occur inside the confined space defined by droplets. This modus operandi for mixing reactants is called one-pot method. Contrary to a postcore method, the chemical reduction of the two metals can take place simultaneously by the one-pot procedure. Once the synthesis is finished, surfactant around the nanoparticles can easily be taken off by addition of Carbon Vulcan. Then, this mixture is stirred. To break the microemulsion, tetrahydrofuran is also added. Solid particles are obtained after decantation, filtration and centrifugation. Finally, particles are washed with a water/ethanol mixture and dried.

For example, Au/Pt nanoparticles have been prepared in an isooctane 75%/Tergitol 20%/water 5% microemulsion by a one-pot method as follows [20]:

Materials: Hydrogen tetrachloroaurate (HAuCl$_4$.3H$_2$O), hexachloroplatinate (H$_2$PtCl$_6$.6H$_2$O), isooctane (2,2,4-trimethylpentane, ≥99%), Tergitol (Type 15-S-5), tetrahydrofuran (≥99%) and hydrazine monohydrate (98%) were supplied by Sigma Aldrich. Carbon Vulcan 72 XR was purchased from Cabot. All aqueous solutions were prepared with deionized water from a Milli-Q system (Millipore).

Method: The microemulsions containing the metal precursor were prepared as follows: aqueous solutions (5 w%) of the appropriate amounts of the metals were added to a mixture of Tergitol (20 wt%) and isooctane (75 wt%) under stirring. It results in a transparent mixture. Another microemulsion with same composition (isooctane 75%, Tergitol 20% and water 5%) was prepared. In this second microemulsion, a hydrazine aqueous solution is used as water phase. The radius of the droplets composing the microemulsion was measured by Dynamic light scattering (DLS) (7.3 ± 0.1 nm).

By mixing the two microemulsions (the one containing metal precursors and the one containing reducing agent) under constant stirring at room temperature for 5 h, the nanoparticles were formed. Then, Carbon Vulcan was added in order to separate particles from the microemulsion and eliminate the surfactant. The appropriate amount of Carbon Vulcan is calculated to be 20% wt loading of Pt in the resulting nanoparticle. After the mixture was stirred overnight, the microemulsion was broken by adding tetrahydrofuran. By filtration and centrifugation, a solid was obtained after decantation. Then particles were washed several times with a mixture composing by water and ethanol, and finally dried at 100 °C for 12 h prior to any further analysis.

4.2.2 Computational model to simulate the synthesis of bimetallic nanoparticles in microemulsions by a one-pot method

The simulation model recreates the kinetic course of the reaction (see reference [46] for details). Briefly, a microemulsion is represented as a set of micelles positioned at random on a three-dimensional space. The one-pot procedure is reproduced by mixing equal volumes of the microemulsions, each one carrying one of the reactants (two metal salts A^+ and B^+ and reductor R). Reactants are initially distributed throughout droplets using a Poisson distribution. The initial average occupancy $\langle c \rangle$ can be modified in order to study the influence of concentration.

Micelles diffuse and collide with each other. To simulate an effective collision, 10% of micelles are chosen at random to collide, and then fuse and redisperse. Only in the case of collision was energetic enough, a water channel may be established between micelles, allowing material intermicellar exchange. At this stage, the material located inside colliding micelles is checked according to the exchange criteria described as follows. Once the composition inside micelle is revised, one Monte Carlo step is considered to be completed.

1. If the material are reactants and non-aggregated metal atoms, a concentration gradient principle is considered to redistribute material between two colliding micelles: material carried by the more occupied droplet can be moved to the less occupied one. The maximum number of reactants (A^+, B^+ and/or R) and atoms (A and/or B) transferred during a collision is quantified by the exchange parameter k_{ex}.

2. If the material is a growing particle (built up by aggregated metal atoms), the intermicellar exchange is limited by the size of the channel through which

fused micelles can interchange their content. Channel size is strongly dependent on the flexibility of the surfactant film, which is included in the model by means of the flexibility parameter (f). It quantifies the maximum particle size for transfer between droplets. In addition, Ostwald ripening is also taken into account. In order to minimize surface energy, Ostwald ripening assumes that smaller particles solubilize more than larger ones, so components diffuse from smaller to larger particles, which will grow by condensation of material on their surface. This phenomenon is included in the model by assuming that, when both colliding micelles contain particle, the material coming from smaller one is transferred to the micelle containing the larger one, provided that the inter-micellar channel would be of sufficient size to let material go through.

Whenever the metal salt (A^+ and/or B^+) and the reductor (R) are located inside the same micelle (as a consequence of redistribution of material between micelles during a collision), chemical reduction can occur. The reduction rate is simulated by the percentage of metal precursor which is reduced during a collision. For example, Au reduction is instantaneous, so 100% of Au salt inside the same micelle is reduced if there is enough reducing agent. In the case of Pt, whose reduction is slower, the reduction of 10% of Pt precursor successfully reproduced experimental results [20]. The remaining slower metal salt and reducing agent stay in the micelle. As the process advances, posterior collisions will give rise to the exchange and reaction of this material. In this way, a metal A is characterized by means of its reduction rate, v_A.

The reduction rate ratio of two oxidized metals A and B, which are reduced simultaneously (as in a one-pot method) to obtain an A/B bimetallic nanoparticle, can be related to the standard potential ($\varepsilon^0{}_A$) by means of the Volmer equation:

$$\frac{j_A}{j_B} = \frac{v_A n_A F}{v_B n_B F} = \frac{n_A F k_{red,A} c_{O,A} \exp\left[\frac{-\beta_A n_A F \varepsilon_A}{RT}\right]}{n_B F k_{red,B} c_{O,B} \exp\left[\frac{-\beta_B n_B F \varepsilon_B}{RT}\right]} = \exp\left[\frac{\beta n F (\varepsilon_B - \varepsilon_A)}{RT}\right] \qquad (4.1)$$

being j_A the current density, n_A is the number of electrons, F is the Faraday constant, $k_{red,A}$ is the chemical rate constant, β_A is the transfer coefficient, $c_{O,A}$ is the concentration of oxidized A, R is the gas constant and T is the temperature. Equation (4.1) can be simplified if the concentration of the two metal precursors A and B are initially equal ($c_{O,A} = c_{O,B}$). Furthermore, taking into account that electrochemical potential is the main factor which determines reduction rates, the following approximations can be assumed: the chemical rate constants ($k_{red,A} = k_{red,B} = k_{red}$), the transfer coefficients ($\beta_A = \beta_A = \beta$) and the number of electrons ($n_A = n_B = n$) are equal. Therefore, the difference between the standard potentials of two metals A and B can be related to the rates of electron transfer ratio as follows,

$$\log\frac{v_A}{v_B} = \frac{1}{2.3}\frac{\beta nF(\varepsilon_B - \varepsilon_A)}{RT} \tag{4.2}$$

This equation obeys the general belief that the more different the standard reduction potentials of the two metals are, the greater the two metals reductions rates ratio is.

The surfactant film flexibility is simulated by the material intermicellar exchange criteria 1 and 2 as follows. The prerequisites for material intermicellar exchange to occur are: On one hand, during a collision fused micelles should be together long enough to material be redistributed. On the second hand, the channel which connects both micelles must become large enough to allow the material exchange. Model assumes that the dimer stability is the key point restricting the exchange of reactants and free metals (isolated species), because these species can travel through the channel one by one. Therefore, a higher dimer stability means that two micelles stay together longer, so more material can be exchanged. Channel size is not influential in this case. Based on this assumption, k_{ex} parameter is directly related to the dimer stability. In contrast, if a nanoparticle (composed by aggregation of a number of metal atoms) is the exchanged specie, the channel size becomes crucial because a particle has to cross the channel as a whole. As a result, the exchange of a particle is restricted by the channel size between micelles (f parameter). Summarizing, the flexibility of the surfactant film is incorporated in the model by the k_{ex} (dimer stability) and the f (intermicellar channel size) parameters [47]. For example, a successful agreement between model and experiments was achieved when a flexible film, such as isooctane/Tergitol/water, was related to a channel size $f = 30$ (only aggregates composed by less than 30 atoms can traverse the channel) and $k_{ex} = 5$ (a maximum of 5 free atoms are allowed to be exchanged during a collision) [20]. These two factors have to decrease together when the film is more rigid. Accordingly, a microemulsion composing of AOT (dioctyl sodium sulfosuccinate)/n-heptane/water was simulated by the values $f = 5$ and $k_{ex} = 1$ [48].

It is believed that nanoparticle formation starts from a nucleus, which grows by building up new layers until it becomes a particle. So the sequential material layer processing of the metals determines the metal segregation in final nanostructure. Each simulation was run until the number and type of species carried by all micelles remains unchanged, that is, both chemical reactions and the intermicellar exchange is end up. Each run provides a set of micelles that can or not contain one particle, whose bimetallic nanoarrangement may be distinct. The order of metal deposition of each nanoparticle is studied as time advances. Finally, the resulting nanostructures are averaged over 1,000 runs. Assuming a spherical ordering, the averaged sequence is divided in ten concentric layers and lastly the averaged composition (%A) and the nanoparticle dispersion are determined layer by layer.

Finally, phase inmiscibility also plays a relevant role in the resulting metal segregation. For example, recent results show that Pd-Pt homogeneous nanoalloys exhibit a tendency to segregate, being this segregation easier at nanoscale [49]. The main driving force for the metal segregation in Pd-Pt is relative surface tension. In opposition, other physicochemical properties such as negative enthalpy of formation and lack of miscibility gap in bimetallic phase diagram favour the formation of homogeneous alloys. These features are not included in the model.

4.3 Characterization methodologies and instrumentation techniques

Model predictions were successfully compared with experimental results of Au/Pt nanoparticles prepared as described in Section 4.2.1, and under similar conditions than those explored by simulation (see reference [20] for details). The particles were characterized by HR-STEM including cross section with Energy Dispersive X-Ray Spectroscopy (EDX) analysis to obtain their nanostructure. For comparison purposes, the structures obtained by simulation were used to calculate the amount of each metal crossed by a beam of 2 Å (which is the estimated size of the experimental EDX beam), so that theoretical STEM profiles can be obtained. Both experimental and simulated profiles were normalized to 1 in order to make a comparison easier. Left column of Figure 4.1 shows the histograms for the theoretical particles at different reactant concentrations. The theoretical and experimental STEM profiles are shown in centre and right columns, respectively. Both STEM profiles show a Pt (the slower reduction metal) enrichment in the outermost layers, while Au (the metal which reduces faster) accumulates in the core. One can observe that a deeper STEM profile is obtained as concentration increases. That is, the inner layers contain less Pt and more Pt is located in the outer ones. This means that a high concentration leads to an improved segregation of metals. The good agreement between experimental and theoretical results proves the ability of the model to predict the nano-arrangement of bimetallic particles synthesized by one-pot microemulsion method.

4.4 General remarks about factors affecting metals distribution

Based on previous studies, the following rules can be established:

Rule 1: The higher the difference between reduction potentials of the two metals composing the bimetallic nanoparticle, the higher the metal segregation. The reason for this is that the reduction potentials are directly related to the tendency of the salt to acquire electrons (and then be reduced). So, it is believed that the metal salt with a higher reduction potential has the priority in chemical reduction. Therefore, taking into account that the reductions of the two metals start simultaneously, one can expect that a large difference in the reduction potentials will give rise to a core-shell arrangement, and a small one leads to a nanoalloy [50].

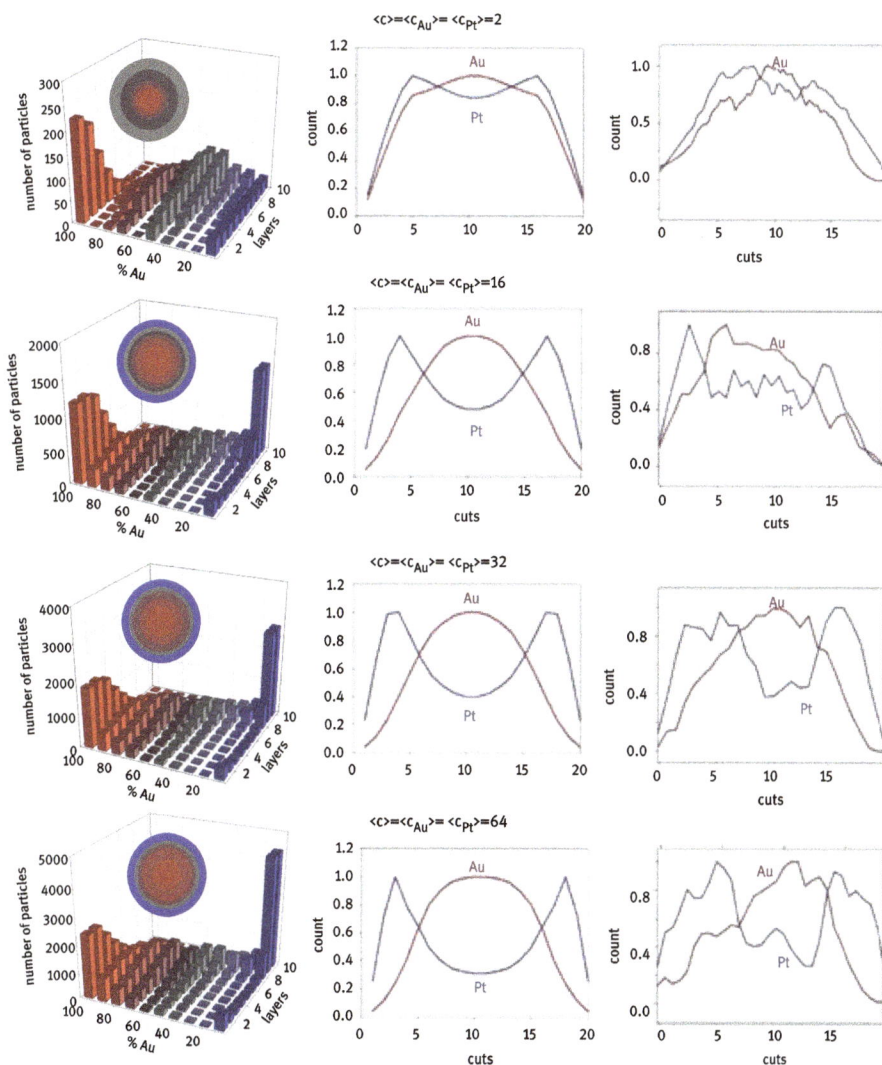

Figure 4.1: Left column: histograms obtained by simulation using different initial concentrations (Au:Pt = 1:1). Lighter colour means a higher proportion of pure metal in the layer (red: 100% Au, blue: 100% Pt, grey: 50% Au-Pt). Centre column: theoretical STEM profiles obtained for the average nanoparticle. Right column: experimental STEM profiles obtained for Au/Pt nanoparticles prepared in a water/Tergitol/isooctane microemulsion. Simulation parameters: flexible film (k_{ex} = 5, f = 30); reduction rate ratio (v_{Au}/v_{Pt} = 100/10); reducing agent concentration $\langle cR \rangle = 10\langle M^+ \rangle$. Adapted with permission from reference [20]. Copyright (2015) American Chemical Society.

For example, during the simultaneous reduction of $AuCl_4^-$ (ε^0 = 1.002 V) and $PtCl_6^{2-}$ (ε^0 ($PtCl_6^{2-}/Pt$) = 0.742 V [51]), the earliest nuclei will be mainly composed of Au, which plays as the seed for the deposition of Pd on their surface. As the slowest reduction starts to take place, the middle layers will show a progressive enrichment in Pt, which builds up the covering shell (surface). This sequence of metal salt reduction leads to a core-shell nanoarrangement. On the contrary, almost equal reduction potentials suggest that both metals are reduced at the same time, which improves the mixture of the two metals at atomic level, and resulting in a nanoalloy. This is the case of Ag-Pd nanoparticles synthesized from Ag^+ ($\varepsilon^0 = 0.80$ V) and Pd^{2+} ($\varepsilon^0 = 0.915$ V), which were always obtained as alloys [8, 52, 53] in microemulsions by a one-pot method.

Rule 2: The more flexible the surfactant film of the microemulsion, the greater the degree of mix in the final nanoparticle. Both experimental and simulation results showed that the metal pairs with higher enough difference in standard potentials ($\Delta\varepsilon^0 = 0.20$ V) **[54]** have been prepared as nanoalloys or core-shell structures using different microemulsions. In all cases, surfactants that provide a rigid film favour the formation of segregated nanoarrangements.

Rule 3: The higher the concentration, the higher the metal segregation. Both experiments and simulation proved that the metal segregation in Au/Pt nanoparticles is better as concentration increases [20] (for a deeper kinetic study, see reference [39]).

Finally, recent simulation results suggest that the ability of the microemulsion to modify the bimetal intrastructure is lost if both reductions are slow [40]. That is, only in case of one of the metals has a fast metal reductions, the bimetallic arrangement can be changed by using a different microemulsion composition and/or changing concentration. Under other conditions, the material intermicellar exchange rate is irrelevant because of a chemically controlled kinetics. It leads to a fixed metal distribution that only depends on the reduction rates of the particular pair of metals, and neither microemulsion composition nor concentration can alter the final metal distribution.

4.5 How to prepare Au/M (M = Ag, Pt, Pd) in a core-shell structure

Table 4.1 shows different Au/M (M = Ag, Pt, Pd, Ru) nanoparticles synthesized in microemulsions taken from literature. Rather than making a review of Au/M nanoparticles, only those Au/M nanoparticles obtained from microemulsions by a one-pot method are under consideration. Table 4.1 provides a summary information on the difference between the reduction potential of the two metals, the intrastructure of the final nanoparticle, microemulsion composition, metal salts and reductor agent, flexibility of surfactant film, metal salts concentration and resulting particle size, if available. Au/M nanoparticles data are arranged in Table 4.1 according to the difference between reduction potential in each couple of metals. This is because, as mentioned in Rule 1, it is assumed that very different reduction potentials

Table 4.1: Different Au/M (M = Ag, Pt, Pd) bimetallic nanoparticles prepared from microemulsions by a one-pot method.

Experiments	Metals	$\varepsilon°$ (V)	Structure	Microemulsion; reactants	Film flex	c (M)	Size (nm)	Ref.
1	Au-Ag	0.20	Au core enriched in Ag shell	Water/AOT/isooctane; N_2H_5OH; Ag^+, $AuCl_4^-$; $\omega = 10$	Rigid	0.1	5.1	[56]
2			Alloy	Water/$C_{11}E_3$-$C_{11}E_5$/ cyclohexane; $NaBH_4$; Ag^+, $AuCl_4^-$	Flex	0.05	6.7	[57]
3			Alloy	Water/TritonX-100/cyclohexane $NaBH_4$; Ag^+, $AuCl_4^-$; $\omega = 3$	Flex	0.05	23	[58]
4	Au-Pt	0.26	Core-shell	Water/AOT/isooctane; N_2H_5OH; $AuCl_4^-$, $PtCl_6^{2-}$; $\omega = 8$	Rigid	0.5	3.8	[34]
5			Core-shell	Water/Tergitol/isooctane; N_2H_5OH; $AuCl_4^-$, $PtCl_6^{2-}$	Flex	0.01–0.40	–	[20]
6			Alloy	Water/Tergitol 15-S-5/isooctane; N_2H_5OH; $AuCl_4^-$, $PtCl_6^{2-}$	Flex	–	–	[59]
7			Alloy	Water/TritonX-100/cyclohexane/1-hexanol; $NaBH_4$; $AuCl_4^-$, $PtCl_6^{2-}$; $\omega = 3$	Flex	0.00027	2.5	[60]
8	Au-Pd	0.39	Core-shell	Water/AOT/isooctane; N_2H_5OH; $AuCl_4^-$, $PdCl_4^{2-}$; $\omega = 6$	Rigid	0.5	2.8	[61]
9			Enriched in Au core/enriched in Pd shell	Water/Brij-30/n-heptane; $NaBH_4$; $AuCl_4^-$, $PdCl_4^{2-}$	Rigid	–	5.0	[62]
10			Alloy	Water/TritonX-100/n-hexane/n-hexanol; N_2H_5OH; $AuCl_4^-$, $PdCl_6^{4-}$	Flex	0.005/0.006	5.1	[63]
11	Au-Ru	0.40	Core-shell	Water/BrijL4/ heptane; $NaBH_4$; $AuCl_4^-$, Ru^{3+}; $\omega = 8$	Rigid	0.0025	4–5	[64]

will give rise to well segregated structures, and quite similar ones lead to alloys [50]. Table 4.1 follows the expected trend, that is, as both reduction rates are more different [55], core-shell bimetallic nanoparticles can be obtained. Au-Ag nanoparticles, whose difference in reduction potentials is small ($\Delta\varepsilon^0 = 0.20$ V), were obtained as homogeneous alloys when flexible films were used. Only using a rigid film Au-Ag nanoparticles with a slight metal segregation (an enriched in Au core surrounded by enriched in Ag outer layers) can be prepared [56]. As the difference between standard potentials of the couple is increased, true core-shell structures have been obtained for Au-Pt, Au-Pd and Au-Ru, even if a flexible film such as water/Tergitol/isooctane microemulsion is used [20].

The second point of interest is related to the influence of microemulsion composition. As established in Rule 2, a clear tendency towards core-shell structures can be observed as the film is more rigid. Thus, when the microemulsion composition includes surfactants with low flexibility such as AOT [34] and Brij-30 [13] nanoparticles showed core-shell structures, contrary to the ones obtained with more flexible surfactants (Tergitol [11] and TritonX-100 [58]).

A further interesting aspect to note in the experiments shown in Table 4.1 is the influence of reactant concentration. As established in Rule 3, the effect of increasing the amount of reactants on final metal segregation was proven to be determinant to obtain core-shell structures. This tendency can be observed in all cases except for Au/Ru, which only be prepared using a value of concentration.

In all couples shown in Table 4.1, Au precursor is $AuCl_4^-$, which has a very fast reduction rate. As a matter of fact, Au reduction is even quick enough to make necessary the use of stopped-flow spectrophotometer, because the detection of the time required to reach final sizes could not be detected by conventional methods [34]. This is in agreement with simulation results, which proved that Au reduction rate was successful simulated by considering that 100% of metal salt within the micelle is reduced at each collision [20]. As mentioned, from a kinetic point of view, it indicates that $AuCl_4^-$ reduction in microemulsion is controlled by the intermicellar exchange rate. The presence of Au is relevant, because the ability to modulate the nanostructure associated to the microemulsion composition and concentration is only proved when Au is one of the metals composing the nanoparticle. In accordance with the current status of researches and to the best of our knowledge, there is no bimetallic nanoparticle without Au whose intrastructure has been changed just by changing the microemulsion composition or reactants concentration. Moreover, simulation results suggest that only if both metal reductions are quite rapid when compared to the exchange rate (dictated by the microemulsion), a change in the microemulsion composition allows to alter the final bimetallic nanostructure. Otherwise, slower metal reductions result in chemically controlled kinetics, in which case neither microemulsion composition nor concentration can change the resulting nanostructure [40].

One can conclude that bimetallic nanoparticles containing Au ($\varepsilon°(\text{AuCl}_4{}^{2-}/\text{Au})$ = 1.00 V) combined with another slower metal M ($\varepsilon°(\text{M}) < 1.00$ V) can be obtained as a core-shell or as an alloy just by changing the reactants concentration and/or microemulsion composition.

In order to get more insight at the influence of reagents concentration on metal segregation, Figure 4.2 shows Au/Pt nanostructures obtained by simulation using a flexible film (such as the one provided by 75% Isooctane/20% Tergitol/5% water microemulsion) and different concentrations. At low concentration (Figure 4.2(a), $\langle c[\text{AuCl}_4]^-\rangle = \langle c[\text{PtCl}_6]^{2-}\rangle = \langle c\rangle = 2$ metal salts in a micelle) most of the particles are formed by a core mainly composed by Au, surrounded by mixed layers, i.e. an alloyed surface is obtained. In Figure 4.2(b) a rather high concentration ($\langle c\rangle = 4$ metals salts in a micelle) results in an outer layer slightly enriched in Pt. By observing blue bars on the right, one can see that this Pt improvement in the surface is more noticeable as increasing concentration (see Figure 4.2(c), $\langle c\rangle = 6$ metals salts in a micelle) until a nearly core-shell structure is obtained at $\langle c\rangle = 8$ metals salts in a micelle, as shown in Figure 4.2(d). That is, a transition from a mixed nanoalloy to a well segregated nanostructure can be observed as concentration is increased. Similar behaviour has been obtained using different film flexibilities. One can conclude that Au/Pt particles can be prepared with a predesigned nanoarrangement just by properly choosing initial reagents concentration and/or materials composing the microemulsion. For example, if a core-shell nanoparticle is our objective, metal salt concentration must be higher than 8 metal ions in a micelle.

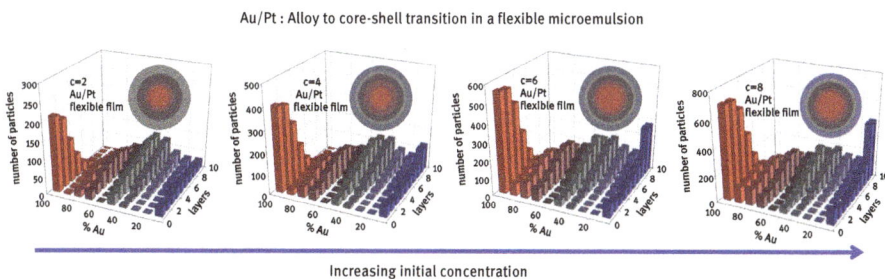

Au/Pt : Alloy to core-shell transition in a flexible microemulsion

Increasing initial concentration

Figure 4.2: Histograms obtained by simulation using a flexible film (k_{ex} = 5, f = 30) and different initial concentrations ($\langle c[\text{AuCl}_4]^-\rangle = \langle c[\text{PtCl}_6]^{2-}\rangle = \langle c\rangle$; $\langle c$ hydrazine$\rangle = 10 \langle c\rangle$). Histograms represent the quantity of nanoparticles with a fixed percentage of Au in each layer, from the inside (core, layer one) to the outside (surface, layer ten). Chemical reduction rates: Au reduction is assumed as an instantaneous one (100% of Au salts within colliding micelles react); in the case of Pt, the amount which reacts in each collision is only 10% of Pt salts colour pattern: blue (0–45% of Au), grey (45–55% of Au), red (55–100% of Au). Less red signifies less Au. Nanoarrangement is also represented by concentric layers, which obeys the same colour pattern. Adapted with permission from reference [46]. Copyright (2018) Elsevier.

For the purpose of providing assistance in the preparation of core-shell Au/M (M = Ag, Pt, Pd) nanoparticles, the minimum concentration to prepare core-shell structures was calculated by simulation for different pairs and in different microemulsions. Due to the segregation between the two metals is progressive as concentration is increased, a criterion is needed to define when a nanostructure can be considered as core-shell. The following criterion was chosen: a minimum of 80% of particles must have a surface whose composition is at least 70% in the slower metal (70–80 criterion). In order to take a close look at the surface of nanoparticles, Figure 4.3 shows simulations results on the distribution of particles with a given percentage of Pt in the surface, that is, the distribution composition of the outer layer in the histograms shown in Figure 4.2. Figure 4.3(a) shows that the outer layer in most of particles is composed by a 50/50 mixture of Au and Pt, and only 27% of particles contain between 70% and 100% Pt in the surface. As concentration increases, this percentage also increases. Thus, 47% of particles meet the Pt requirement at $\langle c \rangle = 4$ (Figure 4.3(b)), and 53% of particles at $\langle c \rangle = 6$ (Figure 4.3(c)). In spite of the quite better metal segregation observed in Figure 4.2(d) ($\langle c \rangle = 8$), only 60% of particles meet the surface requirement. In order for the requirement that 80% particles have at least 70% Pt in the surface to be satisfied, it was necessary to rise concentration until $\langle c \rangle = 18$ metal salts per micelle. Indeed, a true core-shell structure is obtained at $\langle c \rangle = 18$ (see Figure 4.4(a)). The surface composition, shown in Figure 4.4(b), satisfies the required conditions (70–80 criterion).

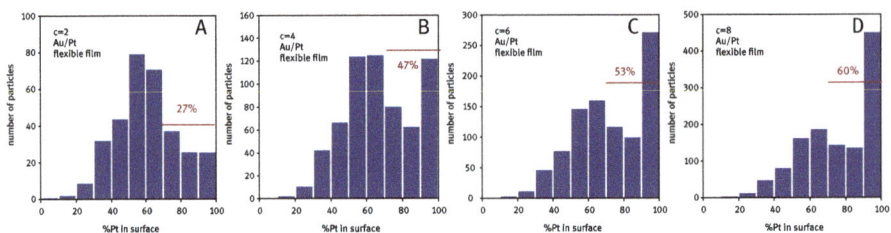

Figure 4.3: Percentage of Pt in Au/Pt nanoparticle surface (100% of Au salts and 10% of Pt salts react in each collision). Flexible film ($k_{ex} = 5$, $f = 30$); reagents concentration: ($\langle c[AuCl_4]^- \rangle = \langle c[PtCl_6]^{2-} \rangle = \langle c \rangle$; $\langle c$ hydrazine$\rangle = 10 \langle c \rangle$).

In case of less matching requirements, lower concentrations are sufficient. For example, if only 70% of particles must be composed by a minimum of 60% Pt, a concentration $\langle c \rangle = 6$ ions per micelle is enough, unlike for "70–80" criterion, which does not rises until $\langle c \rangle = 18$.

A similar qualitative behaviour has been obtained when different microemulsion compositions (changing f and k_{ex} simulation parameters) are studied. A transition from nanoalloy to well-segregated core-shell structures was obtained, but a lower concentration is necessary to meet the requirements as the surfactant film is more rigid, as expected.

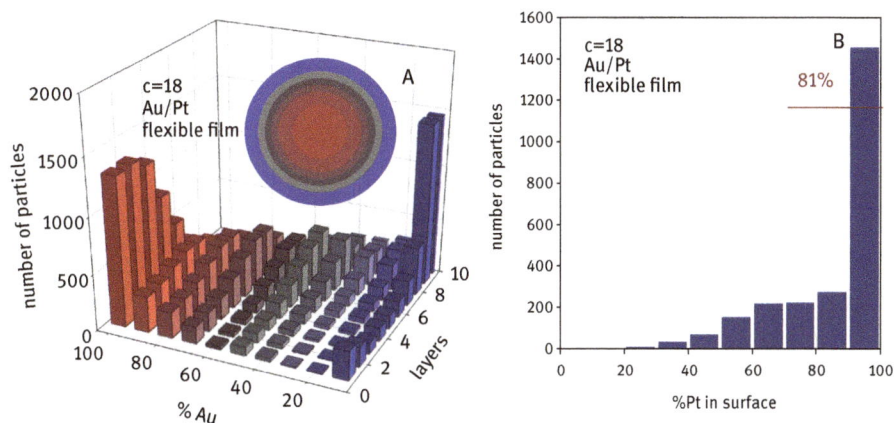

Figure 4.4: (A) Number of particles composing by different percentages of Au in each layer for an initial concentration ($\langle c[AuCl_4]^-\rangle = \langle c[PtCl_6]^{2-}\rangle = \langle c\rangle = 18$; $\langle c$ hydrazine$\rangle = 10 \langle c\rangle$), and employing a flexible film ($k_{ex} = 5$, $f = 30$). Au and Pt chemical rates: 100% of Au salts and 10% of Pt salts react in each collision. Colour pattern: blue (0–45% of Au), grey (45–55% of Au), red (55–100% of Au). Less red signifies less Au. Nanoarrangement is also represented by concentric layers which obeys the same colour pattern. (B) Surface composition (% of Pt).

The calculation of the minimum concentration to get a well-defined core-shell structure that meets the requirement 70–80 was repeated for different pairs of metals. To this end, the reduction of $AuCl_4^-$ (characterized by a $v_\% = 100\%$ reactants are reduced in each collision) was simulated combined with different metals M, whose reduction was slower than that of Au precursor. Metals are characterized by their reduction rate $v_\%$ (ranging from 3% to 25% of reactants that react in a collision). For each pair Au/M, the study was carried out using microemulsion with three different compositions (different f and k_{ex} simulation parameters). Figure 4.5 shows the minimum value of concentration to get a core-shell structure (expressed in initial number of metal ions per micelle) for different slow metals M and different microemulsions. The faster metal was always Au.

In order to guide experiments, the minimum concentration to obtain a core-shell structure is also expressed in molar concentration on the right y-axis. It was calculated from an isooctane 75%/Tergitol 20%/water 5% microemulsion, whose micelle radius was 7.3 nm (obtained by DLS [20]). For example, from this radius, and assuming spherical shape ($V_{micelle} = 4/3\pi r^3$), the molar concentration of a micelle carrying 18 metal salts can be calculated as follows:

$$\langle c\rangle = 18 \frac{\text{ions}}{\text{micelle}} \frac{1}{V_{micelle}(L.\text{micelle}^{-1})} \frac{1}{N_{Av}(\text{ions.mol}^{-1})} = 0.15\,M \qquad (4.3)$$

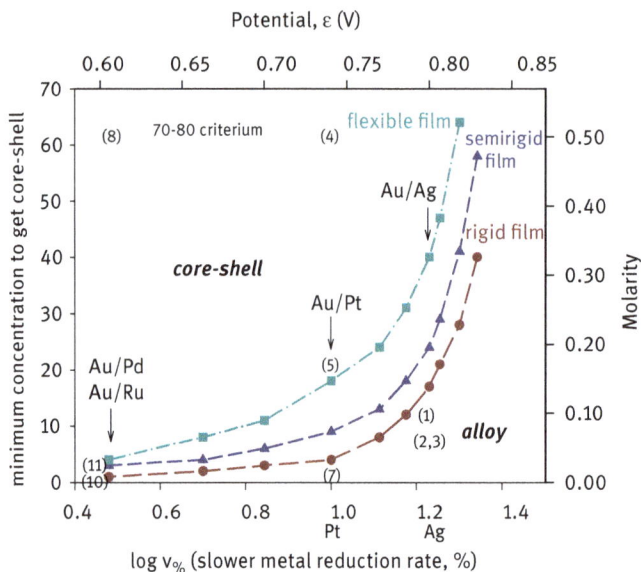

Figure 4.5: Simulation results on the minimum initial concentration required to obtain a core-shell nanostructure (80% of particles have at least 70% of slower metal in the surface) versus log $v_\%$ (reduction rate of the slower metal in percentage of metal salts which are reduced during a collision). Upper x-axis: standard potential ε° (V), based in a linear log $v_\%$ vs. ε° plot. Red, blue and green symbols show results using a rigid ($f = 5$, $k_{ex} = 1$), semi-rigid ($f = 15$, $k_{ex} = 3$) and flexible ($f = 30$, $k_{ex} = 5$) film, respectively. Lines are only guides to the eye. Molarity is calculated for a micelle of radium $r = 7.3$ nm.

where N_{Av} is the Avogadro's number. The concentration of the reducing agent $\langle c_R \rangle$ was always ten times larger than that of metal salts.

This study is restricted to metals whose reduction rate is slower than 25% (25% of reactants inside the micelle can react in each collision), because over this value the concentration to get core-shell structures was always larger than $\langle c \rangle = 64$ ions/ micelle ($\langle c \rangle = 0.52$ M in an isooctane 75%/Tergitol 20%/water 5% microemulsion, micelle radius 7.3 nm). Using high concentration inside micelles results in larger nanoparticles, which lose interest as catalysts. Only for reduction rates slower than 25% (log $v_\%$=1.4, see Figure 4.5), and depending on the microemulsion composition, a transition from alloy to core-shell structures was obtained at concentration smaller than $\langle c \rangle = 64$ ions/micelle.

A key outcome from Figure 4.5 is that, for a particular Au/M pair of metals, the required concentration to prepare a core-shell structure strongly depends on the microemulsion composition. Taking into account that Au is the faster metal in all pairs, the rate of the metal with a slower reduction is the main feature. Therefore, one can conclude that the greater the difference between the two metals reduction rates, the lower the concentration needed to prepare core-shell nanoparticles, as expected.

With the aim of successfully designing experiments to obtain a given nanoparticle architecture, it would be interesting to establish a link between the metal standard potential and its reduction rate, but it is not obvious. Indeed it was proposed that redox-potential of atomic silver is $\varepsilon^{\circ}(Ag_1/Ag^+) = -1.8$ V instead of $\varepsilon^{\circ}(Ag/Ag^+) = +0.799$ V for Ag metal [65, 66]. However, to the best of our knowledge, the relationship between reduction rate and standard reduction potential could obey a log v – ε° relationship (see eq. (4.1)). In this line, Troupis et al. [67] have achieved an accurate redox-control on electron transfer rate for silver nanoparticles synthesized using polyoxometalates: The reaction rate of Ag^+ reduction using different polyoxometalates as reducing agents was measured, and it was found that polyoxometalates with the more negative redox potentials are faster in electron transfer to Ag^+. A linear function of log k (the rate constant of Ag^+ reduction) versus the reduction potential of various polyoxometalates was obtained (see Figure 6 in reference [67]). This linearity of log k versus ε° is in agreement to Volmer equation (see eq. (4.1)).

In accordance with previous experimental approaches on the preparation of Au/Pt nanoparticles, Au and Pt reduction rates can be related to 100% and 10% reaction in a collision, respectively [20]. These simulated reduction rates (in percentage of reaction per collision) are directly related to the rate chemical constants of metal ions reduction, which in turn are related to standard reduction potentials. Therefore, Au is characterized by a $v_\% = 100$ (log $v_\% = 2$) and $\varepsilon^{\circ}(AuCl_4^{2-}/Au) = 1.00$ V, and Pt is associated to $v_\% = 10$ (log $v_\% = 1$) and ε° ($PtCl_6^{2-}/Pt) = 0.742$ V. That is, the more positive the standard reduction potential is the more likely the metal precursor will be reduced. Taking these two values and assuming a linear behaviour of log $v_\%$ versus ε°, the value of $v_\%$ can be calculated for different metals from their standard potential value ε°. In this way, ε° of the slower metal M is also represented on the upper x-axis in Figure 4.5. From this understanding, Au/Ag, Au/Pt and Au/Ru pairs are marked in Figure 4.4 ($v_{\%,Ag} = 17\%$, $v_{\%,Pd} = 3.3\%$ and $v_{\%,Ru} = 2.85\%$).

Knowing the relationship among variables (microemulsion composition, slower metal standard potential and minimum concentration to get core-shell arrangement) can help to guide further experimental studies.

With the aim to check the model predictions, data shown in Figure 4.5 are used to cross-check results from the experiments in Table 4.1.

Au/Ag nanoparticles: Assuming $\varepsilon^{\circ}(Ag/Ag^+) = +0.799$ V and a log v – ε° linear behaviour, the reduction rate of Ag precursor can be approached as $v_{Ag,\%} = 17\%$. AOT surfactant provides a rigid film. Therefore, from Figure 4.5, one can see that an $\varepsilon^{\circ} = 0.799$ V cuts the rigid line at $\langle c_{min}\rangle = 17$ ions per micelle = 0.14 M. This means that concentration must be at least 0.14 M to obtain a well-defined core-shell structure. Due to the concentration used in the experiment was smaller ($c = 0.1$ M[56]), an Ag enrichment in surface is observed, but not a true core-shell structure. This experiment is located in Figure 4.4 as (1).

In experiments 2 and 3, both microemulsions (water/$C_{11}E_3$-$C_{11}E_5$/ cyclohexane and water/TritonX-100/cyclohexane, respectively) can be considered as flexible ones. Under these conditions, one can infer from Figure 4.5 that a core-shell Au/Ag nanoparticle can only be obtained by using a minimum concentration $\langle c_{min} \rangle = 40$ ions per micelle = 0.33 M. Taking into account that the concentration was much lower (0.05 M in both experiments) an alloyed distribution of metals was obtained in both cases (noted in Figure 4.4 as (2, 3)).

Au/Pt nanoparticles: As mentioned, Pt ($\varepsilon°(\mathrm{PtCl_4^{2-}/Pt}) = +0.742$ V) was success-fully simulated with a $v_{Pt,\%} = 10\%$ (log $v_{Pt,\%} = 1$). For a rigid film (water/AOT/isooc-tane microemulsion) the minimum required concentration to obtain a core-shell arrangement is $\langle c_{min} \rangle = 4$ ions per micelle = 0.03 M, much lower than $c = 0.5$ M, the concentration used in experiment 4 [34]. Therefore, a core-shell structure was ob-tained, as predicted (noted in Figure 4.5 as (4)).

In experiment 5 [20], Au/Pt nanoparticles were synthesized from a water/Tergitol/isooctane (flexible film). The simple approach of varying the concentration results in different metal distributions, ranging from pure Au core surrounded by alloyed surfa-ces at $c = 0.01$ M to typical core-shell structures at 0.08 M, 0.16 M and 0.40 M, varying the degree of metal segregation as a function of concentration. Metal distributions at 0.16 M and 0.40 M satisfy the 70–80 criterion used in Figure 4.5 to establish a $\langle c_{min} \rangle = 18$ ions per micelle = 0.15 M as the concentration required for true core-shell Au/Pt nanoparticles synthesized from a flexible microemulsion (in Figure 4.5 as (5)).

Experiment 6 [11] cannot be included in this study because concentration is not available. In relation to experiment 7 [58], a water/TritonX-100/cyclohexane/ 1-hex-anol microemulsion can be associated to a flexible film, so a $\langle c_{min} \rangle = 18$ ions per mi-celle = 0.15 M is required, much higher than $c = 0.00027$ M used in this experiment, which justifies the obtained alloyed nanoparticle (in Figure 4.5 as (7)).

Au/Pd nanoparticles: Based on standard potential ($\varepsilon°(\mathrm{PdCl_6^{4-}/Pd}) = 0.615$ V), the reduction rate of Pd can be related to $v_{Pd,\%} \sim 3\%$. The microemulsion used in experiment 8 [61] was water/AOT/isooctane, so the film can be characterized as a rigid one. From Figure 4.5, the minimum concentration to obtain a core-shell arrangement is $\langle c_{min} \rangle = 0.008$ M, much lower than the one used in the experiment ($c_{exp8} = 0.5$ M). Therefore, a core-shell nanoarrangement was obtained, as ex-pected (in Figure 4.5 as (8)).

In experiment 9, 1:1 Au/Pd nanoparticles have a palladium-rich surface [62]. These data can not be verified because concentration is not available. But taking into account the small value of $\varepsilon°$ (small $v_{Pd,\%}$) combined with the rigidity of the film (water/Brij-30/n-heptane microemulsion), a concentration $\langle c_{min} \rangle = 0.008$ M is enough to justify the metal segregation. On the contrary, the higher flexibility of the water/TritonX-100/n-hexane/n-hexanol microemulsion employed in experiment 10 [63] leads to a $\langle c_{min} \rangle = 0.03$ M, that is higher than experimental one $c_{exp10} = 0.005$ M. As a consequence, an alloy is obtained (in Figure 4.5 marked as (10)).

Au/Ru nanoparticles ($\varepsilon°(\mathrm{Ru^{3+}/Ru}) = 0.60$ V, $v_{Ru,\%} \sim 3\%$): The water/BrijL4/ hep-tane employed in experiment 11 [64] can be associated with a rigid film, so a $\langle c_{min} \rangle = 0.008$ M is needed to get core-shell structures. Due to the concentration used was higher ($c_{exp11} = 0.025\,M$), the resulting nanoparticle shows a core-shell metal distribution (in Figure 4.5 as (11)).

Summarizing, the intrastructure of Au/M nanoparticles taken from literature was compared with the ones predicted by simulation. This holds true in all studied cases.

4.6 Conclusions and future perspective

The minimum reactant concentration required to obtain true core-shell Au/M nanoparticles by the one-pot microemulsion route has been calculated by a simulation model under different synthesis conditions. This minimum concentration was proved to be dependent on the standard reduction potential of the metal whose reduction is slower and on the flexibility of the surfactant composing the microemulsion. Simulation results were tested by comparing to Au/M nanoparticles taken from literature. In all cases, experimental data obey model predictions. From this agreement, one can conclude that the smaller the standard potential of the slower reduction metal, the lower the minimum concentration needed to prepare well-defined core-shell nanoparticles. In addition, the higher the surfactant flexibility, the higher the minimum concentration to get core-shell nanoparticles.

Model prediction results allow to propose which is the best value of concentration in order to prepare a core-shell nanoparticle as a function of standard potential and microemulsion composition. This outlook may become an advanced tool for fine-tuning Au/M nanostructures. Thus, for a given Au/M pair (characterized by the reduction potential of the metal M) the specific combination of reactant concentration and microemulsion composition will determine the final arrangement. Therefore, two strategies are available to synthesize well-segregated Au/M nanoparticles by the one-pot method: either a change in concentration or a change in the microemulsion composition. Thus, just by changing one of these two parameters in accordance with the guide provided by model predictions, the desired nanoparticle can be prepared.

References

[1]   Sinfelt JH. Catalysis by alloys and bimetallic clusters. Acc Chem Res. 1977;10:15–20.
[2]   Sinfelt JH. Structure of bimetallic clusters. Acc Chem Res. 1987;20:134–9.
[3]   Gu X, Lu Z-H, Jiang H-L, Akita T, Xu Q. Synergistic catalysis of metal-organic framework-immobilized Au-Pd nanoparticles in dehydrogenation of formic acid for chemical hydrogen storage. J Am Chem Soc. 2011;133:11822–5.
[4]   Bandarenka AS, Varela AS, Karamad M, Calle-Vallejo F, Bech L, Pérez-Alonso FJ, et al. Design of an active site towards optimal electrocatalysis: overlayers, surface alloys and near-surface alloys of Cu/Pt(111), Angewandte Chemie. Int Ed. 2012;51:11845–8.

[5] Spanos I, Dideriksen K, Kirkensgaard JJ, Jelavic S, Arenz M. Structural disordering of de-alloyed Pt bimetallic nanocatalysts: the effect on oxygen reduction reaction activity and stability. Phys Chem Chem Phys. 2015;17:28044–53.

[6] König RY, Schwarze M, Schomäcker R, Stubenrauch C. Catalytic activity of mono- and bi-metallic nanoparticles synthesized via microemulsions. Catalysts. 2014;4:256–75.

[7] Zielinska-Jurek A, Kowalska E, Sobczak JW, Lisowski W, Ohtani B, Zaleska A. Preparation and characterization of monometallic (Au) and bimetallic (Ag/Au) modified-titania photocatalysts activated by visible light. Appl Catal B. 2011;101:504–14.

[8] Heshmatpour F, Abazari R, Balalaie S. Preparation of monometallic (Pd, Ag) and bimetallic (Pd/Ag, Pd/Ni, Pd/Cu) nanoparticles via reversed micelles and their use in the Heck reaction. Tetrahedron. 2012;68:3001–11.

[9] Jiang H-L, Xu Q. Recent progress in synergistic catalysis over heterometallic nanoparticles. J Mater Chem. 2011;21:13705–25.

[10] Notar Francesco I, Fontaine-Vive F, Antoniotti S. Synergy in the catalytic activity of bimetallic nanoparticles and new synthetic methods for the preparation of fine chemicals. Chem Cat Chem. 2014;6:2784–91.

[11] Hernández-Fernández P, Rojas S, Ocón P, Gómez de la Fuente JL, San Fabián J, Sanza J, et al. Influence of the preparation route of bimetallic Pt-Au nanoparticle electrocatalyst for the oxygen reduction reaction. J Phys Chem B. 2007;111:2913–23.

[12] Boutonnet M, Lögdberg S, Svensson EE. Recent developments in the aplication of nanoparticles prepared from w/o microemulsions in heterogeneous catalysis. Curr Opin Colloid Interface Sci. 2008;13:270–86.

[13] Habrioux A, Vogel W, Guinel M, Guetaz L, Servat K, Kokoh B, et al. Structural and electrochemical studies of Au-Pt nanoalloys. Phys Chem Chem Phys. 2009;11:3573–9.

[14] Sun S, Murray CB, Weller D, Folks L, Moser A. Monodisperse FePt nanoparticles and ferromagnetic FePt nanocrystal superlattices. Science. 2000;287:1989–92.

[15] Ferrando R, Jellinek J, Johnston RL. Nanoalloys: from theory to applications of alloy clusters and nanoparticles. Chem Rev. 2008;108:845–910.

[16] Hrubovcak P, Zelenakova A, Zelenak V, Kovac J. The study of magnetic properties and relaxation processes in Co/Au bimetallic nanoparticles. J Alloys Compd. 2015;649:104–11.

[17] Weber I, Solla-Gullon J, Brimaud S, Feliu JM, Juergen Behm R. Structure, surface chemistry and electrochemical de-alloying of bimetallic PtxAg100-x nanoparticles: quantifying the changes in the surface properties for adsorption and electrocatalytic transformation upon selective Ag removal. J Electroanal Chem. 2017;793:164–73.

[18] Beygi H, Babakhani A. Microemulsion synthesis and magnetic properties of $Fe_xNi_{(1-x)}$ alloy nanoparticles. J Magn Magn Mater. 2017;421:177–83.

[19] Felix-Navarro RM, Beltran-Gastelum M, Salazar-Gastelum MI, Silva-Carrillo C, Reynoso-Soto EA, Pérez-Sicairos S, et al. Pt-Pd bimetallic nanoparticles on MWCNTs: catalyst for hydrogen peroxide electrosynthesis. J Nanopart Res. 2013;15:1802/1–/11.

[20] Buceta D, Tojo C, Vukmirovik M, Deepak FL, López-Quintela MA. Controlling bimetallic nanostructures by the microemulsion method with sub-nanometer resolution using a prediction model. Langmuir. 2015;31:7435–9.

[21] Zhou S, McIlwrath K, Jackson G, Eichhorn B. Enhanced co tolerance for hydrogen activation in Au-Pt dendritic heteroaggregate nanostructures. J Am Chem Soc. 2006;128:1780–1.

[22] Luo J, Njoki PN, Lin Y, Mott D, Wang L, Zhong C-J. Characterization of carbon-supported AuPt nanoparticles for electrocatalytic methanol oxidation reaction. Langmuir. 2006;22:2892–8.

[23] Zhang W, Li L, Du Y, Wang X, Yang P. Gold/platinum bimetallic core/shell nanoparticles stabilized by a fréchet-type dendrimer: preparation and catalytic hydrogenations of phenylaldehydes and nitrobenzenes. Catal Lett. 2009;127:429–36.

[24] Suntivich J, Xu Z, Carlton CE, Kim J, Han B, Lee SW, et al. Surface composition tuning of Au-Pt bimetallic nanoparticles for enhanced carbon monoxide and methanol electro-oxidation. J Am Chem Soc. 2013;135:7985–91.

[25] Chen Y, Wang WG, Zhou S. Size effect of Au seeds on structure of Au-Pt bimetallic nanoparticles. Mater Lett. 2011;65:2649–51.

[26] Liang H-P, Jones TG, Lawrence NS, Jiang L, Barnard JS. Understanding the role of nanoparticle synthesis on their underlying electrocatalytic activity. J Phys Chem C. 2008;112:4327–32.

[27] Chen HM, Peng H-C, Liu RS, Hu SF, Jang L-Y. Local structural characterization of Au/Pt bimetallic nanoparticles. Chem Phys Lett. 2006;420:484–8.

[28] Zhang H, Toshima N. Synthesis of Au/Pt bimetallic nanoparticles with a Pt-rich shell and their catalytic activities for aerobic glucose oxidation. J Colloid Interface Sci. 2013;394:166–76.

[29] Yin Z, Ma D, Bao X. Emulsion-assisted synthesis of monodisperse binary metal nanoparticles. Chem Commun. 2010;46:1344–6.

[30] Wang L, Qi B, Sun L, Sun Y, Guo C, Li Z. Synthesis and assembly of Au-Pt bimetallic nanoparticles. Mater Lett. 2008;62:1279–82.

[31] Khalid M, Wasio N, Chase T, Bandyopadhyay K. In situ generation of two-dimensional Au-Pt core-shell nanoparticle assemblies. Nanoscale Res Lett. 2010;5:61–7.

[32] Luo J, Maye MM, Petkov V, Kariuki NN, Wang L, Njoki P, et al. Phase properties of carbon-supported gold-platinum nanoparticles with different bimetallic compositions. Chem Mater. 2005;17:3086–91.

[33] Wanjala B, Luo J, Fang B, Mott D, Zhong CJ. Gold-platinum nanoparticles: alloying and phase segregation. J Mater Chem. 2011;21:4012–20.

[34] Wu M, Chen D, Huang T. Preparation of Au/Pt bimetallic nanoparticles in water-in-oil microemulsions. Chem Mater. 2001;13:599–606.

[35] Zhang G-R, Zhao D, Feng Y-Y, Zhang B, Su DS, Liu G, et al. Catalytic Pt-on-Au nanostructures: why Pt becomes more active on smaller Au particles. ACS Nano. 2012;6:2226–36.

[36] Hartl K, Mayrhofer KJ, Lopez M, Goia D, Arenz M. AuPt core-shell nanocatalysts with bulk Pt activity. Electrochem Commun. 2010;12:1487–9.

[37] Shao M, Peles A, Shoemaker K, Gummalla M, Njoki PN, Luo J, et al. Enhanced oxygen reduction activity of platinum monolayer on gold nanoparticles. J Phys Chem Lett. 2011;2:67–72.

[38] Zhao L, Thomas JP, Heinig NF, Abd-Ellah M, Wang X, Leung KT. Au-Pt alloy nanocatalysts for electro-oxidation of methanol and their application for fast-response non-enzymatic alcohol sensing. J Mater Chem. 2014;2:2707–14.

[39] Tojo C, de Dios M, Buceta D, López-Quintela MA. Cage-like effect in Au-Pt nanoparticle synthesis in microemulsions: a simulation study. Phys Chem Chem Phys. 2014;16:19720–31.

[40] Tojo C, Buceta D, López-Quintela MA. Slowing down kinetics in microemulsions for nanosegregation control: a simulation study. J Phys Chem C. 2018;122:20006–18.

[41] Yu W, Porosoff MD, Chen JG. Review of Pt-based bimetallic catalysis: from model surfaces to supported catalysts. Chem Rev. 2012;112:5780–817.

[42] Bracey CL, Ellis PR, Hutchings GJ. Application of copper-gold alloys in catalysis: current status and future perspectives. Chem Soc Rev. 2009;38:2231–43.

[43] Toshima N, Yonezawa T. Bimetallic nanoparticles - Novel materials for chemical and physical applications. New J Chem. 1998;22:1179–201.

[44] Muñoz-Flores BM, Kharisov BI, Jiménez-Pérez VM, Elizondo Martínez P, López ST. Recent advances in the synthesis and main applications of metallic nanoalloys. Ind Eng Chem Res. 2011;50:7705–21.

[45] Bönnemann H, Richards RM. Nanoscopic metal particles - Synthetic methods and potential applications. Eur J Inorg Chem. 2001;10:2455–80. DOI: 10.1002/1099-0682(200109)2001.

[46] Tojo C, Buceta D, López-Quintela MA. Bimetallic nanoparticles synthesized in microemulsions: a computer simulation study on relationship between kinetics and metal segregation. J Colloid Interface Sci. 2018;510:152–61.

[47] Quintillán S, Tojo C, Blanco MC, López-Quintela MA. Effects of the intermicellar exchange on the size control of nanoparticles synthsized in microemulsions. Langmuir. 2001;17:7251–4.

[48] Tojo C, Blanco MC, López-Quintela MA. Preparation of nanoparticles in microemulsions: a Monte Carlo study of the influence of the synthesis variables. Langmuir. 1997;13:4527–34.

[49] Januszewska A, Dercz G, Lewera A, Jurczakowski R. Spontaneous chemical ordering in bimetallic nanoparticles. J Phys Chem C. 2015;119:19817–25.

[50] Feng J, Zhang C. Preparation of Cu-Ni alloy nanocrystallites in water-in-oil microemulsions. J Colloid Interface Sci. 2006;293:414–20.

[51] Li Y, Jiang Y, Chen M, Liao H, Huang R, Zhou Z, et al. Electrochemically shape-controlled synthesis of trapezohedral platinum nanocrystals with high electrocatalytic activity. Chem Commun. 2012;48:9531–9531.

[52] Wei G, Dai W, Qian L, Cao W, Zhang J. Reverse microemulsions synthesis and characterization of Pd-Ag bimetallic alloy catalysts supported on Al_2O_3 for acetylene hydrogenation. China Pet Process Petrochem Technol. 2012;14:59–67.

[53] Ström L, Ström H, Carlsson P, Skoglundh M, Härelind H. Catalytically active Pd-Ag alloy nanoparticles synthesized in microemulsion template. Langmuir. 2018;34:9754–61.

[54] Tojo C, Buceta D, López-Quintela MA. On metal segregation of bimetallic nanocatalysts prepared by a one-pot method in microemulsions. Catalysts. 2017;7:68–86.

[55] Tojo C, de Dios M, López-Quintela MA. On the structure of bimetallic nanoparticles synthesized in microemulsions. J Phys Chem C. 2009;113:19145–54.

[56] Chen D, Chen C. Formation and characterization of Au-Ag bimetallic nanoparticles in water-in-oil microemulsions. J Mater Chem. 2002;12:1557–62.

[57] Cheng J, Bordes R, Olsson E, Holmberg K. One-pot synthesis of porous gold nanoparticles by preparation of Ag/Au nanoparticles followed by dealloying. Colloids Surf A. 2013;436:823–9.

[58] Pal A, Shah S, Devi S. Preparation of silver, gold and silver-gold bimetallic nanoparticles in w/o microemulsion containing Triton X-100. Colloids Surf A. 2007;302:483–7.

[59] Hernández-Fernández P, Rojas S, Ocón P, Gómez de la Fuente JL, San Fabián J, Sanza J, et al. Influence of the preparation route of bimetallic Pt-Au nanoparticle electrocatalysts for the oxygen reduction reaction. J Phys Chem C. 2007;111:2913–23.

[60] Pal A. Gold–platinum alloy nanoparticles through water-in-oil microemulsion. J Nanostruct Chem. 2015;5:65–69.

[61] Wu M, Chen D, Huang T. Synthesis of Au/Pd bimetallic nanoparticles in reverse micelles. Langmuir. 2001;17:3877–83.

[62] Simoes M, Baranton S, Coutanceau C. Electrooxidation of sodium borohydride at Pd, Au, and Pd_xAu_{1-x} carbon-supported nanocatalysts. J Phys Chem C. 2009;113:13369–76.

[63] Li T, Zhou H, Huang J, Yin J, Chen Z, Liu D, et al. Facile preparation of Pd-Au bimetallic nanoparticles via in-situ self-assembly in reverse microemulsion and their electrocatalytic properties. Colloids Surf A. 2014;463:55–62.

[64] Ayed D, Laubender E, Souiri M, Yurchenko O, Marmouch H, Urban G, et al. Carbon nanotubes supported Ru-Au nanoparticles with core-shell structure for glucose detection with high resistance against chloride poisoning. J Electrochem Soc. 2017;164:B767–B75.

[65] Pomogailo AD, Dzhardimalieva GI. Reduction of metal ions in polymer matrices as a condensation method of nanocomposite synthesis. In: Nanostructured materials preparation via condensation ways. Dordrecht: Springer Science, 2014.

[66] Sviridov VV, Vorob'eva TN, Gaevskaya TV, Stepanova LI. Khimicheskoi osazhdenie metallov iz vodnykh rastvorov. Minsk: Izd. Universitetskoe, 1987:270.
[67] Troupis A, Triantis T, Hiskia A, Papaconstantinou E. Rate-redox-controlled size-selective synthesis of silver nanoparticles using polyoxometalates. Eur J Inorg Chem. 2008;2008:5579–86.

Bionotes

Concha Tojo is Professor of Physical Chemistry at the University of Vigo (Spain). She has published 56 publications with more than 900 citations (h-index 16). Her research fields include the synthesis of metallic and bimetallic particles in microemulsions, nanostructured bimetallic materials and chemical kinetics in microemulsions.

David Buceta defended his PhD in Chemistry in 2011 in the University of Santiago de Compostela (USC, Spain). He is currently a Postdoc of Physical Chemistry at same university. He was 3 years as a Postdoc in Germany, at the Technische Universität Berlin, working in photocatalysis. Before, he spent 3 years in USA, at the Brookhaven National Laboratory (NY), working in fuel cells and bimetallic particles. He has more than 25 papers with more than 450 citations (h-index 12). Youngest Scientific Advisor of the company NANOGAP (www.nanogap.es), a spin-off from the USC founded in 2006 and dedicated to the production of nanomaterials and sub-nm metal (0) clusters. His research interests are the synthesis of metallic and bimetallic particles and their catalytic and electrocatalytic properties; synthesis of "ligand-free" subnanometric clusters and the study of their properties (for instance photocatalysis, electrocatalysis, heterogeneous and homogeneous catalysis and biomedical properties).

M. Arturo López-Quintela is Full Professor of Physical Chemistry at the University of Santiago de Compostela (USC, Spain). Postdoc in Germany at MPI für Biophysikalishe Chemie, Göttingen and University of Bielefeld. Visiting Professor at MPI für Metallforschung, Stuttgart, Germany; Centre for Magnetic Recording Research, UCLA, USA; Yokohama Natl. University, Japan and Research Centre for Materials Science, Nagoya, Japan. Solvay Award in Chemistry and Burdinola Award in the Field of New Nanotechnologies in Chemistry. Co-founder and Principal Scientific Advisor of the company NANOGAP (www.nanogap.es), a spin-off from the USC founded in 2006 and dedicated to the production of nanomaterials and sub-nm metal (0) clusters. Since 2005 Co-editor of the Journal of Colloid and Interface Science. He has published more than 300 publications (h-index 53) and is co-author of 26 patents (most of them under exploitation). His current research interests are synthesis and properties of nanomaterials and "ligand-free" metal clusters by soft chemical techniques; synthesis of anisotropic nanomaterials and nanocomposites; catalytic, electrocatalytic, photocatalytic and therapeutic properties of "ligand-free" clusters.

Sumana Kundu and Vijayamohanan K. Pillai

5 Synthesis and characterization of graphene quantum dots

Abstract: Conventional inorganic semiconductor quantum dots (QDs) have numerous applications ranging from energy harvesting to optoelectronic and bio-sensing devices primarily due to their unique size and shape tunable band-gap and also surface functionalization capability and consequently, have received significant interest in the last few decades. However, the high market cost of these QDs, on the order of thousands of USD/g and toxicity limit their practical utility in many industrial applications. In this context, graphene quantum dot (GQD), a nanocarbon material and a new entrant in the quantum-confined semiconductors could be a promising alternative to the conventional toxic QDs due to its potential tunability in optical and electronic properties and film processing capability for realizing many of the applications. Variation in optical as well as electronic properties as a function of size, shape, doping and functionalization would be discussed with relevant theoretical backgrounds along with available experimental results and limitations. The review deals with various methods available so far towards the synthesis of GQDs along with special emphasis on characterization techniques starting from spectroscopic, optical and microscopic techniques along with their the working principles, and advantages and limitations. Finally, we will comment on the environmental impact and toxicity limitations of these GQDs and their hybrid nanomaterials to facilitate their future prospects.

This article has previously been published in the journal *Physical Sciences Reviews*. Please cite as: Kundu, S., Pillai, V. K. Synthesis and characterization of graphene quantum dots. *Physical Sciences Reviews* [Online] **2020**, 5. DOI: 10.1515/psr-2019-0013.

https://doi.org/10.1515/9783110345001-005

Graphical Abstract:

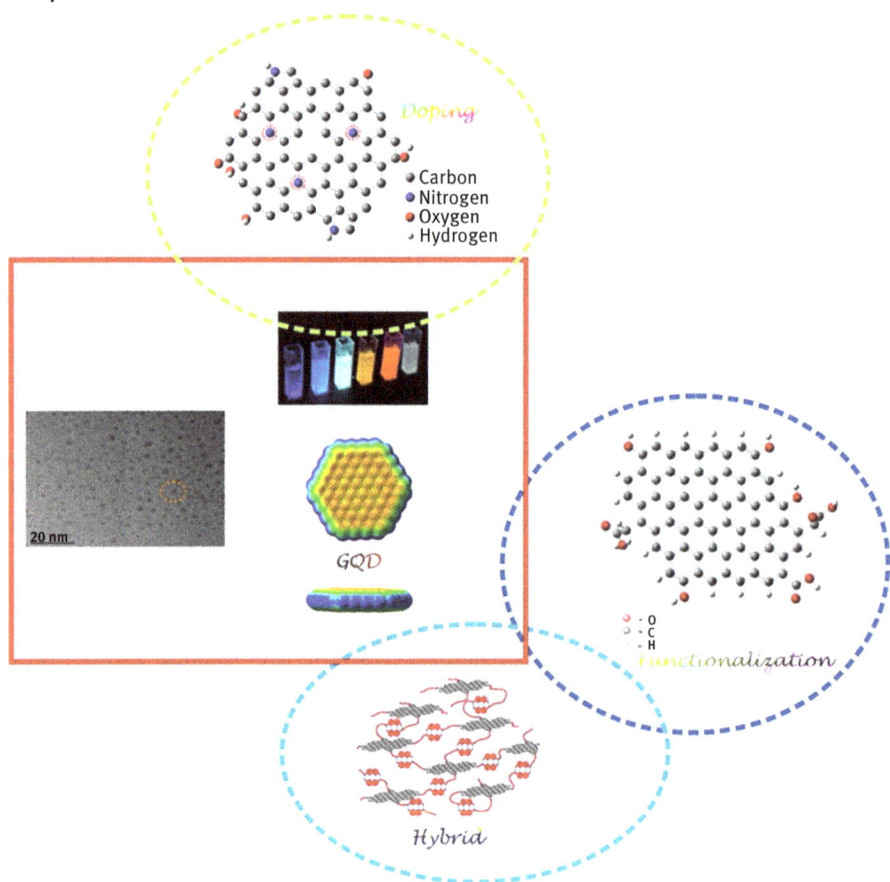

Structure of doped, functionalized and hybrid GQDs

Keywords: Graphene quantum dots (GQDs), Photoluminescence (PL), Electrochemical luminescence (ECL)

5.1 Introduction

5.1.1 The background of graphene

Carbon, the "king of the elements", [1] forms the backbone of organic chemistry and many consider this also as the basis of life. In fact, human beings have made this friendly and essential element to an "enemy" by breaking the natural carbon cycle [2] leading to global warming and related anthropogenic climate issues [2] and people often talk about "low carbon", technologies and efforts for "decarbonization" to lower the carbon footprint, especially while generating and storing energy. Deployment of "low carbon" to "carbon neutral" and "carbon-negative" energy technologies has made a profound impact on the quality of life in many emerging economies. The catenation property of carbon allows it to produce several chains and morphologies and we see various dimensional forms after interconnection carbon with other carbon atoms. Among the three dimensional (3-D) forms, diamond is the most precious and hardest material known so far and graphite is known since antiquity. Another popular allotrope is graphene ([2-D] planar sp^2 hybridized, honeycomb lattice of carbon atoms), nanotubes (quasi 1-D) [3] and fullerene (0-D). The first theoretical approach of studying the properties of graphene in the pretext of graphite was presented by Wallace in 1947 [4, 5]. Later in 1962, a transmission electron microscopic image of graphene has been reported [5]. Graphene is a zero-bandgap semiconductor as the valence and conduction bands of graphene overlap slightly. The room temperature electron mobility of graphene is measured to be as high as $15{,}000 \, cm^2 \, V^{-1} \, s^{-1}$ with very little temperature dependency and zero effective mass for the charge carriers [6]. Thus, unlike in most materials, the electron mobility in graphene is only hindered by defect scattering and not by phonon scattering and is the highest conductive material at room temperature with a resistivity of $10^{-6} \, \Omega \, cm$ [5]. Moreover, a single layer graphene could be seen by the naked eye as it absorbs a relatively high fraction of visible light with little dependence of its wavelength. Thus, graphene has drawn huge attention due to its above-mentioned exciting properties like electronic conductivity, carrier mobility, thermodynamic stability, flexibility, thermal and mechanical strength. Many recent studies provide more exotic properties even for two layers especially after twisting them for a magic angle [7].

5.1.2 Emergence of graphene quantum dots (GQDs)

Nevertheless, ideal 2-D graphene is a zero bandgap material which limits its utility for niche electronic applications despite the other unique properties. Hence, engineering of bandgap is of paramount importance for achieving many technological breakthroughs related to carbon-based electronic devices. Graphene has a Bohr-exciton radius of infinity [8] and so the so-called "three dimensionally quantum confinement" can be introduced to graphene structure to tune it from a semiconducting to an insulating state [8]. Bandgap engineering in graphene is intriguing and can be introduced mainly by reducing the size of graphene to nanoscale regions having one or a few number of layers with a small lateral diameter. Dimensionality regulates most of the physical and chemical properties and QDs stand for the confinement of electrons in all three spatial dimensions, that leads to quantization of the energy spectrum and graphene quantum dots (0-D, GQDs) derived from graphene structure are one such class of nanomaterials, well known for their variation of bandgap with size. To understand the size-dependent photoluminescence (PL) properties of GQD, Mahasin et al. have calculated theoretically the emission wavelength of pristine zig-zag-edged GQDs with different diameters which shows PL emission from deep UV to near infrared on varying the size from 0.46 to 2.31 nm [8]. However, the bandgap and size of GQDs are inversely proportional (i. e. the smaller the GQDs, the higher is the bandgap) and theoretical calculations show that a maximum of 7 eV band gap can be achieved from the smallest possible graphene quantum dot (GQD) which is supposed to have a benzene like structure [5]. On the other hand, different functionalization strategies could be employed systematically to tailor the bandgap structures of GQDs by creating intermediate energy states in between the valence and conduction band keeping the size of GQDs smaller below 10 nm whereby making them suitable for particular applications as shown in Figure 5.1 [9]. Flexibility in the band gap engineering as well as good optical absorption has made these materials promising in the area of optical and optoelectronic applications. By virtue of the quantum confinement, GQDs show fluorescence and single electron transfer properties observed more clearly in ultra-small (3 nm) GQDs despite reproducibility concerns [10]. For example, Ponomarenko et al. has demonstrated GQDs allowing a range of operational regimes from conventional single-electron detectors to Dirac billiards, in which size effects are exceptionally strong and chaos develops easily [11]. In a recent study, Jairo et al. have explored the single-electron charging of localized states that arise

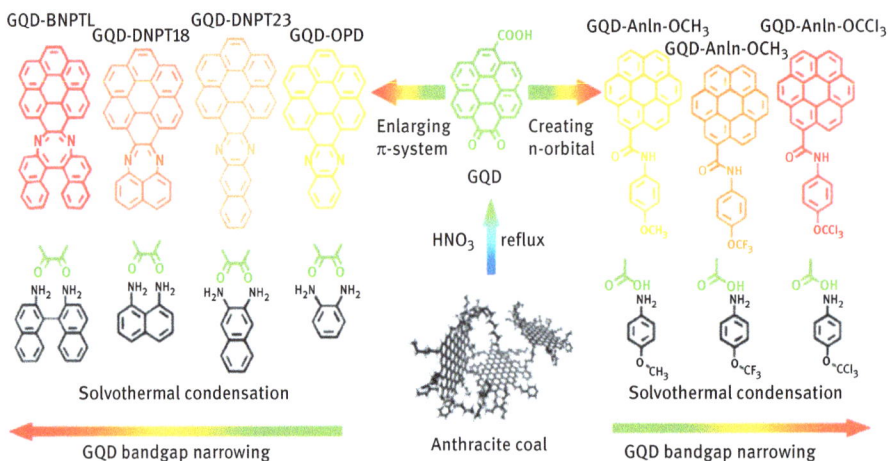

Figure 5.1: Illustration of bandgap narrowing by enlarging π-conjugated system via conjugating GQD with poly aromatic rings or by introducing intermediate n-orbital via conjugating with electron-donating groups. For clarity, GQDs (contains ~ 60 carbon rings and various chemical groups) are oversimplified to a seven-ring structure. And for the same reason, only one conjugated moiety is illustrated for modified GQDs. ["Reprinted with permission from (Ref [9]). Copyright (2018) American Chemical Society."].

from the quantum confinement of massive Dirac Fermions within their exposed bilayer graphene quantum dots [12]. However, a poor quantum yield limits the performance of pristine GQDs in some potential optical applications that can be avoided by encapsulating GQDs with some organic ligands or doping to tune the quantum yield [13]. Moreover, the ease of dispersion in several solvents unlike graphene, doping with hetero atoms like N, S, B, etc. and its ease of functionalization make the researchers to play with the structural features to unravel their exciting properties [14]. Moreover, the recent discovery of a one-step multi-gram synthesis of GQDs from cheaper materials like coal and other carbon sources opens the possibility of their large-scale industrial production [9, 15]. In order to make the surface grafting for various application, it is very crucial to learn about the basic structure of GQDs and proper synthetic route to tailor the desired product because the properties of the material are highly dependent on the synthetic route. There are several review articles and book chapters recently published that discuss about the various methods of synthesis, properties and characterization, some are focused on their potential application with future aspect. For instance, Ju et al. [16] have focused their article on the recent development of GQDs as fluorescence probes, including their synthesis and application in metal ions detection specifically. Du et al. [17], specifically have reviewed about doped fluorescent carbon dots (DFCDs) including their synthesis, dopant categories, formation mechanism and the wide range of applications during the year 2012–2015. In another report, Zhu et al. [18] have reviewed the current status and future prospect of GQDs and carbon dots (CDs) mainly focusing on the PL mechanism of them. Similarly, Gao et al. [19] have reviewed on CDs as nanoprobes for metal ions detection and their potential usage in drug delivery and biomedical applications. However, these articles do not discuss in detail about the synthetic routes of preparation. They either have overviewed the synthesis and characterization without the experimental details or some of them are mainly focused on some particular property like optical or electronic. On the other hand, Musselman et al. [20] discusses the other QD systems beyond graphene like 2D-QDs, including MoS_2-QDs, GQDs, g-C_3N_4-QDs and BP-QDs, and their application in solar cells as charge transporting layers, co-sensitizers, additives for the active layer, etc. Similarly, Xu et al. [21] have also made a comprehensive review on categories, synthetic routes, properties, functionalization and applications on all the extant 2D-QDs based on graphene, phosphorene, silicene, carbides, nitrides, transition metal dichalcogenide, transition metal oxides, MXenes, etc. In this article, we will discuss the synthetic routes for preparation of GQDs with some experimental elaboration along with basic structure, characterization of the GQDs and properties of GQDs along with potential, additional applications of GQDs. In the following section we will briefly discuss the structural features of GQDs and their properties starting from its parent material (graphene) before explaining the possible synthetic routes as the structure, properties and synthetic routes are highly correlated.

5.2 Structure, defects and properties of GQDs

5.2.1 Structure of graphene and GQDs

Electronic structure of a material plays a critical role in controlling all other properties and hence before discussing the applications based on unique properties of GQDs, we should look into the electronic structure and properties of graphene. The electronic structure of 3-D graphite has been extracted from the wave functions of the lower dimensional graphene lattice (2-D) using perturbation calculation [22]. Using tight binding model, the electronic properties can be understood. The honeycomb lattice structure of graphene has two atoms per unit cell (Figure 5.2 and Figure 5.3 left hand side image), (ABAB) sub-lattices of graphite unit cell with different C-C bond distances and the right hand side image describes the first Brillouin zone (BZ) in reciprocal space for graphite unit cell., which results in two "conical" points per Brillouin zone (Dirac Cone), where band crossing occurs, K and K' (Figure 5.3 right hand side image) and near these crossing points, the electron energy is linearly dependent on the wave vector [16]. K and K' can be expressed as below in eq. (5.1).

$$K = (2\pi/3a, 2\pi/3\sqrt{3}a), K' = (2\pi/3a, -2\pi/3\sqrt{3}a) \qquad (5.1)$$

where a = carbon–carbon distance; a ≈ 1.42 Å.

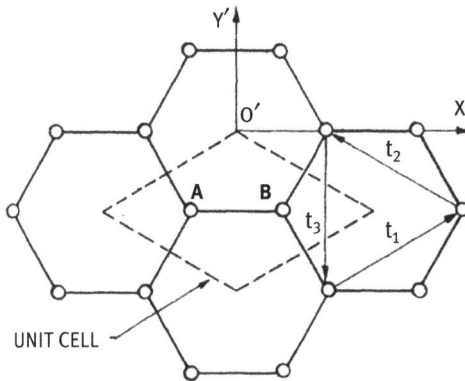

Figure 5.2: A layer plane of the graphite lattice. The circles represent carbon atoms. "Reprinted (Figure 5.1) with permission from [(J. C. Slonczewski and P. R. Weiss. *Phys. Rev.*, *330*, 272–279, 1985.)] Copyright (1985) by the American Physical Society."].

The unique electronic structure of σ bonded carbon atoms in graphene is responsible for its flexibility and the robustness but stacking of layers in graphene can change the electronic properties considerably. The full band electronic dispersion

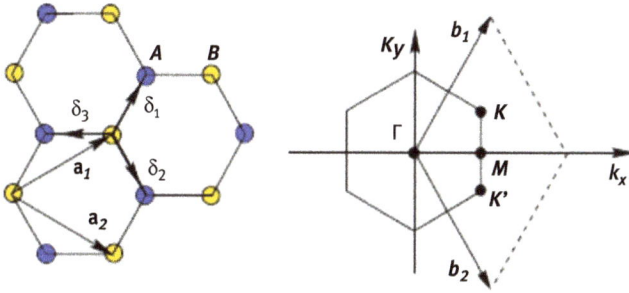

Figure 5.3: Honeycomb lattice and its Brillouin zone. Left: lattice structure of graphene, made out of two interpenetrating triangular lattices (a1 and a2 are the lattice unit vectors, and δi, i = 1,2,3 are the nearest-neighbour vectors). Right: corresponding Brillouin zone. The Dirac cones are located at the K and K' points. "Reprinted (Figure 5.2) with permission from [(Castro Neto, A. H.; Guinea, F.; Peres, N. M. R.; Novoselov, K. S.; Geim, A. K., *Rev. Mod. Phys.* 2009.)] Copyright (2009) by the American Physical Society."].

energy of graphene has been shown in Figure 5.4 with both t and t′ where "t" is the nearest-neighbour hopping energy (hopping between different sublattices), and t′ is the next nearest-neighbour hopping energy (hopping in the same sublattice). Herein, the zoomed in of the band structure close to one of the Dirac points (at the K or K′ point in the BZ). The electronic band structure with stacking of layers is displayed in Figure 5.5 to reveal that the graphene layers can be rotated relative to each other owing to its weak Van der Waals' bonding between the planes which permits for the various quasi-degenerate orientational states [8]. For

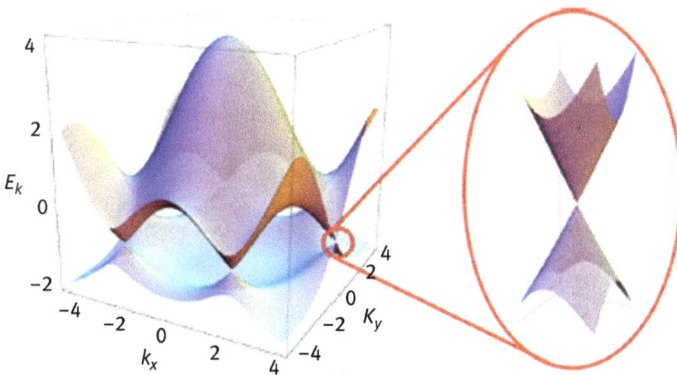

Figure 5.4: Electronic dispersion in the honeycomb lattice. Left: energy spectrum (in units of t) for finite values of t and t′, with t = 2.7 eV and t′ = − 0.2t. Right: zoom in of the energy bands close to one of the Dirac points. "Reprinted (Figure 5.3) with permission from [(Castro Neto, A. H.; Guinea, F.; Peres, N. M. R.; Novoselov, K. S.; Geim, A. K., *Rev. Mod. Phys.* 2009.)] Copyright (2009) by the American Physical Society."].

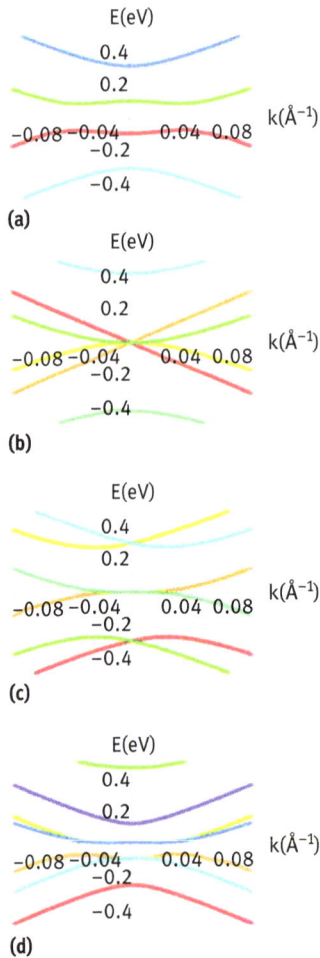

Figure 5.5: Electronic bands of graphene multilayers. (a) Biased bilayer. (b) Tri-layer with Bernal stacking. (c) Tri-layer with orthorhombic stacking. (d) Stack with four layers where the top and bottom layers are shifted in energy with respect to the two middle layers by + 0.1 eV. "Reprinted (Figure 5.13) with permission from [(Castro Neto, A. H.; Guinea, F.; Peres, N. M. R.; Novoselov, K. S.; Geim, A. K., *Rev. Mod. Phys.* 2009.)] Copyright (2009) by the American Physical Society."].

certain angles, the graphene layers become commensurate with each other thus lowering the electronic energy. The superlubricity and flexibility in graphite comes from the electronic structure dependence on the relative rotation angle between layers of graphene and hence its extensive utility. However, major limitations of graphene include, difficulties of scale-up, poor dispersion in common solvents due to aggregation and its zero optical bandgap preventing its utility for the fabrication of electronic devices.

Nevertheless, a tuneable bandgap could be created when the ideal graphene with zero gap is transformed into a nanometre sized (quasi-zero dimensional) GQDs thus introducing the way for many spectacular applications for graphene. GQDs could be considered as nanographene fragments that are small enough to cause exciton confinement in all three dimension by reducing the size (diameter below 20 nm) of the graphene moiety [5]. The exact structure of the GQDs depends upon the method used for the synthesis, and also the possibility of functionalization or doping carried out intentionally to tailor the bandgap. For instance, Pan et al. [18] have proposed a structure of GQDs synthesized by oxidative cutting of oxidized graphene sheets by hydrothermal method to show a carbene-like zig-zag and carbyne-like armchair surface structure along with the carbonyl and carboxylic acid groups as demonstrated in Figure 5.6. In a similar manner, the electronic structure of hexagonal zigzag-edged and armchair GQDs has been investigated theoretically using

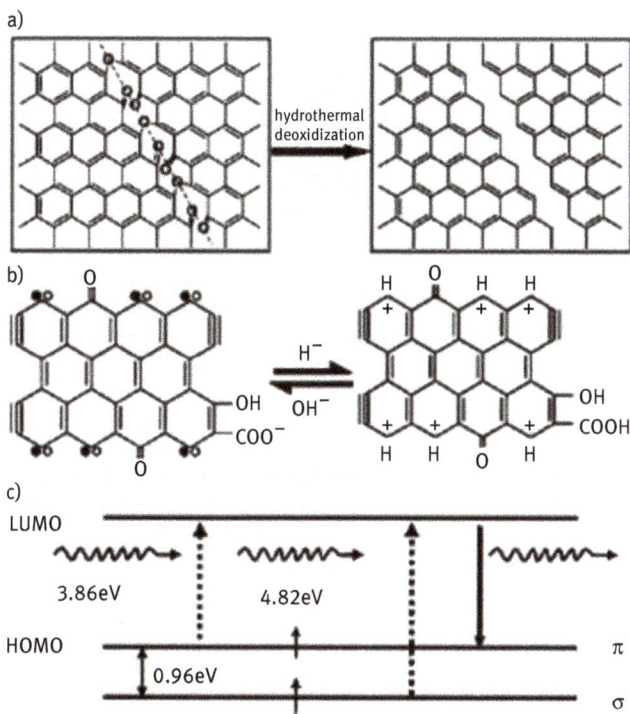

Figure 5.6: (a) Mechanism for the hydrothermal cutting of oxidized GSs into GQDs: a mixed epoxy chain composed of epoxy and carbonyl pair groups (left) is converted into a complete cut (right) under the hydrothermal treatment. (b) Models of the GQDs in acidic (right) and alkali (left) media. The two models can be converted reversibly depending on the pH. The pairing of σ (°) and Π (·) localized electrons at carbene-like zigzag sites and the presence of triple bonds at the carbyne-like armchair sites are represented. (c) Typical electronic transitions of triplet carbenes at zig-zag sites. [(Reprinted with permission from Ref [23] Copyright 2010 Wiley.)].

(e. g. electron-electron and electron-hole interaction which are responsible in determining the electronic structure and optical properties) perturbation theory where
many-body effects are predominant owing to quantum confinement and reduced
screening [24]. Herein, the quasi-particle corrections and exciton binding energies
are much larger than those of other carbon allotropes with higher dimensionality.
Although, tight-binding calculations have revealed that the HOMO of GQDs can be
altered by the geometry shape more strongly, although the edge effect and corner
termination contribute little in such cases, experimental evidences have indicated
that the bandgap can be independently controlled by either size tuning, or doping
or functionalization or both. Apart from this, the bandgap can be tuned by applying
external electric field [24]. Hence, a thorough knowledge of their electronic structure is indispensable for effective applications. Accordingly, Figure 5.7 demonstrates the effect of size reduction on bandgap and interestingly, in line with these,
time-dependent density functional theory (TDDFT) calculations show that the low-
est bright spin-singlet states in GQDs are two-fold degenerate and that dark singlet
states do exist below the bright one. Their work has a strong implication on revealing the importance of the lateral size and the edge geometry in determining the

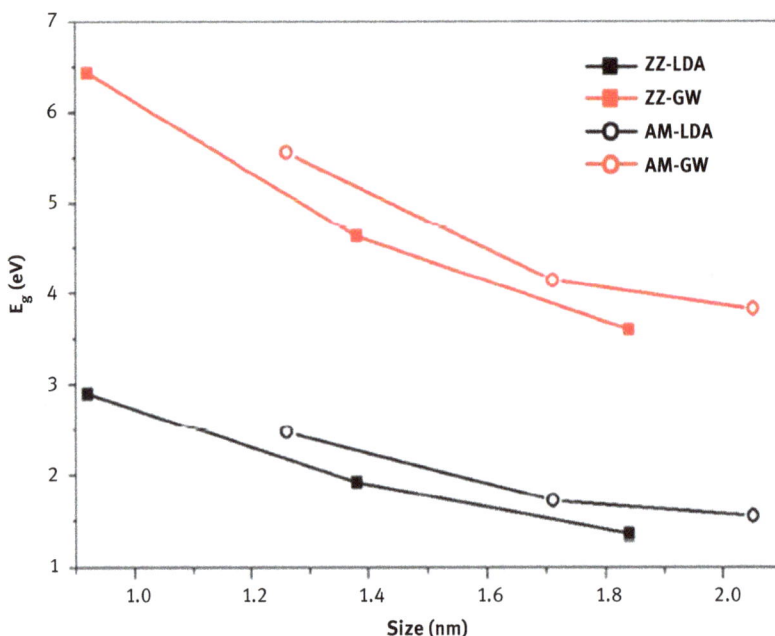

Figure 5.7: Variation of HOMO-LUMO gap with the lateral size. ZZ represents zig-zag edges, AM
represents arm chair edges; GW approach comes from (Green's function G and screened Coulomb
interaction W) and LDA approach arises from local-density approximation LDA based functionals
["Reprinted with permission from (Ref [24]). Copyright (2015) American Chemical Society."].

band gap and exciton binding energy of GQDs because both band gap and binding energy of the first bright exciton decrease as the lateral size of GQD increases.

The electronic structure and optical properties of edge-functionalized GQDs have also been investigated by density functional and many-particle perturbation theories where the mechanism has been explained based on the competition and collaboration between frontier orbital hybridization and charge transfer [124]. They explore the frontier orbital hybridization and functional groups of GQDs reducing the energy gap between HOMO and LUMO while the charge transfer from GQD moiety to functional group enlarges it. It also describes how the functional groups affect the bandgap more illustrating using carbon–oxygen double bond (C = O), aldehyde (–CHO), ketone (–COCH$_3$) and carboxyl (–COOH), which are perhaps more responsible for tailoring the electronic and optical properties of pristine GQD compared to amino group (–NH$_2$), which indeed has a much weaker influence on electronic structure since the large charge transfer cancels out the effect of frontier orbital hybridization [124].

In a recent study using low-temperature scanning tunnelling spectroscopy, the local density of states of GQDs supported on Ir (111) could be mapped due to a band gap in the projected Ir band structure around the graphene K point to reveal that the electronic properties of the QDs are dominantly graphene like [25]. Moreover, these results could be compared favourably with tight binding calculations on the honeycomb lattice based on parameters derived from density functional theory. This clearly suggests that the interaction with the substrate near the edge of the island gradually opens a gap in the Dirac cone, which implies softwall confinement. Interestingly, this confinement results in highly symmetric wave functions [21]. Another group study the symmetry classes of graphene quantum dots, both open and closed, through the conductance and energy level statistics [26]. Here they study universalities in the spectrum and conductance of GQDs in both the closed Coulomb blockade and the open ballistic regime, respectively. Moreover, they demonstrate the universality class of the conductance can be different from that of the spectrum and the main reason behind this paradox is the separation of time scales characterizing the conductance (escape time) and the spectrum (Heisenberg time, i. e. inverse level spacing), allowing scattering times to be smaller than one but larger than the other [26]. Recently, Okamoto et al. have investigated real-space characterization of surface plasmonpolariton (SPP) in GQDs on SiC substrate by utilizing a scattering-type scanning near-field optical microscope (s-SNOM) in the mid-infrared (MIR) band. Their study revealed that the SPP in the GQDs was excited at the wavenumber of 929.02 cm^{-1}, where the SPP in 2-D graphene has not been reported. The present findings provide not only a deeper understanding of the SPP in graphene nanostructures, but also a first step for exploring new plasmonic functionalities impossible with conventional 2-D graphene [27].

5.2.2 Defects in graphene-based nanomaterials

The properties of the material are highly sensitive towards the defect chemistry. We would discuss here the general defects in graphene. They are usually categorized into two different groups: one is intrinsic defects – arises due to non-sp^2 orbital hybrid carbon atoms in graphene and the second defects are extrinsic defects which could arise due to various chemical treatment, different experimental parameter used for synthesis [28]. The general five different kinds of defects which may arise in graphene-based nanomaterials are shown in Figure 5.8 [29]. The formation energy for these intrinsic defects is tabulated below in Table 5.1.

1. Structural Defects: Appears mainly due to significant structural changes caused by the presence of pentagons or heptagons within the hexagonal sp^2 hybridized carbon lattice.
2. Topological Defects/Bond rotations or grain boundaries: Mainly occurs on graphene surfaces, which do not result in large curvature distortions of the sheet.
3. Non-sp^2 hybridized carbon defects: Include edges, vacancies, sp^3 carbon centres due to functionalizations, ad-atoms, interstitials, carbon chains, etc.
4. Doping-Induced Defects: Doping of hetero atoms like B, N, S, etc. can create defect in the organized hexagonal graphene moiety.
5. Folding-induced defects: Arise mainly from significant deformation of the graphene sheet, thus altering the orbitals [29].

Graphene can have either zigzag or armchair edges, which exhibit distinct quantum confinement properties. For example, graphene nanoribbons (GNRs) with dominant zigzag edges have a smaller band gap (0.14 eV) as compared to the similarly sized GNRs with dominant armchair edges (0.38 eV) because of localized states on zigzag edges [29]. In a similar manner, the localized states in zigzag-edged GQD are pushed to the edge sites while similarly sized armchair-edged GQD have localized states scattered in the centre [29]. Unlike armchair-edged counterparts, the localized states at zigzag edge sites lower the energy of the conduction band and thus reduce the band gap and so it is expected that the armchair edge would widen the band gap of GQD and consequently blue-shift the emission. Moreover, calculations show that 1.27 and 2.06 nm armchair-edged GQDs emit at 450.5 and 678.2 nm, respectively, while the predicted emission wavelengths of their zigzag-edged counterparts are ~ 551 and ~ 872 nm, respectively [7]. In brief, the spectrum of realistic graphene quantum dots in the presence of disorder (edge roughness or defect) reveal unique features which differ from both Schrödinger or Dirac billiards of confined massive or massless free particles and the graphene band structure near the K points leaves clear imprints. They include interference structures in the wave functions, enhanced confinement effects, and a delayed

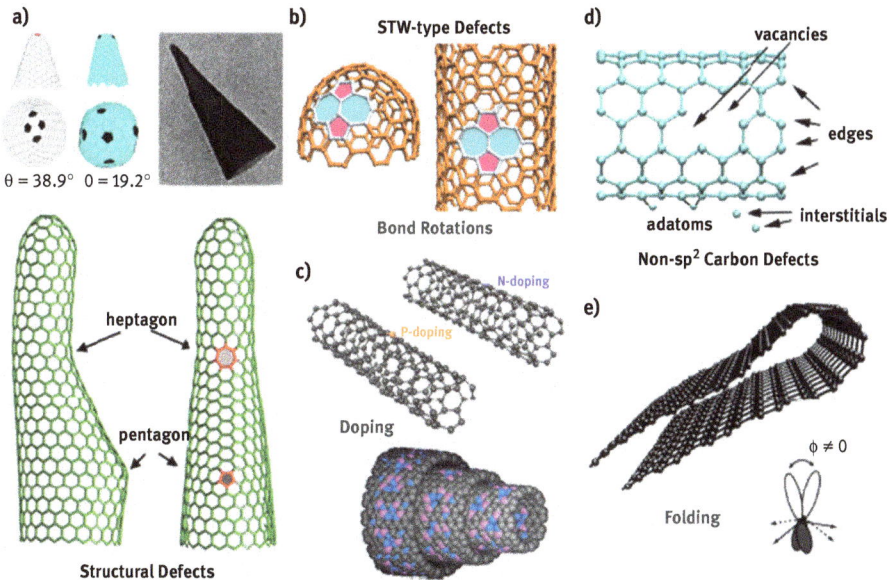

Figure 5.8: Schematic models representing different types of defects in graphene-like materials. (a) Structural defects induce significant structural changes caused by the presence of pentagons or heptagons within the hexagonal sp^2 hybridized carbon lattice (Image of cones courtesy of M. Endo and T.W. Ebbesen); (b) Topological defects, also termed Stone-Thrower-Wales defects, do not result in big structural changes. Shown here is the formation of 5-7-7-5 pairs created by rotating an individual carbon-carbon bond 90°; (c) Doping consists of replacing a carbon atom with another element within the hexagonal lattice (here, N and P) or a CNT randomly doped with B and N; (d) Non-sp^2 hybridized carbon defects, including vacancies, edges, adatoms, interstitials, carbon chains, etc.; (e) Folding-induced defects, which result from significant deformation of the graphene sheet, thus altering the π orbitals. The direction of the π orbital is then called the π orbital axis vector (POAV). The angle $\theta_{\sigma\pi}$ between the POAV and a σ direction (i. e. a bond) indicates the degree of "pyramidalization" and the hybridization. For $\theta_{\sigma\pi} = 90°$ (planar system), the σ orbitals are in a sp^2 hybridization and the π orbital is a pure pz orbital. For a folded graphene sheet, $\theta_{\sigma\pi}$ has an intermediate value which decreases as the inverse of the radius of the curvature of the folding, and reaches 90° at the limit $R \rightarrow \infty$. ["Reprinted from Nano Today, 5, Mauricio Terronesa, Andrés R. Botello-Méndezb, Jessica Campos-Delgadoc, Florentino López-Uríasd, Yadira I. Vega-Cantúd, Fernando J. Rodríguez-Macíasd, Ana Laura Elíase, Emilio Mu˜noz-Sandovald, Abraham G. Cano-Márquezd, Jean-Christophe Charlierb, Humberto Terronesb, Graphene and Graphite Nanoribbons: Morphology, Properties, Synthesis, Defects and Applications, 351–372, Copyright (2010), with permission from Elsevier.]".

transition from Poisson to Wigner-Dyson nearest-neighbour distributions. However, one still "cannot hear the imperfect shape of the drum", the size and roughness of graphene quantum dots can be, indeed, inferred from the spectral properties [30]. The recent report on investigation on the magnetic properties and type of edges depicts the appearance of π-electron-based

Table 5.1: Defect and energy of formation correlation.

Name of the defect	Energy of formation of the defect (eV)	Reference
Stone-Wales Defects	5	[28]
Single vacancy defects	7.5	[28]
Multiple vacancy defects	7	[28]

magnetism at the armchair edges under consideration [31]. The chemical and physical aspects of graphene edges may have strong implications for the research field of graphene-based nanoscale systems. It is widely accepted that the zigzag edge of graphene supports the localized π state on its boundary and gives rise to *intravalley* scattering of extended electronic states while its armchair counterpart does not possess the edge state and leads to *intervalley* scattering of charge carriers [32]. However, the recent studies show that the distinctions between zigzag and armchair edges of graphene can be completely eliminated by proper modification of the zigzag edges with certain chemical species (e. g. hydrogen) [32]. Thus, the properties could be tailored with defect chemistry accordingly to get the desired performance. The properties of GQDs will be discussed in the next section before discussing the synthetic routes of various GQDs (Section 5.3).

5.2.3 Properties of GQDs

5.2.3.1 Optical properties of GQDs

The optical properties of GQDs are intriguing due to a tuneable energy bandgap. Although, theoretically, all GQDs show similar absorption spectra, with a prominent peak emerging below the absorption onset of the non-interacting spectrum when electron-hole interaction is included [24]. The electron-hole overlap relates to the spin singlet-triplet splitting, which can be approximately measured by the overlap between frontier orbitals involved in the optical transitions [24]. In this regard, the strong many-body effects in GQDs should be of great importance for predicting the optical properties of the material. The experimental UV absorption peak of GQDs appears around 230 nm owing to the Π-$\Pi^\star$ transition from the aromatic sp^2 domain of GQDs with a long-tail extending into the visible range due to n-$\Pi^\star$ transition [5]. However, another shoulder absorption peak could be observed in the wavelength region 270–390 nm (n-$\Pi^\star$) due to $-C = O$. The peak position is more sensitive to the preparation method rather than the size of the dots and also depends on the solvent used for measurement. For instance, GQDs prepared from hydrothermal/solvothermal methods show the same absorption wavelength (320 nm, n-$\Pi^\star$), although the average size is different (~ 9.6 and 5.3 nm, respectively) [18, 23, 83]. Another appealing optical feature is PL due to the semiconductor nature of GQDs with a certain bandgap. The observed PL spectrum is usually broad and both excitation dependent and independent PL spectra could be seen based on the synthetic route. For example, electrochemically synthesized GQDs show excitation dependent [29] PL which is due to the presence of different surface and energy states whereas GQDs prepared by ultrasonic method shows excitation independent [30] PL spectra. Also, size dependent PL variation of the GQDs as shown have been found in the literature and presented in the schematic (Figure 5.9). The authors explain the level of the n-states is not disturbed by a change in the particle size, whereas the levels of the π, π^* and σ^* orbitals vary considerably with the size [33]. They also describe the excitation-wavelength-independent PL in the GOQD specimens arises due to the $\pi^* \to n$ recombination, which involves phonon scattering and an electron transition and the relaxation of electrons from the σ^* orbital to the π^* orbital is essential for PL emissions induced by short-wavelength excitation [33]. In addition, evidences show the nature of the PL spectrum depends on the pH, solvent and counter ions even if the size is similar [34]. Apart from this, up conversion PL (Figure 5.10), that is emitting shorter wavelength upon simultaneous absorption of two or sequential absorption of multiple longer-wavelength photons, could be observed for some of the GQDs [35–37]. This phenomenon could be attributed to the following reasons like multi-photon active process and the anti-Stokes photoluminescence (ASPL), where the energy difference between the Π and σ orbitals is near 1.1 eV [35, 36]. However, Shen et al. [35] speculate that ASPL is a more relevant phenomenon

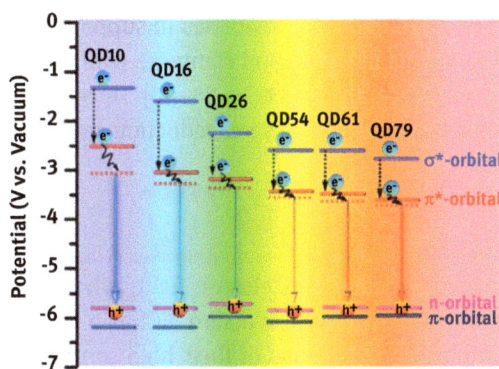

Figure 5.9: A Schematic diagram of energy level diagram for GOQD specimens. A schematic of the energy levels associated with size-dependent PL emissions from the GOQDs. The PL colour varies from orange-red (for QD79) to blue (for QD10). QD10 denotes 1.0 nm diameter, QD16 (1.6 nm diameter GQDs), QD26 (2.6 nm diameter), (d) QD54 (5.4 nm diameter), (e) QD61 (6.1 nm diameter), and (f) QD79 (7.9 nm diameter). ["Reprinted with permission from (Ref [33]). Copyright (2015) American Chemical Society."].

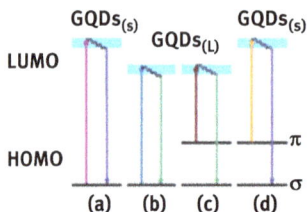

Figure 5.10: A schematic illustration of various typical electronic transitions processes of GQDs. Normal PL mechanisms in GQDs for small size (a) and large size (b); Upconverted PL mechanisms in GQDs for large size (c) and small size (d). [(Reprinted with permission from Ref [35] Copyright 2011 Royal Chemical Society.)].

compared to the mostly believed multi-photon active process to explain the upconverted PL property of GQDs. However, the exact reason is still not known.

5.2.3.2 Electrochemical properties of GQDs

The electrochemistry is highly sensitive to the surface and defect of a material and the tunable size, and in these quantum confined structures more surface area of GQDs has been found to reveal several interesting and important electrochemical properties like quantized double layer (QDL) charging behaviour depending upon the core size and the nature of the surface passivating molecules [10]. The inherent charge storage properties of GQDs arise due to graphitic domain for the use of GQDs as electrode material for micro-supercapacitor [54][119]. The electrocatalytic behaviour of GQDs, doped and interconnected GQDs also has been demonstrated for oxygen reduction, oxygen evolution and hydrogen evolution reactions in support with conductive materials like GNR and noble metals like gold and competes with its 1-D analogous material, GNRs as metal-free electrocatalysts [33–36, 38–40]. However, a critical comparison of literature demonstrates that for oxygen reduction reaction, GNRs are far more superior to the GQDs [41].

5.2.3.3 Electrochemiluminiscnce (ECL)

ECL is a smart combination of chemiluminescence and electric field, where the light-emitting reaction is preceded by an electrochemical reaction. ECL is a fundamental study with several analytical applications because of their high sensitivity and wide range of working concentrations [42]. The excited state can either be produced through the reaction of cation and anion radicals generated from the same luminophore, or from two different precursors (luminophore and co-reactant), via co-reactant mechanism. A typical mechanism of ECL in GQDs is schematically shown in Figure 5.11. To replace the conventional toxic semiconductors for biological application of ECL, low cost and compatible materials, several new materials

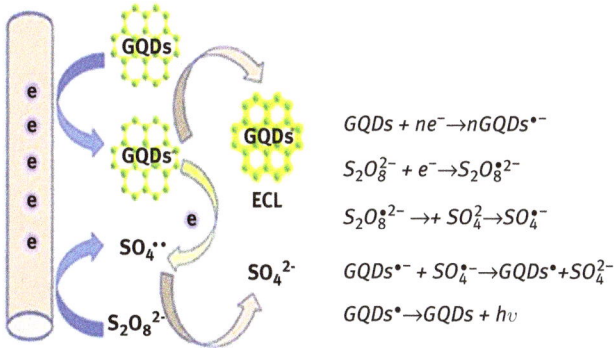

The equations shown in the figure:

$$GQDs + ne^- \rightarrow nGQDs^{\bullet-}$$

$$S_2O_8^{2-} + e^- \rightarrow S_2O_8^{\bullet2-}$$

$$S_2O_8^{\bullet2-} \rightarrow + SO_4^{2-} \rightarrow SO_4^{\bullet-}$$

$$GQDs^{\bullet-} + SO_4^{\bullet-} \rightarrow GQDs^\bullet + SO_4^{2-}$$

$$GQDs^\bullet \rightarrow GQDs + h\upsilon$$

Figure 5.11: A schematic illustration of the ECL mechanism of GQDs on the left and at the right equations of the mechanism explaining the respective ECL of GQDs [(Reprinted with permission from Ref [43] Copy-right 2012 Wiley.)].

are being developed and GQD could be a potential candidate with excellent physical and optical properties. The ECL emission could be originated from the GQDs⁻ reacted with SO_4^- radicals and explained by the equations on the right image of Figure 5.11 [43].

5.2.3.4 Cytotoxicity

Cytotoxicity is very crucial factor in biological applications such as *in vitro* and *in vivo* imaging studies. However, most semiconductors are derived from heavy metals and suffer from their high toxicity. In this context, GQDs are promising alternatives due to their excellent semiconductor-like properties. GQDs have been tested for cytotoxicty by several groups till date. For instance the toxicity of GQDs has been tested for male mouse sexual behaviours, reproductive and off spring health by Zhang et al. [44] whereas Wang et al. have demonstrated the use of GQDs as a potent inhibitor against the in vivo aggregation and toxicity of human islet amyloid poly-peptide (IAPP), a hallmark of type 2 diabetes [45]. Jin et al. have shown S-GQDs, as the effective cell-imaging material, would easily penetrated into the cell membranes of Hela cell and exhibit relatively low cytotoxicity and when treated with the typical bacteria medium (Staphylococcus aureus LZ-01 and Escherichia coli DH5α), with no conspicuous antiseptic qualities. Therefore, the S-GQDs would have potential applications in the bio-imaging [46]. In a recent study, Xie et al. have systematically studied the toxicity of functionalized GQDs [47]. They have employed a lung carcinoma A549 cells as the model to investigate the cytotoxicity and autophagy induction of three types GQDs, including cGQDs (COOH-GQDs), hGQDs (OH-GQDs) and aGQDs (NH₂-GQDs). The results showed hGQDs was the most toxic, as significant cell death is induced at the concentration of 100 μg/mL, determining by WST-1 assay as well as Annexin-V-FITC/PI apoptosis analysis, whereas cGQDs and aGQDs are non-cytotoxic within the measured concentration. Further, polymer modification of GQDs can be adapted to reduce cytotoxity by encapsulating well-defined GQDs in a PEG nanoparticle [48].

5.3 Preparation strategies

Size-tunable GQDs could be synthesized using both top-down and bottom-up methods like other nanomaterials. A graphical representation is illustrated in Figure 5.12 in order to illustrate the difference between the two methods as a change in the synthetic routes could result in the different surface states and structure nature of GQDs.

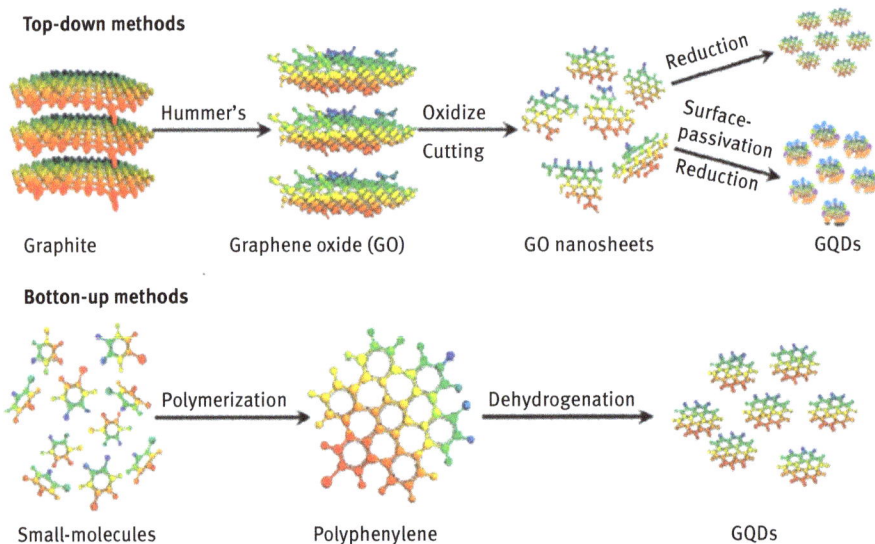

Figure 5.12: Top-down and Bottom-up Method Graphical Representation [(Reprinted with permission from Ref [49] Copyright 2012 Royal Chemical Society.)].

5.3.1 Top-down methods

Generally, top-down approaches use simple precursors and are employed in bulk production. But the GQDs will involve different physical characteristics with high surface defects due to the aromatic framework destroyed during the process. Hence, most of the synthetic approaches for GQDs follow top-down process which involve the breaking down of a bigger structure to a smaller one chemically, by a hydrothermal ultrasonic or microwave synthesis. A simple hydrothermal route can be followed for the cutting of pre-oxidized micrometre-size rippled graphene sheets into ultrafine GQDs with a diameter from 5 to 13 nm [23]. However, these GQDs show bright blue PL (quantum yield ca. 6.9 %) only in alkaline medium. In a similar manner, another group has reported [50], the reduction of graphene nanosheets (GSs) in presence of ploy ethylene glycol (PEG) to produce PEG-surface passivated GQDs by hydrothermal method. In most of the cases, both hydrothermal and electrochemical methods end up producing only one emission colour either green or blue luminescence depending on the size. In a typical chemical method, graphitic oxide nanosheets are produced first by acidic oxidation followed by are reduction to produce GQDs by using hydrazine-vapour exposure (from 20 s up to 60 min), or hydrazine hydrate or sodium borohydrate [35, 51, 52].

Previously mentioned methods produce GQDs which show excitation-dependent PL behaviour while GQDs synthesized by other routes can exhibit excitation-independent PL. For instance, GQDs produced by ultrasound method, where ultrasound can generate alternating low-pressure and high-pressure waves in liquid, leading to the formation and violent collapse of small vacuum bubbles and these cavitations cause high-speed impinging liquid jets, deagglomeration, and strong hydrodynamic shear-forces, exhibit excitation-independent PL [53]. Microwave synthesized GQDs display not only monodispersity in particle size distribution but also excitation-independent PL behaviour [13].

Moreover, combining both high energy microwave and hydrothermal techniques, is more convenient to prepare monodispersed samples using shorter time for the reduction of GO [54]. Reduced graphite oxide materials could readily be prepared in the few minutes with a very high yield. Furthermore, a facile microwave-assisted approach for the preparation of stabilizer-free two-colour GQDs from GO nanosheets under acid conditions is also possible as per the schematic representation in Figure 5.13 [43].

The desired geometry of GQDs can be produced ranging from 10 to 240 nm diametre reliably by Electron-beam lithography [11, 55]. Ponomarenko et al. have first prepared graphene by the scotch tape method after placing graphene on to a silicon wafer with a thickness of 300 nm SiO_2 top layer, followed by electron beam lithography (EBL) carving of the pattern-protected graphene to form GQDs with desired geometries. Moreover, still smaller diameter GQDs are also possible to fabricate by this method and the prepared GQDs are undoubtedly of very high quality and defect free with desired surface or edges. Unfortunately, EBL needs special equipment, expensive raw materials and special skill in operating the instruments.

Another popular electrochemical method, involves synthesizing the material using a higher potential, (± 1.5 V to ± 3 V), which is higher enough to either oxidize the $C = C$ bonds or oxidize water to generate hydroxyl and oxygen radicals playing the role of an electrochemical "scissors" in its oxidative cleavage reaction [56]. Mostly, graphite and multiwalled carbon nanotubes (MWCNTs) are used as starting

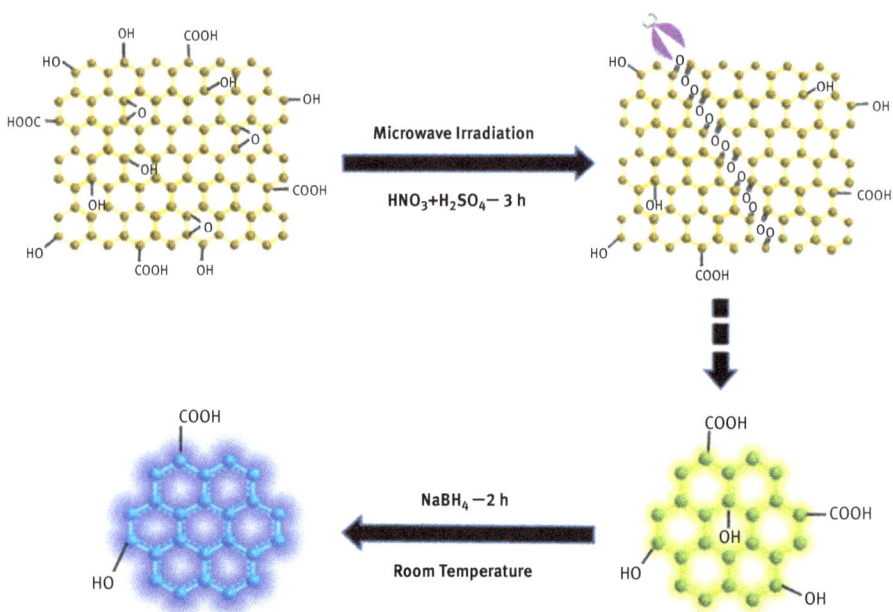

Figure 5.13: A schematic representation of the preparation of GQDs by microwave method [(Reprinted with permission from Ref [43] Copyright 2012 Wiley.)].

materials for this kind of techniques [56]. Longitudinal unzipping of MWCNT produces nanoribbon while lateral or longitudinal or any other kind of scissoring of MWCNT gives GQDs along with many irregular types of debris. Similar kind of electrochemical synthesis of C-dots, which do not contain graphitic moiety [57], has been reported by oxidizing a graphite electrode at 3 V using a saturated calomel reference electrode and a Pt wire counter-electrode in 0.1 M NaH_2PO_4 aqueous solution [58]. Other example of this kind of synthesis includes electrochemical oxidation of a graphene electrode in phosphate buffer solution to produce green-luminescent functional GQDs with a uniform size of 3–5 nm with topographic heights between 1 and 2 nm [59]. The oxygen-containing groups on the surface of the GQDs are responsible for their aqueous solubility and promote further surface functionalization if required. A sustained oxidation (15 h) of MWCNTs in non-aqueous solvent like propylene carbonate with salt $LiClO_4$ at 1 V followed by reduction has yielded discrete spherical particles of GQDs presumably due to lateral unzipping (Figure 5.14) [60]. However, these electrochemical methods suffer from drawbacks like poor yield and excitation-dependent PL behaviour especially towards certain application perspective.

In order to alleviate such problems, several other methods have been introduced to obtain GQDs with a higher yield and stronger PL quantum yield. Oxygen plasma assisted method is such kind of a technique where strong PL could be induced in a single-layer graphene (SLG) to produce GQDs [61]. Herein, graphene

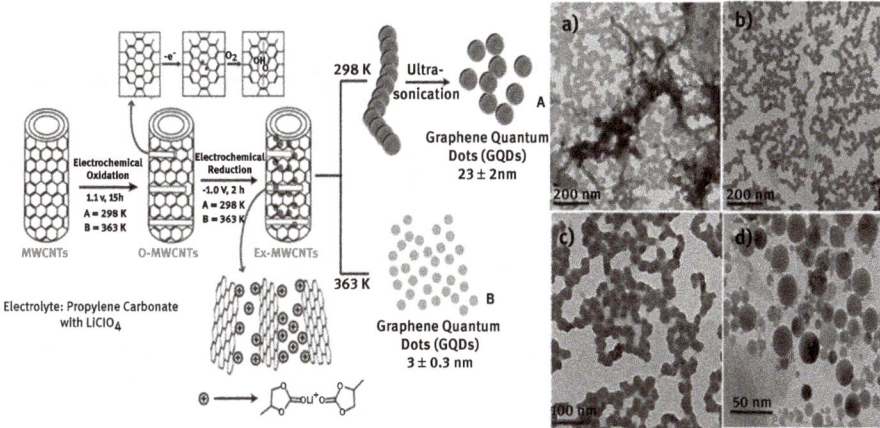

Figure 5.14: Schematic representation (left) of various processing stages involved in the preparation of photo-luminescent GQDs from MWCNTs by the electrochemical approach [60] and in right hand side (a), (b) Typical TEM images of GQDs synthesized at 30 °C by using oxidation at 1 V for 7 h followed by reduction at −1 V; (c) GQDs prepared by oxidation at 1 V for 11 h, followed by reduction at −1 V; (d) GQDs with different levels of sonication prepared at 1 V for 15 h oxidation followed by reduction at −1 V [(Reprinted with permission from Ref [60] Copyright 2012 Royal Chemical Society.)].

samples produced by micro-cleavage of graphite on a silicon substrate are exposed to oxygen: argon (1: 2) RF plasma (0.04 mbar, 10 W) for increasing time (1–6 s) for which the structural and optical changes have been monitored by Raman spectros-copy and elastic light scattering [62, 63]. Another group has also demonstrated that GO nanosheets possessed visible and near-infrared (Vis-NIR) fluorescene despite poor stability [64].

5.3.1.1 Experimental details of a few top-down methods
In this sub-section, few synthetic experimental details will be mentioned related to the top-down routes as discussed in the preceding paragraph. At first we will dis-cuss about the hydrothermal route for the synthesis of GQDs reported by Pan et al. [23].

5.3.1.1.1 Hydrothermal method
GO sheets were prepared from natural graphite powder by oxidation using modified Hummers method in the first step. In the second step, Graphene sheets (GSs) were obtained by thermal reduction of GO sheets in a tube furnace at 200–300 °C for 2 h with a heating rate of 5 °C/min in a nitrogen atmosphere. Next, in another step, GSs (0.05 g) were oxidized in concentrated H_2SO_4 (10 mL) and HNO_3 (30 mL) for 15–20 h

under mild ultrasonication (500 W, 40 kHz) followed by dilution of the mixture with deionized (DI) water (250 mL) and filtered through a 0.22-mm microporous membrane to remove the acids. Again, the purified oxidized GSs (0.2 g) were re-dispersed in DI water (40 mL) and the pH was tuned to 8 with NaOH and the suspension was transferred to a Teflon-lined autoclave (50 mL) and heated at 200 °C for 10 h for hydrothermal treatment. Finally, after cooling down the autoclave to room temperature, the resulting black suspension was filtered through a 0.22-mm microporous membrane and a brown filter solution (yield ca. 22 %) was separated. So, the colloidal solution which still contained some large graphene nanoparticles (50–200 nm) (emitted weak blue fluorescence) was further dialyzed in a dialysis bag (retained molecular weight: 3500 Da) overnight and GQDs that were strongly fluorescent through the bag were obtained with a yield of ca. 5 %.

5.3.1.1.2 Oxidation method
In another simple method, oxidative scissoring of GO sheets to GQDs has been reported by Tang et al. later [65]. In this method also GO was synthesized from purified natural graphite powder by a modified Hummers method like the earlier report [23]. In order to prepare the GQDs, GO (0.200 g) was added into a mixture of concentrated H_2SO_4 (30 mL) and HNO_3 (10 mL). Next, the mixture was sonicated for 2 h and stirred for 3 h at 100 °C. After cooling down to room temperature, the product was diluted with ultra pure water and the pH was tuned to seven with Na_2CO_3 in an ice-bath. A light yellow solution was obtained after removing the residual GO from the mixture by filtration through the PTFE membrane (pore size of 0.22 μm). Then, the solution was dialyzed in a dialysis bag (MWCO: 100 Da) for 2 days to remove the $NaNO_3$ and Na_2SO_4 formed during the neutralization process. Finally, the GQDs water dispersions were separated by further dialysis (MWCO: 1000 Da) for another 2 days.

5.3.1.1.3 Microwave method
Another faster technique reported by Shin et al. [84] involves microwave treatment of graphite powders. Herein, GQDs were synthesized from natural graphite powder (200 mg, 2–15 micron Alfa Aesar) by microwave irradiation. At first, graphite powder was sonicated in conc. H_2SO_4 (100 mL) for one followed by the addition of $KMnO_4$ (1 g) slowly to keep the temperature of the suspension lower than 25 °C. Next, the mixture was heated and atmospheric refluxed with a MAS-II microwave operating (Tsingtao Unicom-optics instruments Co., Ltd, China) at a power of 600 W for 1 h. The microwave treated, transparent suspension was then cooled to room temperature and diluted with DI water (700 mL). The solution was neutralized to pH seven with NaOH in an ice bath. In the next step, the diluted suspension was filtered through a 200 nm nanoporous membrane to separate large graphene nanoparticles and a brown filtrate. The final product solution was further dialyzed through a dialysis bag (retained molecular weight: 2000 Da) for 7 days to obtain the product.

5.3.1.1.4 Ultrasonic method

In ultrasonic method, Zhou et al. prepared GQDs from graphene by the following steps [36]. At first, graphene (0.05 g) was oxidized in a mixture of conc. H_2SO_4 (10 mL) and HNO_3 (30 mL) at room temperature for 12 h. Then, the mixture was subsequently treated ultrasonically for 12 h with an ultrasonic instrument (model KQ-300 TDE, 300 W, 80 kHz) followed by calcination in a furnace installed with an exhaust gas recovery at 350 °C for 20 min to remove the concentrated H_2SO_4 (boiling point, 338 °C) and HNO_3 (boiling point, 83 °C). Then the product was re-dispersed in DI water (40 mL) and the resulting black suspension was filtered through a 0.22 μm microporous membrane to get a brown filter solution. In the final stage, the solution was further dialyzed in a dialysis bag (retained molecular weight: 3500 Da) overnight to obtain GQDs.

5.3.1.1.5 Electrochemical method

The electrochemical preparation of functional GQDs was carried out by Li et al. [59]. The experimental route involved a CV scan within ± 3.0 V at a scan rate of 0.5 V/s in 0.1 M PBS using a graphene film (5 mm × 10 mm) as working electrode, Pt wire as counter and Ag/AgCl used reference electrodes, respectively. Graphene film was treated with O_2 plasma for seconds prior to the preparation of GQDs in order to enhance its hydrophilicity. The water-soluble GQDs were collected after filtering followed by a dialysis with a cellulose ester membrane bag [MD77(8000–14,000)] and the purified GQDs were re-dispersed in DI water for further characterization.

5.3.1.1.6 Photo-fenton reaction method

The photo-Fenton reaction preparation of GQDs by Zhou et al. [66] is as follows: 5 mL of 0.5 mg/mL GO aqueous suspension, 20 mL of 20 mM H_2O_2 and 100 μL of 1.0 mM of $FeCl_3$ were mixed in a quartz tube under vigorous stirring and the pH of the mixture was adjusted to four. The reaction was initiated by exposing the quartz tube to a mercury lamp (365 nm, 1000 W) and finally the reaction products were dialyzed in ultrapure water for 2 days to remove iron ions, trace H_2O_2, and other small molecular reaction products. The photo-Fenton reactions were carried out in a photoreactor equipped with the irradiation lamps having different emission wave- lengths and powers (Bilon, Shanghai).

5.3.1.1.7 K-intercalation method

In this method, Lin et al. [67] has used MWCNTs for K-intercalation to exfoliate and cut down graphene moiety. In a typical reaction, 0.0820 g MWCNTs were put into a Pyrex tube with a stop and a side vacuum connection, followed by adding ~ 0.6 g of K and the two materials were mixed gently by shaking the Pyrex tube. Next, the tube was initially heated in an oil bath to ~ 110 °C (the temperature of the oil bath) under vacuum condition (0.05 Torr) and held for 10 min to remove any evaporable phases and then further heated to 190–200 °C and held for more than 3 h until the mixture in the tube changed to bronze colour (a few silver grey segments were seen on the surface of the mixture), after which the mixture was held for another 1 h and then after cooling down the tube to room temperature, the vacuum pump was turned off. The following process needs to be carefully carried out since the intercalated MWCNTs are highly reactive in air, ethanol and water. After taking all the security measures, the bottom of the Pyrex tube was placed in a room temperature water bath in an ultrasonic vibrator (Bandelin Sonorex RK-100 H) and then air was introduced slowly into the Pyrex tube by carefully controlling the valve (Caution: the rapid introduction of air into the tube through the side valve or by opening the stop cover will make violent/explosive burning of the mixture). After light and burning in the Pyrex tube were not visible, the ultrasonic vibrator was turned on and then the stop cover of the tube was opened and 50 mL ethanol was added into the tube followed by addition of 50 mL deionized water (do not add water until the residual silver K suspension on the surface of EtOH disappears). The tube was sonicated for 2 h. Next, a deep yellow solution containing GQD was separated from the residual solid carbon by using a Hettich Zentrifugen EBA 21 centrifuge. The remaining K-ions in the solution were absorbed/removed with the cation exchange resin and the resultant faint yellow solution was centrifuged at least 3 times (each time for 15 min) at 6000 RPM to further remove any residual solid carbon, before it was concentrated to a higher concentration (deep yellow) solution via distillation at 120 °C for requisite times.

5.3.1.1.8 Precursor pyrolysis

Herein, Tang et al. [68] used a series of glucose solutions with various concentrations (2.2, 4.4, 6.7, 8.9, 11.1 wt %) (by using glucose and DI water as source and solvent, respectively). Then, glucose solutions of 2–2.5 mL were siphoned to a glass bottle (4 mL) with tightened cover. The glass bottle was heated with a conventional microwave oven at a certain power (280, 336, 462, 595 and 700 W) for a period of time (1, 3, 5, 7, 9 and 11 min) and they observed that the experimental parameters such as microwave power, heating time, source concentration, as well as solution volume have a distinct effect on the growth of GQDs. Hence, these parameters can be tuned to prepare GQDs in a controllable way. In the process of microwave heating, the solution changes colour (e. g. from transparent to pale yellow) as a result of formation of GQDs and subsequently, the glass bottle was cooled to room temperature for characterizations. Besides glucose, they also investigated sucrose and fructose.

5.3.1.1.9 Thermal plasma method

Kim et al. [69] reported preparation of GQDs from Carbon tubes. CNTs of various lengths (5–20 cm; 2 cm diameter) were attached to the anode and thermal plasma of

Ar (99.999 %, at the injection flow rate of 13.5 L per min) was generated by applying a high voltage of ~3 kV between a zirconium-containing tungsten cathode and a copper anode. Then a thermal plasma jet for generating a carbon atomic beam was operated by a dc of ~200 A and 60 V. A plasma jet with a value close to sound velocity flowed into a Cu nozzle (6 mm in inner diameter), then continued through an attached carbon tube and ethylene gas was introduced continuously (2.5 L per min) as a carbon source into the torch using a gas flow meter. Carbon soot produced in the process was dispersed in ethanol by stirring with a stirring rod. Herein, isolated GQDs were dispersed in ethanol.

5.3.1.1.10 E-beam lithography method

Massabeau et al. [55] reported about the synthesis of GQDs using E-beam lithography insitu for a THz device. For the device fabrication, first they exfoliated large single layer graphene flakes. Next, hBN/graphene/hBN sandwiches were realized by a dry transfer technique of the graphene onto hBN layers. Then, the encapsulated graphene was transferred onto a high resistivity silicon substrate. For patterning the graphene quantum dot, they used a standard e-beam lithography process. Then a 30 nm Al mask was deposited followed by a lift-off. A subsequent Fluorine-reactive ion etching was performed to etch away the exposed area, leaving the pattern of quantum dot.

5.3.2 Bottom-up methods

The major disadvantages of such a top–down approach are that an undefined mixture of all possible molecules and structures is produced at the end, with each of them having its own properties, and more significantly the difficulty in controlling the size and defective surface structure. On the contrary, the bottom-up approaches using solution chemistry and pyrolysis or carbonization of organic molecules to form bulk materials by cage opening of fullerene and GQDs derived from organic molecules are simpler. Hence, it is possible to module the structure size, shape and surface state during synthesis. A seven-step syntheses from benzene derivative ending up with three different graphene molecules, two of which are deliberatively doped with N to produce GQDs have been reported using solution chemistry [70]. In a similar manner, a mechanistic approach to the synthesis of GQDs by metal-catalysed cage-opening of C_{60} [71] shows the fragmentation of the embedded molecules at elevated temperatures producing carbon clusters that undergo diffusion and aggregation to form GQDs.

Moreover, solution-phase methods by oxidative condensation of aryl groups have also been successfully implemented to synthesize GQDs [72, 73]. In this, synthesis of stabilized GQDs has been demonstrated with uniform and tunable size by multiple 2',4',6'-trialkyl phenyl groups covalently attached to the edges of the graphene moieties which consist of graphene moieties containing 168, 132 or 170 conjugated carbon atoms, respectively [73]. The formation of 3-D protection is attributed to the strong covalent binding and twisted phenyl groups owing to the steric hindrance at the edge of the graphene moieties. Apart from this, multicolour GQDs have been prepared with a uniform size of approx. 60 nm diameter and 2–3 nm thickness by using unsubstituted hexa-peri-hexabenzocoronene (HBC) as the precursor [74]. However, the main challenges associated with this bottom up technique are multiple steps and a poor yield.

Different synthetic top down techniques for synthesizing GQDs have been summarized at a glance in Table 5.2. Most of the starting materials are graphite, MWCNTs or coal and the process involves mainly oxidative cutting, hydrothermal/solvothermal breakdown of the starting materials or in some of the cases microwave heating or ultrasonication to achieve the desired goal for faster synthesis.

Table 5.2: Different top-down techniques for GQDs using oxidation, hydrothermal, microwave and ultrasonication.

Methods	Starting material	Size (nm)	Height (nm)	Colour	Yield (%)	Ref
Top Down	GO	~ 3–4	~ 1.5	Blue, Green	N/A	[65]
Acidic oxidation	Carbon black	15	0.5	Green	44.5	[75]
	Carbon fibres	1–4	0.4–2	Blue, Green, Yellow	N/A	[76, 77]
	Lignin biomass	2–6	0.4–2	Cyan	Gram scale	[78]
	GO	1–8	1–2	Orange-red to Blue	N/A	[33]
	Coal	1–7	1.2	N/A	N/A	[79]
	Coal	2.30 ± 0.78	1.5–3	Yellow, Green, Blue	20	[15]
	MWCNTs	4–26	N/A	Green	N/A	[80]
Top Down Hydrothermal/ Solvothermal	GO	5–13	1–2	Blue	5	[23]
		1.5–5	1.5–1.9	Green	N/A	[81]
		2.5	1.13	Blue to Yellow	N/A	[82]
		3–5	0.95	Green	N/A	[83]
Microwave	Graphite	2–8	0.7–3	Blue to Green	70 ± 5	[84]
	1,3,6-trinitropyrene	4.12–3.33	0.4 – 2	Yellow Cyan	35 %	[85]
	MWCNTs	2	1	Green	85 %	[13]
Ultrasonic	Graphene	3–5	N/A	Blue	N/A	[35]
	GO	3.0	0.7–3	Cyan	N/A	[86]

Apart from the above mentioned top-down techniques, electrochemical scissoring has been used to synthesize GQDs and are listed below in Table 5.3.

Table 5.3: Synthesis of GQDs using electrochemical techniques (Top-Down).

Methods	Starting material	Size (nm)	Height (nm)	Colour	Yield (%)	Ref
Electrochemical	Graphene	3–5	1–2	Green	N/A	[59]
	Graphite rods	5–10	<0.5	Yellow	N/A	[87]
		3 ± 0.3	1–2	Green	N/A	[60]
	MWCNTs	5 ± 0.3	3	Green	N/A	[60]
		8.2 ± 0.3	5	Green	N/A	[60]
		23 ± 2		Green	N/A	[60]

Table 5.4 includes other different top-down techniques used for synthesis of GQDs excluding the methods described in the earlier tables. In Table 5.5, the synthesis of doped GQDs has been tabulated using common techniques like acidic oxidation, electrochemical scissoring, hydrothermal or solvothermal cutting of the precursors.

Table 5.4: Synthesis of GQDs using other top-down methods.

Methods	Starting material	Size (nm)	Height (nm)	Colour	Yield (%)	Ref
Photo-Fenton Reaction	GO	40	1.2	Blue	45	[66]
K-intercalation	MWCNTs	~20	<1	Blue	22.96	[67]
	Graphite flakes	~20	0.9	Blue	9.9	[67]
Precursor pyrolysis	Glucose	1.65–2.1	3.2	DUV, blue	N/A	[68]
	Citric acid	~15	0.5–2	Blue	N/A	[88]
Pyrolysis and Exfoliation	Unsubstituted HBC	~60	2–3	Blue	N/A	[74]
Thermal plasma	Ethylene gas, CNTs	10–19	<1	N/A	N/A	[69]
E-beam lithography	Graphene	240	N/A	N/A	N/A	[55]

From Table 5.2 to Table 5.5, we have tabulated the popular top-down methods. However, there are very few reports on synthesis of GQDs via bottom-up approach (Table 5.6). They involve mainly organic precursors, but the multistep process makes this approach least viable in terms of bulk synthesis for practical applications.

Table 5.5: Different top-down techniques for doped GQDs.

Methods	Starting material	Size (nm)	Height (nm)	Colour	Yield (%)	Ref
Hydrothermal Cutting (Boron-doped GQDS)	Boron-doped graphene	2–4	0.5–0.8	Blue	N/A	[89]
Electrolysis (Sulphur-doped GQDs)	Graphite	2–4	0.7	Blue to Green	N/A	[90]
Electrochemical method (N-doped GQDs)	Graphene and Tetrabutyl ammonium perchlorate	2–5	1–2.5	Blue	N/A	[91]
Hydrothermal (N-doped GQDs)	Citric acid	1.8–3.8	0.5–2	Blue	N/A	[92]
Hydrothermal (S, N co-doped GQDs)	Citric acid, Urea	2.69 ± 0.42	0.5–2	Yellow Green	N/A	[93]
Hydrothermal (S, N co-doped GQDs)	Oxidized graphene, ammonia, powder S	2.4–5.35		Blue		[94]
Hydrothermal	Pyrene	2.4–4.8	0.9–1.9	Yellow, Blue, Green, Cyan	63	[95]
Acidic oxidation (Amino functionalized GQDs)	GO, ammonia	5–7	N/A	Blue, Green, Cyan	N/A	[96]
Microwave (S, N and F co-doped GQDs)	MWCNTs	~ 2	1	Green	~ 85	[13]

Table 5.6: Bottom-up techniques for the synthesis of GQDs.

Methods	Starting material	Size (nm)	Height (nm)	Colour	Yield (%)	Ref
Bottom up Catalysed cage-opening	C_{60}	2.7–10		N/A	15–30	[71]
Bottom up Stepwise solution Chemistry	Organic precursors	~ 2.5–5		Red	N/A	[72, 73]

5.3.2.1 Experimental details of a few bottom-up methods
In a similar manner to Section 5.3.1.1, now we will discuss about some experimental details to synthesize GQDs by bottom-up technique. Only a few bottom-up

approaches have been reported till date either due to complication of multiple step procedure or difficult experimental set up. [71–73] Here we will discuss the following bottom-up synthetic method described below.

5.3.2.1.1 Bottom-up catalysed cage-opening method

Lu et al. [71] derived GQDs from the surface-catalysed decomposition of C_{60} adlayers on reactive transition metals. They have illustrated the transformation of a high coverage of C_{60} ($\theta > 0.7$ monolayer (ML)) to a single layer of graphene covering a Ru (0001) surface after annealing the sample at 1,200 K for 5 min. They have shown that within a coverage range of 0.2 ML $< \theta < 0.7$ ML C_{60}, the short diffusion distance between the fragmented molecules favours the aggregation of these carbon clusters, resulting in the growth of larger-sized and irregularly shaped graphene nano-islands. Also, they found that to grow GQDs, the interparticle diffusion length must be sufficiently long (mean distance between C_{60} molecules, 15 ± 3 nm) or the interparticle diffusion velocity must be sufficiently low to limit diffusional aggregation of the decomposed C_{60} fragments. By controlling these factors, Lu et al. has synthesized GQDs with well-defined geometrical shapes can be assembled from the carbon clusters derived from C_{60}.

5.3.2.1.2 Stepwise solution chemistry method

Yan et al. [72] reported a stepwise organic synthesis of GQDs from small starting molecule. They have carried out the experiment as stated below: Compound 1 is final product GQDs.

4-bromo-3-iodoaniline (S2) 4-bromo-3-iodobenzoic acid (**S1**; 1.9 g, 5.8 mmol), diphenylphosphoryl azide (1.9 g, 6.9 mmol), triethylamine (1.3 g, 13 mmol) and t-butanol (100 mL) were heated to reflux for 5 h under argon atmosphere. After cooling, the solvent was removed under reduced pressure. The crude product was purified by column chromatography on silica gel with CH_2Cl_2/hexanes (1:4) to afford t-butyl 4-bromo-3-iodophenylcarbamate (2.2 g, 95 %) as a white powder. t-butyl 4-bromo-3-iodophenylcarbamate (2.2 g, 5.5 mmol) was then dissolved in CH_2Cl_2 (50 mL). A solution of trifluoroacetic acid (15 mL) in CH_2Cl_2 (50 mL) was added slowly. The reaction was stirred at room temperature until no starting materials was detected by TLC. After evaporating the solvent and excess trifluoroacetic acid in a vacuum, the crude product was purified by column chromatography on silica gel with ethyl acetate/hexanes (1:4) to afford **S2** (1.4 g, 81 %) as a brown solid: 1 H NMR (400 MHz, CDCl$_3$): $\delta = 7.31$ (d, J = 8.4 Hz,1 H), 7.19 (d, J = 2.8 Hz, 1 H), 6.51 (dd, J = 2.8, 8.8 Hz, 1 H), 3.67 (br, 2 H)); 13C NMR (100 MHz, CDCl3): $\delta = 146.25$, 132.59, 126.05, 117.11, 116.47, 101.34. 1-bromo-3-(phenylethynyl)benzene 1-bromo-3-iodobenzene (2.26 g, 8 mmol), Pd(PPh$_3$)$_4$ (92 mg, 0.08 mmol), CuI (15.2 mg, 0.08 mmol) and piperidine (20 ml) were mixed together under argon atmosphere. The mixture was degassed by two "freeze-pump-thaw" cycles. Then phenylacetylene (817 mg, 8 mmol) was then added slowly.

The reaction mixture was stirred for 4 h at room temperature. Then 2 M aqueous NH_4Cl was added to quench the reaction. The mixture was extracted twice with dichloromethane, and the organic layer was washed twice with water and dried over Na2SO4. After evaporating the solvent, the product was purified by column chromatography on silica gel with hexanes to afford 1-bromo-3- (phenylethynyl)benzene (2.0 g, 97 %) as colourless oil: 1 H NMR (400 MHz, $CDCl_3$): δ = 7.70 (t, J = 1.6 Hz, 1 H), 7.53 (m, 2 H), 7.46 (m, 2 H), 7.36 (m, 3 H), 7.22 (t, J = 8 Hz, 1 H); 13C NMR (100 MHz, $CDCl_3$): δ = 134.29, 131.65, 131.35, 130.12, 129.75, 128.61, 128.39, 125.30, 122.73, 122.15, 90.66, 87.78.

3-(phenylethynyl)phenylboronic acid (S3) 1-bromo-3-(phenylethynyl)benzene (2.1 g, 8.2 mmol) was dissolved in anhydrous THF (10 mL) and cooled down to −78 °C in a dry ice/acetone bath. n-BuLi (9.8 mmol in 4 ml Hexanes) was slowly added into the reaction mixture under argon atmosphere. The resulting mixture was stirred for 1 h at −78 °C, and then B(OMe)3 (1.37 mL, 12.3 mmol) was added. The reaction mixture was allowed to warm up to room temperature gradually. The reaction was quenched with 1 M HCl and extracted twice with ethyl acetate. The organic layer was washed twice with 10 % HCl and dried over Na2SO4. Evaporating the solvent afforded a white solid (1.9 g). The crude product was used in the following steps without further purification. **6-Bromo-3'-(phenylethynyl)biphenyl-3-amine (S4)** Freshly made 3-(phenylethynyl)phenylboronic acid (S3; 1.9 g), 4-bromo-3-iodoaniline (S2; 2.2 g, 7.4 mmol), $Pd(PPh_3)_4$ (170 mg, 0.15 mmol), toluene (20 mL), ethanol (5 mL) and 2 M K_2CO_3 (10 mL) were mixed together. The resulting mixture was degassed by three "freeze-pump-thaw" cycles and then heated at 60 °C and stirred for 3 d. After cooling, the reaction mixture was extracted twice with ethyl acetate. The organic phase was dried over Na_2SO_4 and concentrated under reduced pressure. The product was purified by column chromatography on silica gel with CH2Cl2/hexanes (1:1) to afford **S4** (2.4 g, 94 %) as a yellow waxy solid: 1 H NMR (400 MHz, $CDCl_3$): δ = 7.54 (m, 4 H), 7.42–7.33 (m, 6 H), 6.66 (d, J = 2.8 Hz, 1 H), 6.56 (dd, J = 2.8, 8.4 Hz, 1 H), 3.71 (br, 2 H); 13C NMR (100 MHz, CDCl3): δ = 145.72, 142.15, 141.44, 133.54, 132.29, 131.58, 130.56, 129.29, 128.31, 128.26, 127.89, 123.15, 122.91, 117.59, 115.79, 110.157, 89.53, 89.23. 2-bromo-5-iodo-3'-(phenylethynyl)biphenyl **(S5)** 6-bromo-3'-(phenylethynyl)biphenyl-3-amine (**S4;** 2.4 g, 6.9 mmol) and iodine (2.6 g, 10.4 mmol) were dissolved in dry benzene (35 mL) and cooled down to 5 °C under argon atmosphere. t-butyl nitrite (1.3 g, 12.5 mmol) was added dropwise over 30 min via a syringe. The resulting mixture was allowed to warm up to room temperature and stirred for 3 h. About 1 M NaOH (15 mL) was added to quench the reaction. The reaction mixture was extracted twice with ethyl acetate. The organic layers were combined and dried over Na_2SO_4. After evaporating the solvent, the crude product was purified by column chromatography on silica gel with hexanes to afford S5 (1.7 g, 54 %) as a yellow waxy solid: 1 H NMR (400 MHz, $CDCl_3$): δ = 7.68 (d, J = 2.4 Hz, 1 H), 7.54 (m, 5 H), 7.42 (t, J = 7.6 Hz, 1 H), 7.39 (d, J = 8.4 Hz, 1 H), 7.35 (m, 4 H); 13C NMR (75 MHz, CDCl3): δ = 143.75, 139.83, 139.68, 137.87, 134.66, 132.18, 131.63, 131.15, 129.12, 128.35, 128.15, 123.32, 123.05, 122.44, 92.33, 89.92, 88.90. 1,3,5-tris

(3,7,11,15-tetramethylhexadecyl)benzene **(S6)**. A mixture of 1,3,5-trichlorobenzene (1.21 g, 6.67 mmol) and NiCl$_2$(dppe) (51.8 mg, 0.1 mmol,) in dry diethyl ether (10 mL) was stirred under argon atmosphere at 0 °C. To this solution freshly made 3,7,11,15-tetramethylhexadecyl-1-magnesiumbromide (30 mmol in 30 mL dry diethyl ether) was slowly added. The resulting mixture was allowed to warm up to room temperature and stirred overnight. The mixture was refluxed for another 2 h before the reaction was quenched with ice/water. The organic phase was separated and dried over Na$_2$SO$_4$. After evaporating the solvent, the yellow crude product was purified by column chromatography on silica gel with hexanes to afford **S6** (yield: 92 %) as colourless oil with 2,6,10,14-tetramethylhexadecane as byproduct (generated from excess 3,7,11,15-tetramethylhexadecyl-1-magnesiumbromide): 1 H NMR (400 MHz, CDCl$_3$): δ = 6.81 (s, 3 H), 2.55 (m, 6 H), 1.65–0.80 (m, 117 H); 13C NMR (75 MHz, CDCl$_3$): δ = 143.04, 125.71, 39.43, 39.21, 39.12, 37.51, 37.45, 37.35, 33.58, 32.83, 32.74, 31.65, 28.02, 27.50, 24.86, 24.53, 22.76, 22.66, 19.80, 19.73. 2-bromo-1,3,5-tris(3,7,11,15-tetramethylhexadecyl)benzene **(S7)** 1,3,5-tris(3,7,11,15-tetramethylhexadecyl)benzene **(S6**; 5.87 mmol) was dissolved in CCl$_4$ (3.3 mL) under argon atmosphere. To this solution Br$_2$ (0.3 mL, 5.87 mmol) was added drop-wise over 30 min and the reaction mixture was stirred overnight at room temperature. The reaction was quenched with 1 M NaOH and extracted twice with CH$_2$Cl$_2$. The organic phases were combined and dried over Na$_2$SO$_4$. After evaporating the solvent, the crude product was purified by column chromatography on silica gel with hexanes to give **S7** (yield: 95 %) as colourless oil: 1 H NMR (400 MHz, CDCl$_3$): δ = 6.86 (s, 2 H), 2.72 (m, 4 H), 2.51 (m, 2 H), 1.65–0.80 (m, 117 H); 13C NMR (100 MHz, CDCl$_3$): δ = 142.73, 141.86, 127.72, 123.33, 39.39, 37.47, 37.41, 37.31, 37.24, 34.83, 33.01, 32.79, 31.60, 27.99, 27.50, 24.82, 24.49, 24.44, 22.73, 22.64, 19.77, 19.71. trimethyl(2',4',6'-tris(3,7,11,15-tetramethylhexadecyl)biphenyl-4-yl)silane **(S9)** A round bottom flask was charged with 2-bromo-1,3,5-tris(3,7,11,15-tetramethylhexadecyl)benzene (**S7**; 3.60 g, 3.61 mmol), 4- (trimethylsilyl)phenylboronic acid (**S8**; 1.05 g, 5.41 mmol), Pd(OAc)$_2$ (20 mg, 0.09 mmol), 2-(2', 6'- dimethoxybiphenyl)-dicyclohexylphosphine (74 mg, 0.18 mmol), and K$_3$PO$_4$ (2.30 g, 10.8 mmol). Toluene (15 mL) was added via a syringe under argon atmosphere and the reaction mixture was stirred for 10 min at room temperature to allow the catalyst to be dissolved. Then the resulting mixture was heated to 80 °C overnight. After cooling, the reaction solution was filtered and concentrated under reduced pressure. The crude product was purified by column chromatography on silica gel with hexanes to afford **S9** (yield: 94 %) as colourless oil: 1 H NMR (400 MHz, CDCl$_3$): δ = 7.52 (d, J = 8 Hz, 2 H), 7.16 (d, J = 8 Hz, 2 H), 6.93 (s, 2 H), 2.62 (m, 2 H), 2.28 (m, 4 H), 1.65–0.65 (m, 117 H), 0.30 (s, 9 H); 13C NMR (100 MHz, CDCl3): δ = 141.90, 141.24, 140.80, 138.44, 138.03, 132.72, 129.36, 126.37, 39.43, 39.19, 39.08, 38.97, 37.52, 37.47, 37.36, 37.07, 33.42, 32.82, 31.46, 28.02, 27.50, 24.87, 24.53, 22.77, 22.68, 19.81, 19.74, 19.42, 19.37, −0.974. 4,4,5,5-tetramethyl-2-(2',4',6'-tris(3,7,11,15-tetramethylhexadecyl)biphenyl-4-yl)-1,3,2-dioxaborolane **(S10)** trimethyl(2',4',6'-tris(3,7,11,15-tetramethylhexadecyl)biphenyl-4-yl)silane (**S9**; 1.54 g, 1.44 mmol) was dissolved in anhydrous dichloromethane (4.5 mL) under argon atmosphere. BBr$_3$ (433 mg, 1.73 mmol) was slowly added. The resulting solution was stirred for 5 h at room temperature before water was added to quench the reaction. The reaction mixture was extracted twice with CH2Cl2 and the organic layer was dried over Na$_2$SO$_4$ and concentrated under reduced pressure. The crude product was dissolved in anhydrous CH$_2$Cl$_2$ (4.5 mL) again. Pinacol (340 mg, 2.88 mmol) was then added. The reaction mixture was stirred overnight under argon atmosphere. After evaporating the solvent, the crude product was purified by column chromatography on silica gel with CH$_2$Cl$_2$/hexanes (1:10) to afford **S10** (yield: 75%) as colourless oil: 1H NMR (400 MHz, CDCl$_3$): δ = 7.81 (d, J = 7.4 Hz, 2 H), 7.18 (d, J = 7.4 Hz, 2 H), 6.91 (s, 2 H), 2.60 (m, 2 H), 2.26 (m, 4 H), 1.65–0.68 (m, 129 H); 13C NMR (100 MHz, CDCl3): δ = 143.67, 141.94, 140.98, 138.29, 134.33, 129.40, 126.27, 83.67, 39.40, 39.06, 38.97, 37.49, 37.44, 37.33, 37.08, 34.69, 33.39, 32.81, 32.76, 31.62, 31.39, 27.99, 25.29, 24.96, 24.83, 24.51, 24.35, 22.74, 22.64, 19.78, 19.71, 19.35, 19.47, 14.14.

Compound **S11** 4,4,5,5-tetramethyl-2-(2',4',6'-tris(3,7,11,15-tetramethylhexadecyl) biphenyl-4-yl)-1,3,2- dioxaborolane (S10; 827 mg, 0.74 mmol), 2-bromo-5-iodo-3'-(phenylethynyl)biphenyl (S5; 313 mg, 0.68 mmol), Pd(PPh$_3$)$_4$ (40 mg, 0.035 mmol), toluene (5 mL), ethanol (2 mL) and 2 M aqueous K$_2$CO$_3$ (2.5 mL) were mixed together. The resulting mixture was degassed via three "freeze-pump-thaw" cycles and then heated at 80 °C and stirred overnight. After cooling, the reaction mixture was extracted twice with ethyl acetate and dried over Na$_2$SO$_4$. After evaporating the solvent, the crude product was purified by column chromatography on silica gel with CH$_2$Cl$_2$/hexanes (1:20) to afford **S11** (830 mg, 92%) as yellow oil: 1H NMR (400 MHz, CDCl$_3$): δ = 7.75 (d, J = 8.4 Hz, 1 H), 7.65 (m, 4 H), 7.60–7.50 (m, 4 H), 7.45 (m, 2 H), 7.35(m, 3 H), 7.26 (d, J = 6.8 Hz, 2 H), 6.94 (s, 2 H), 2.61 (m, 2 H), 2.31 (m, 4 H), 1.70–0.60 (m, 117 H); 13C NMR (100 MHz, CDCl3): δ = 142.16, 142.10, 141.32, 141.24, 140.43, 140.19, 137.71, 137.57, 133.49, 132.47, 131.65, 130.85, 130.65, 129.61, 129.46, 128.33, 128.05, 127.43, 126.42, 126.33, 123.25, 123.22, 121.35, 89.73, 89.19, 39.39, 39.10, 39.04, 39.01, 38.97, 37.46, 37.40, 37.36, 37.31, 37.04, 33.39, 32.80, 32.73, 31.46, 27.98, 24.82, 24.47, 24.35, 22.74, 22.65, 19.76, 19.69, 19.62, 19.35, 19.47.

Compound S12 Compound **S11** (275 mg, 0.21 mmol) was dissolved in anhydrous THF (2 mL) under argon atmosphere and cooled down to −78 °C in a dry ice/acetone bath. n-BuLi (0.25 mmol in 0.1 mL hexanes) was added slowly into the reaction mixture resulting a yellow solution. The reaction mixture was stirred for 1H at −78 °C, and then B(i-Pr)$_3$ (0.46 mL, 2 mmol) was added. The resulting mixture was allowed to warm up to room temperature gradually overnight. The reaction was quenched with 1M HCl and extracted twice with ethyl acetate. The combined organic layers were dried over Na$_2$SO$_4$ and concentrated under reduced pressure. The crude boronic acid (yield: 60%, estimated from 1H NMR) was used directly without further purification. The fresh made boronic acid (0.13 mmol, 4 equiv.), 1,3,5-triiodobenzene (14.6 mg, 0.032 mmol, 1 equiv.), Pd(PPh$_3$)$_4$ (3 mg, 0.0025 mmol), toluene (1 mL) and 2 M K$_2$CO$_3$ (0.5 mL) were mixed together. The resulting mixture was degassed by three "freeze-pump-thaw" cycles and then heated at 80 °C and stirred overnight. After cooling, the reaction was extracted twice with dichloromethane and dried over Na$_2$SO$_4$. After evaporating the solvent, the product was purified by column chromatography on silica gel with CH$_2$Cl$_2$/hexanes (1:10) to afford **S12** (102 mg, 84 %): 1 H NMR (400 MHz, CDCl$_3$): δ = 7.73 (d, J = 1.6 Hz, 3 H), 7.70 (d, J = 8 Hz, 6 H), 7.65 (dd, J = 1.6, 8 Hz, 3 H), 7.60 (s, 3 H), 7.55 (m, 6 H), 7.48 (d, J = 8 Hz, 3 H), 7.40–7.30 (m, 12 H), 7.26 (d, J = 8 Hz, 6 H), 7.04 (d, J = 8 Hz, 3 H), 7.00 (d, J = 8 Hz, 3 H), 6.95 (s, 9 H), 2.62 (m, 6 H), 2.33 (m, 12 H), 1.70–0.65 (m, 351 H); 13C NMR (100 MHz, CDCl3): δ = 142.06, 141.98, 141.32, 140.73, 140.40, 140.15, 140.01, 139.76, 139.26, 139.02, 138.25, 137.92, 132.71, 131.62, 131.14, 130.71, 130.48, 129.98, 128.84, 128.33, 127.95, 126.45, 126.37, 123.34, 123.24, 89.60, 89.50, 39.36, 39.06, 37.42, 37.30, 37.14, 33.40, 32.79, 31.47, 29.71, 27.96, 24.80, 24.46, 24.35, 22.72, 22.63, 19.74, 19.68, 19.47, 19.40. HR-MS (MALDI-TOF) calculated. for C282H426 3816.34 [M +] found 3816.80.

Compound 2 In a round bottom flask equipped with a condenser, a mixture of compound **S12** (177 mg, 0.046 mmol), tetraphenylcyclopentadienones (214 mg, 0.56 mmol) and diphenyl ether (1 mL) was heated to reflux under argon atmosphere for 17 h. After cooling, the reaction mixture was quenched with methanol (40 mL). The precipitate was filtered and purified by column chromatography on silica gel with CH$_2$Cl$_2$/hexanes (3:7) to afford 2 (85 mg, 40 %) as a white solid: 1 H NMR (400 MHz, CDCl$_3$): δ = 7.63 (m, 9 H), 7.27 (m, 12 H), 7.10–6.40 (m, 87 H), 5.77 (d, J = 6.4 Hz, 3 H), 5.63 (s, 3 H), 5.19 (s, 3 H), 2.63 (m, 6 H), 2.35 (m, 12 H), 1.70–0.60 (m, 351 H); 13C NMR (100 MHz, CDCl3): δ = 142.18, 142.04, 141.21, 141.00, 140.68, 140.61, 140.52, 140.42, 140.31, 140.06, 139.72, 139.50, 138.62, 138.58, 137.90, 132.52, 131.69, 131.41, 131.06, 130.49, 130.19, 129.58, 129.50, 129.30, 129,22, 127.13, 126.96, 126.62, 126.50, 126.34, 125.83, 125.54, 125.30, 125.08, 124.67, 39.38, 39.09, 39.00, 37.46, 37.43, 37.30, 33.40, 32.81, 31.33, 29.70, 27.98, 24.81, 24.50, 22.74, 22.64, 19.77, 19.71, 19.52, 19.45. HR-MS (MALDI-TOF) calculated for C366H486 4884.81 [M + H] + found 4885.72.

Compound 1 Compound 2 (20 mg, 0.0041 mmol) was dissolved in dry dichloromethane (10 mL). The solution was degassed via bubbling argon for 10 min. Then FeCl$_3$ (144 mg, 0.88 mmol) in dry nitromethane (0.6 mL) was added slowly via a syringe. The resulting mixture was kept under argon flow during the entire reaction. After 3 h, the reaction was quenched with a large volume of methanol. The dark precipitate was collected and washed with methanol. The crude product was dissolved in CH2Cl2 (0.1 mL) again and precipitated in methanol (20 mL). The black precipitate was then collected and washed with methanol and dried in a vacuum to afford crude product 1 (17 mg, 86 %) as a black powder, which was subsequently purified with reverse-phase HPLC.

5.3.3 Mechanism of oxidation of carbon skeleton

Commonly used method for the synthesis of GQDs or GNRs involve oxidative cleavage of MWCNTs/graphite/graphene in a suitable medium. For example, Kosynkin et al. have proposed the same mechanism of cutting of carbon fibre to micrometre sized pitch-based carbon fibres as described in the chemical oxidation route to unzip the CNTs (Figure 5.15) [97]. A graphical representation of this proposed mechanism of oxidative cleavage is demonstrated below.

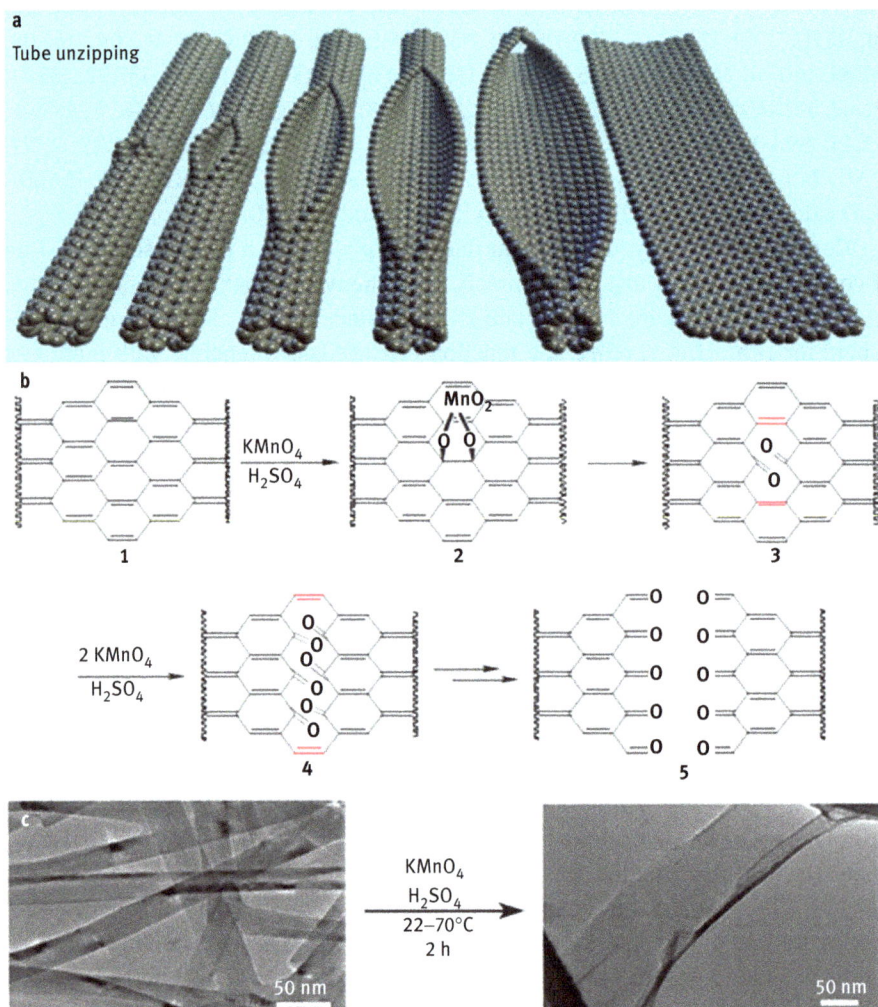

Figure 5.15: (a). A Schematic diagram of unzipping of CNTs. (b). Proposed $KMnO_4$ oxidative cutting mechanism. (c). TEM image showing the transformation of GNRs from CNTs on oxidation in acidic medium. [(Reprinted with permission from Ref [97] Copyright 2009 Nature Publishing Group.)].

In a similar manner, exfoliation method of graphene through oxygen binding [98] has also been described – an epoxy bridge is formed on the graphite surface, where a single oxygen atom bonds to two adjacent carbon atoms, forming a triangle and rest of the mechanism is exactly what followed by Li et al. [99] where these epoxy bridges do generate stress that leads to rupturing of the graphite lattice. DFT calculations [99] show that side-by-side parallel positioning of the epoxy bridges energetically favours to line up on the graphite surface and they jointly induce enough strain on the underlying lattice to break the basic carbon skeleton. However, all these mechanisms are theoretical speculation and experimental confirmation is awaited.

5.3.4 Effect of functionalization and doping of GQDs

Functionalization of graphene and graphene based nanomaterials precludes the challenges of zero bandgap and poor dispersion in several solvents and hence it is very crucial to control both the nature and degree of functionalization from the application point of view. In the earlier section, we have seen that functionalization plays an important role in controlling the cytotoxicity and also some solvent-assisted techniques like spin-coating, sputter coating, and filtration require high dispersion in a solvent which is only possible for graphene by means of selective functionalization [35][129][37]. Generally, both covalent and non-covalent techniques could be used for this purpose. The former processes include simultaneous oxidation of graphene to GO followed by reduction of GO to generate selected groups. However, this route potentially brings down the electrical conductivity of the modified graphene compared to pristine one due to harsh chemical treatment and alters the band structure especially if the degree of functionalization is high [37, 38]. On the other hand, the organic covalent functionalization involves two general routes: (1) the formation of covalent bonds between free radicals or dienophiles and $C = C$ of pristine graphene and (2) the formation of covalent bonds between organic functional groups and the oxygen groups of GO [35]. In contrast, the non- covalent technique involves functionalization by π-π interactions and thus is more preferred over the former one as it does not disturb the electronic network of pristine graphene [35]. In a similar manner GQDs are also functionalized accordingly for particular purposes. For instance, GQDs have been functionalized with thymine for the detection of mercury (II) [100].

Similar to functionalization, doping is another important technique to tailor the electronic arrangement of GQDs. Theoretical reports demonstrate doping with heteroatoms like nitrogen and boron changes the local bonding environment of carbon atoms, and induces the charge trapping states into the energy gaps of boron doped graphene quantum dot (BGQD)and nitrogen doped graphene quantum dot (NGQD) [101]. The high electron mobility of BGQD and high hole mobility of NGQD suggest the slow and asymmetric electron and hole relaxations of BGQDs to be beneficial for both oxidation and reduction reactions for water splitting, while the slower electron relaxation than hole relaxation promises NGQDs for catalysing the water splitting. Another group studied the effect of carrier doping and external electric field on the optical properties of graphene quantum dots, where the optical properties of finite-sized graphene quantum dots have been effectively controlled by doping it with different types of charge carriers (electron/hole) [102]. Moreover, this reveals that the energy band-gap increases when the diamond-shaped DQD (used for the computational study) is doped with holes while it decreases on doping it with electrons. It demonstrates that the application of external transverse electric field results in a substantial blue-shift of the optical spectrum for charge-doped DQD but the influence of charge-doping is more prominent in tuning the optical properties of finite-sized GQDs as compared to externally applied electric field. Hence, undoubtedly, synthetic route has a huge impact on the optical and other properties and thus need to be tailored according to the requirement of the particular application.

5.4 Characterization methodologies

Material characterization is an essential part of nanomaterial synthesis in order to understand the structure, morphology and composition as a function of both size and shape irrespective of the application. Although GQDs possess unique optical, physical and chemical features these properties are required to be measured and correlated using different characterization techniques. The characterizations techniques including X-ray Diffraction (XRD), UV–Vis spectroscopy, Raman spectroscopy, PL, transmission electron microscopy (TEM), atomic force microscopy (AFM), Fourier transform infrared spectroscopy (FT-IR) and X-ray photoelectron spectroscopy (XPS) are required to analyse their crystal structure, electron state, fluorescence property, functional group composition, surface morphology and vibrational patterns are discussed below with respect to GQDs.

5.4.1 X-ray diffraction (XRD)

XRD is an important technique for phase identification (crystalline/amorphous), and to understand the extent of crystallinity, particle size distribution of many nanomaterials and more significantly, for gathering valuable insight about the unit cell dimensions by determining the precise arrangements of the atoms. When an incident X-ray hits the surface of a crystal, the X-ray will be diffracted in a specific direction by the plane of the crystal and the intensity of the diffraction signal is plotted against the angle of the detector. A picture of 3-D electron density within the crystal can be acquired along with important information like chemical bonding, disorder. However, there are few limitations associated with this techniques are: (i) XRD of macromolecules (large repetitive unit cells) appear less resolved compared to that for small molecules (less than 100 atoms in the asymmetric unit). (ii) Material containing a mixture of crystallites below 5%, it could not be detected by this technique. (iii) Smaller particle size causes broadening in the XRD pattern giving an impression of amorphous nature of the sample although the small particles may retain their proper crystal structure. For GQDs, a broad XRD peak appears around at $2\theta \sim 25$ corresponding to (002) plane of graphite [76, 85]. The peak value shifts depending on the inter-planar distance of the graphene layers in GQDs.

5.4.2 Raman spectroscopy

Raman spectroscopy is a fundamental tool to acquire information regarding the molecular vibrations. A monochromatic light beam is illuminated on a sample and the scattered light is detected where the majority of the scattered light is of the same frequency as the excitation source, known as Rayleigh or elastic scattering. However, a very small amount of the scattered light is shifted from the laser frequency due to interactions between the incident light and the vibrational energy levels of the molecules in the sample. The shifted light intensity is plotted against the shifted energy known as Raman shift in Raman spectrum. It could be used widely to analyse the materials, especially graphene-based nanomaterials. It can unveil information on the crystal structure, electronic structure and lattice vibration of material. The typical Raman bands appear when there is a change in polarizability and this can be used to study the surface defects and electronic structure of carbon based materials like nanotubes, GQDs where the tangential mode (G-band) and defect mode (D-bands) of vibrations change systematically after CNTs to graphene transformation. The G band in GQDs appears near $1590 \, cm^{-1}$, representing the E_{2g} vibrational mode of aromatic domains in the 2-D hexagonal lattice structure [13, 102, 104]. D band (near $1350 \, cm^{-1}$) is usually observed due to the defect generated by surface functional groups or destroyed by synthesis process and the I_D/I_G ratio is generally used to estimate the defect extent of GQDs. Raman scattering of GQDs varies with varying sizes from 5 to 35 nm and reports suggest that the frequency of the G band increases as the size increases as long as GQDs keep their shape circular for $d \leq \sim 17$ nm [105]. The D band peak frequency is almost independent of size, whilst the G band shows clear size dependence. The second order of the D band (2D band, $\sim 2800 \, cm^{-1}$), is a single, sharp peak in monolayer graphene, while this peak broadens and shifts to higher frequency with increasing no. of layers, reflecting the evolution of the band structure. Hence, it is expected that the 2-D band of GQDs will show a blue shift with a larger FWHM with increasing size due to the higher average thickness. However, it is not true for GQDs, the 2D band of GQDs is much broader than that of graphene sheets, irrespective of thickness and the probable reason has been given that the defective nature of GQDs could dominantly contribute to deter-mining the frequency of the 2D band, thereby deactivating the effect of the thick-ness variation [105]. The broadening in the 2D band could be attributed to the relaxation of the double-resonance Raman selection rules associated with the ran-dom orientation of GQDs with respect to each other [105]. Hence, this technique is quite useful in determining several structural features also. However, the disadvan-tage associated with this technique is that intense laser beam can destroy the sam-ple to provide Raman spectrum.

5.4.3 Ultraviolet-visible spectroscopy (UV–VIS)

UV–Vis spectroscopy is a kind of absorption or reflectance spectroscopy in the UV and visible region and sometimes it extends up to near infrared (NIR) for investigat-ing the electronic transition of materials. The electromagnetic spectrum is gener-ated when the atoms or molecules undergo electronic transitions from ground to excited state after absorbing in the UV region. The relationship between the energy absorbed and an electronic transition in terms of wavelength (λ) or frequency (υ) for a transition is given below:

$$\Delta E = h\upsilon = hc/\lambda$$

Where, h = Planck's constant, ΔE = energy absorbed during electronic transition in a molecule from ground state to excited state. Lambert-Beer law is an important and more convenient expression for UV absorption which relates absorption (A) di-rectly to the concentration (c) of the absorbing species.

$$A = \text{€}cl$$

where, € is the molar extinsion co-efficient and l is the cell-path length.

This is a basic tool to identify the quantitative determination of different analytes, highly conjugated organic compounds and optically active quantum dots. Hence, this technique is widely used to investigate the excitation wavelength of GQDs. The UV–Vis spectrum analysis of GQDs generally shows two types of peak, the peak below 300 nm is attributed to the $\Pi \rightarrow \Pi^*$, appears due to aromatic C = C structure and the peak between 300 and 390 nm is attributed to the electrons transition from n $\rightarrow \Pi^*$, due to the oxygen-containing groups on GQDs surface [76, 106]. The absorption peaks have been shown in Figure 5.16 (black line). However, different functional groups on GQDs surface or different surface structure may give rise to shift in the wavelength which is clearly visible in Figure 5.17a and Figure 5.17b where the absorption wave-length of GQDs changes with the boronization [107]. UV–visible spectrum analysis helps us to get the excitation range of GQDs and thus determine the possibilities for optoelectronic applications. Mostly, sample preparation involved in solution mode for UV–visible analysis by choosing the desired solvent and calibrated using the blank

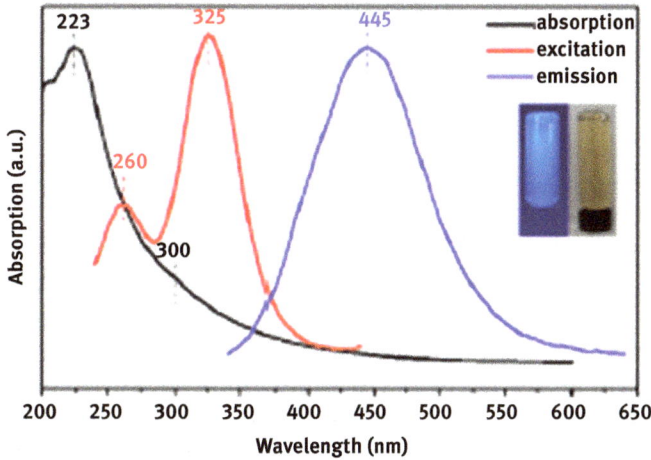

Figure 5.16: Optical characterization of GQDs-1 (GQD 1 prepared from coal tar pitch (CTP)). Combined UV-Vis absorption (black line), PLE spectrum with detection wavelength of 445 nm (blue line) and PL spectrum excited at 325 nm (red line) of the GQDs-1 dispersed in water. Inset of panel: the left is a photograph of the corresponding GQDs-1 aqueous solution under UV light with 365 nm excitation; the right is a photograph of the corresponding GQDs-1 aqueous solution taken under visible light. [Adapted from Ref [106] copyright free from MDPI publishers].

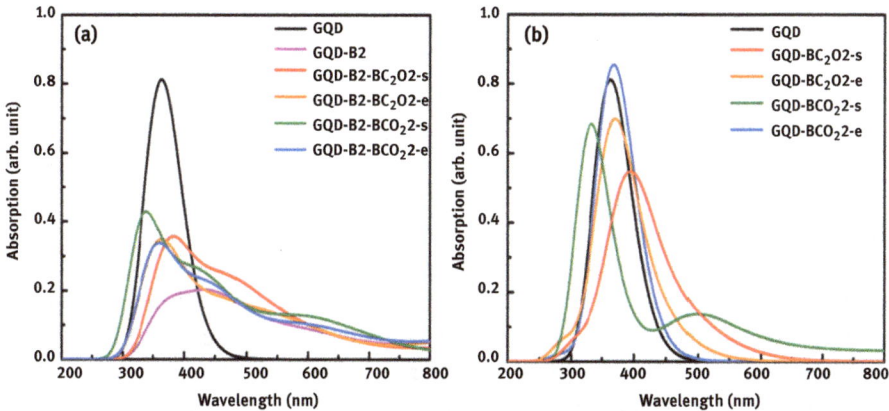

Figure 5.17: The calculated absorption spectra of the GQD–B2 (a) and pristine GQDs (b) with BC_2O and BCO_2 groups on various positions. Herein, GQD–B2 represents $C_{40}B_2H_{18}$, those of edge doped GQD–B2 with including GQD–BCO22-e, GQD–BC$_2$O2-e, GQD–B$_2$–BCO22-e and GQD–B$_2$–BC$_2$O2-e, are $C_{42}B_2O_4H_{20}$,$C_{42}B_2O_2H_{16}$,$C_{40}B_4O_4H_{20}$ and $C_{40}B_4O_2H_{16}$; those of surface-functionalized GQDs, for GQD – BCO22-s, GQD–BC$_2$O2-s, GQD–B2–BCO22-s and GQD–B2–BC$_2$O2-s, are $C_{42}B_2O_4H_{22}$, $C_{42}B_2O_2H_{20}$,$C_{40}B_4O_4H_{22}$, and $C_{40}B_4O_2H_{20}$, respectively. [(Reprinted with permission from Ref [107] Copyright 2019 Royal Chemical Society.)].

solvent. However, the solvent and substrate dependence also should be taken into account in order to analyse the data for an accurate understanding of the properties of materials.

5.4.4 Fluorescence spectroscopy

Fluorescence spectroscopy analyses fluorescence or PL from a specimen. It is a complementary phenomenon of UV–VIS spectroscopy where, a beam of UV light excites the electrons in molecules of certain compounds and causes them to emit light. It is primarily associated with the electronic and vibrational states and the sample preparation involves in solution mode using the desired solvent. PL spectrum measurement for GQDs is very important due to their strong fluorescence properties. For a typical measurement, the GQDs sample is excited with certain excitation wavelength and the GQDs emit corresponding emission peaks. Literature reveals GQDs show either excitation dependent or independent emission in the different colour region or emission light such as blue [89, 91], green [13], yellow [93, 95], cyan [96] and red luminescence [72, 73].

In addition to this, another interesting investigation is to analyse their electron transition via measuring PL excitation (PLE). PLE measurement to investigate the electrons transition situation of their GQDs shown in Figure 5.18. Each PLE spectrum can be de-convoluted into three peaks that correspond to electron transitions of $n \rightarrow \sigma^*$ (blue), $\pi \rightarrow \pi^*$ (green) and $n \rightarrow \pi^*$ (red), respectively.

5.4.5 Fourier transform infrared spectroscopy (FT-IR)

FT-IR is an important characterization technique to identify the surface component and functional groups on materials qualitatively. When IR radiation is passed through a sample (usually a small pellet sample), some of the IR radiation is absorbed by the molecular vibration of sample and some of it is transmitting to the sample. It detects the molecular vibration mode and create a particular signal for each molecule in the finger print region and the researchers take the advantage of this phenomenon to investigate the functional groups on the GQDs surface [76]. These functional groups are very important in tailoring the energy states as discussed earlier and thus having an important impact on the optical properties. In general, FT-IR signals of GQDs appear near 3400, 1600 and 1350 cm^{-1}, representing the stretching vibrations of O–H, aromatic C = C stretching and C = O of carboxy, respectively, for GQDs functionalized with carboxylic group and or –OH group. However, the presence of functional groups may vary according to the need of application and on different synthesis methods and precursors.

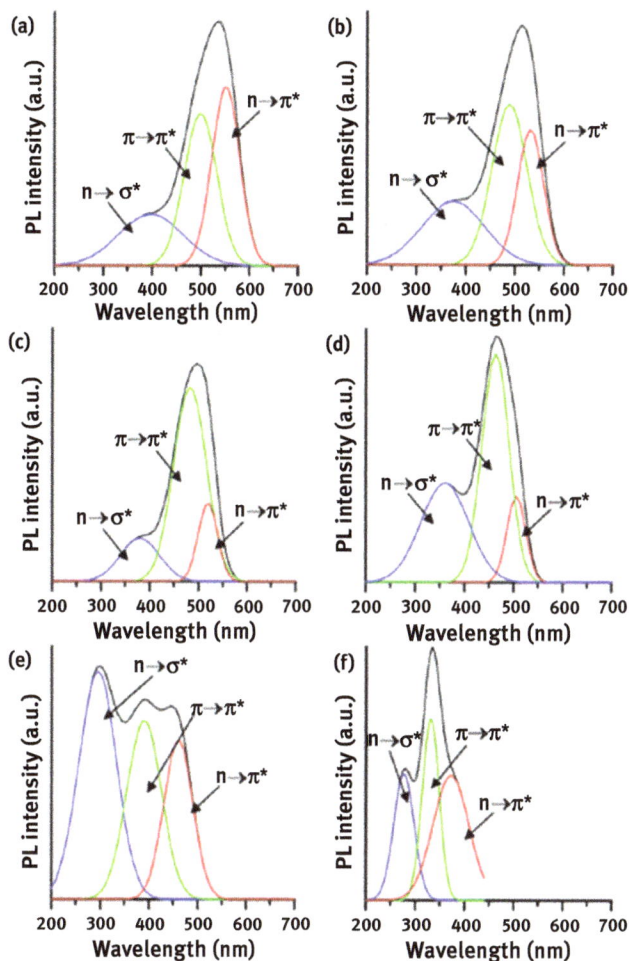

Figure 5.18: The PLE spectra of the aqueous GOQD dispersions. (a) QD79, (b) QD61, (c) QD54, (d) QD26, (e) QD16, and (f) QD10 where QD10 denotes 1.0 nm diameter, QD16 (1.6 nm diameter GQDs), QD26 (2.6 nm diameter), (d) QD54 (5.4 nm diameter), (e) QD61 (6.1 nm diameter), and (f) QD79 (7.9 nm diameter). Each PLE spectrum can be deconvoluted into three peaks that correspond to electron transitions: n → σ* (blue), π → π* (green), and n → π* (red). ["Reprinted with permission from (Ref [33]). Copyright (2015) American Chemical Society."].

5.4.6 TEM and high-resolution transmission electron microscopy (HRTEM)

TEM/HRTEM is one of the most powerful microscopic techniques for nanomaterial characterization which works on the principle of electron diffraction. It produces a high resolution image of the sample in atomic level by the interaction between ultra-thin sample and high energy electron beam emitted from the filament which transmits through the sample along with multiple electromagnetic lenses and then finally make contact with screen. Then the electrons are converted into light to produce image and the image is magnified and then formed onto a phosphor screen via collecting these electrons leaving specimen. According to this working principle, information about the sample morphology could be resolved well. TEM/HRTEM works in two standard imaging modes namely, bright field and dark field. TEM/HRTEM is composed of an illumination system, condenser lens, objective lens, magnification, and the data recording system. The resolution of the instrument depends on the speed of the electrons because the wavelength of the electron is directly proportional to its speed. To increase the resolution of TEM, a monochromator and an aberration corrector are used to reach point resolutions below 0.5 Å, such kind of TEM are called HRTEM. The desired sample could be dispersed by sonication in ethanol/DMF/water (suitable solvent) to prepare a suspension. A drop of this suspension is put on perforated carbon film-coated copper micro grids and after drying, these grids could then be used for structural investigation using a transmission electron microscope (HRTEM). GQDs could easily be visualized using this technique. Particle size of GQDs, crystallinity, morphology etc. could easily be extracted from HRTEM. Figure 5.19a is representative TEM image of GQDs showing uniform dispersion of particle distribution between ca. 2 and 6 nm in diameter (summarized in Figure 5.19d). The corresponding atomic level resolution structure of lattice fringes is visible in HR-TEM (Figure 5.19b), indicative of high crystallinity with a lattice parameter of 0.21 nm (fast Fourier transform (FFT) pattern shown in the inset), similar to the hexagonal pattern of graphene with the (1100) lattice fringes. Even with these extreme advantages, this technique too bears a few drawbacks such as high cost, maintenance, artefacts from sample preparation, high energy beam tolerance needed for the sample for analysis, high vacuum, sensitivity of the instrument with electrical and magnetic field.

5.4.7 Atomic force microscopy (AFM)

AFM is a versatile and powerful microscopic technique which gives insights about the surface topography of the sample. It uses a cantilever with a very sharp tip to scan over a sample surface and operates either by contact or non-contact mode. The cantilever is brought very closer to the surface, increasingly repulsive force takes over and causes the cantilever to deflect away from the surface helping us to learn about the topography of solid surface through the tip scanning the surface. It is usually used to investigate the thickness of GQDs and further estimate the number of graphene layers. For instance, Figure 5.19c demonstrates the topography of the GQDs, the heights of which are 0.4 to 2 nm (Figure 5.19d), corresponding to 1–3 graphene layers [78].

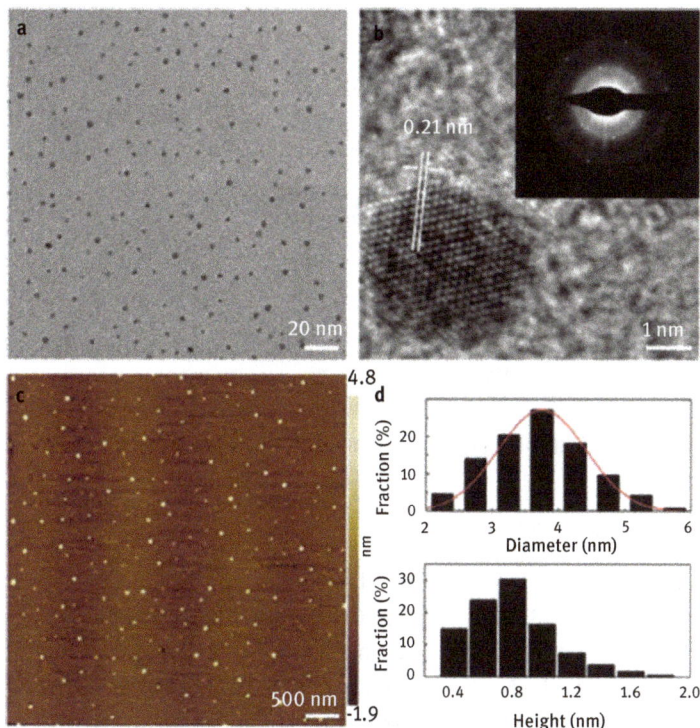

Figure 5.19: Morphology and size distribution of GQDs. (a) TEM image and (b) HR-TEM image of the GQDs (inset: FFT pattern). (c) AFM image of the GQDs. (d) Lateral size and height distribution of GQDs. [(Reprinted with permission from Ref [78] Copy-right 2018 Royal Chemical Society.)].

5.4.8 X-ray photoelectron spectroscopy (XPS)

XPS is a quantitative technique that is useful to analyse surface (0–10 nm surface depth) elemental chemistry of a sample and is based on the well-known photoelectric effect first. Elemental composition, oxidation state of the element present at the surface of the sample could be extracted from XPS analysis. It is always used along with FT-IR to study the material surface components, functional groups. A low energy X-ray irradiates on the sample surface to excite the electrons of the sample atoms and a photo electron is generated if their binding energy is lower than the X-ray energy. By measuring kinetic energy, the binding energy could be derived and the corresponding characteristic peaks could be obtained. The overall process is given below:

$$A + h\upsilon = A^+ + e$$

A = surface of the species, A^+ is the oxidation state of the surface species, $h\upsilon$ is the energy of photon then, according to conservation of energy:

$$E(A) + h\upsilon = E(A^+) + E(e)$$

E(A) and E(A$^+$) are respective energy of the species A and A$^+$ and E(e) represents energy of the electron

since, the electronic energy is totally kinetic, we get:

$$KE = h\upsilon - (E(A^+) - E(A))$$

$$KE = h\upsilon - BE;$$

KE = kinetic energy, BE = binding energy

The BE of solids are conventionally measured with respect to the Fermi level of the solid instead of vacuum. It involves a small correction term called work function (φ) of the solid,

$$KE = h\upsilon - BE - \phi$$

There are several studies for investigating the elemental composition, doping of GQDs by XPS spectroscopy [76, 78, 108]. Pristine GQDs show typical carbon signal which could be de-convoluted into various peaks such as C–C, C–O and C = C bonding. The other elemental peaks like N, P, etc. appear in case of a doped or functionalized GQD. Also, the atomic percentage of the elements could be extracted from XPS analysis accurately. However, the limitations involved with this techniques are: hydrogen and helium cannot be detected, high cost, ultra high vacuum, slow processing (up to 8 h), surface sensitivity (10–100 Å).

5.5 Applications of GQDs

The strong proverb "necessity is the mother of invention" relates to the recent development of GQDs for a variety of applications based on the size distribution which regulate their properties and hence could be thus tailored by doping/functionalization for the necessity of specific applications. The lateral size may be as big as 100–200 nm but their height could be 2–3 nm which essentially brings in interesting optical and electronic properties. The dispersion problems could be obviated using suitable functionalization strategies and it is also possible to regulate the degree of functionalization by several methods. However, research on site-specific functionalization protocols should be focused and there is a dire need to develop these for many biological applications. For example, Figure 5.20 represents some of the diverse range of applications at a glance where GQDs can be efficiently used and few specific examples are discussed below.

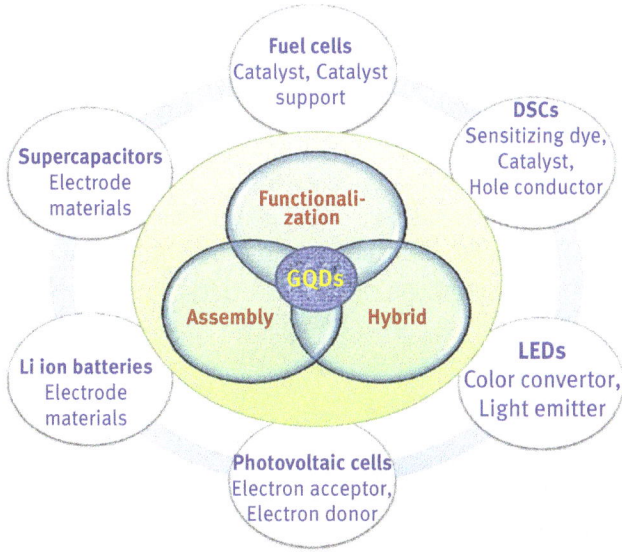

Figure 5.20: Diagram representation of GQDs applications in diverse field from energy applications to catalysis and possible surface modification, hybrid, etc. [(Reprinted with permission from ref. [109] Copyright 2012 Royal Chemical Society.)].

5.5.1 Energy-related applications of GQDs

The PL of GQDs makes this material to be implemented in several green energy generation and storage related applications such as, batteries, light emitting diode (LED) and dye sensitized solar cells (DSSCs) [14, 85, 110, 111]. The broad PL excitation peak matches with the LED application for producing white LED. On the other hand GQDs act as co-sensitizer for DSSCs and harvest the solar energy in the UV-region, which is not possible by using the dye alone. Hence, GQDs provide highly luminous, durable white light for LED and also for efficient light harvesting elements in DSSCs owing to its superior optical properties [37, 112–115]. GQDs have also been used in storage system and coated on VO_2 for Li/Na – ion batteries as efficient electrode material [80, 116–118]. Another energy storage application for GQDs is supercapacitors [119, 120]. Doped GQDs have been utilized as electrocatalysts, especially in the oxygen reduction reactions [70, 109, 121, 122], oxygen evolution [38] and hydrogen evolution [39, 123] reactions which are very crucial in energy conversion systems like fuel cells, and metal air batteries [41]. Also, GQDs are promising in some applications like flash memory [124], thermoelectric energy conversion [125] and light absorbers for photovoltaic devices [126][127–129].

5.5.2 Bio-medical applications of GQDs

The low cytoxicity of GQDs thus make them good candidates for several biological applications from bio-marker to drug delivery system [76, 87, 130–132]. For instance, Figure 5.21 represents GQDs as bio-marker for detection of breast cancer cell. GQDs can also be used in protein analysis by fluorescence resonance energy transfer (FRET), cell tracking, isolation of biomolecules, gene technology owing to their tunable PL, photostability and ease of functionalizations [5]. GQDs are able to penetrate through stem cells easily with a low cytoxicity while providing a clear image [37]. The antibacterial properties of GQDs may further help to develop GQD based band-aids [37].

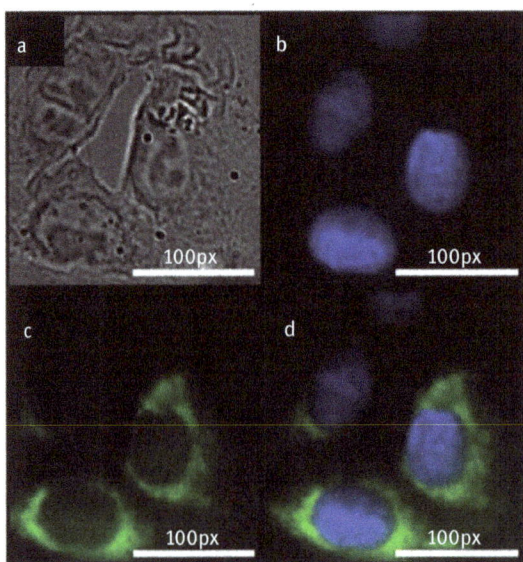

Figure 5.21: Fluorescent images of human breast cancer cell T47D after incubation with green GQDs for 4 h (a) phase contrast picture of T47D cells. (b) Individual nucleus stained blue with DAPI. (c) Agglomerated green GQDs surrounding each nucleus. (d) The overlay high contrast image of nucleolus stained with blue DAPI and GQDs (green) staining. [(Reprinted with permission from ref. [76] Copyright 2012 American Chemical Society.")].

5.5.3 GQDs for sensor applications

GQDs could be used for optical, electrical and electrochemical detection of a variety of analytes and coupled with appropriate functionalization strategies highly selective and sensitive chemical and biosensors could be fabricated. The PL quenching in GQDs may easily be used to sense some of the elements in both qualitative and quantitative manner such as ultrasensitive detection of TNT [133], Electrochemical biosensors [89][134], Autophagy-inducing photodynamic agents [135], Signal amplifiers in stripping voltammetric detection of epithelial cell adhesion molecule (EpCAM) antigen biomarkers [136], Cd^{2+}detection [19], etc. have been demonstrated recently as few examples [131, 137–143]. Ju et al. [144] employed Au-nanoparticle-N-GQDs hybrid as efficient catalyst for efficient H_2O_2 reduction as well as sensing. Although, in their study, they have used N-GQDs as reducing agent and support for the formation of gold nanoparticles by the reduction of $HAuCl_4$ without additional reductant. Zhang et al. [145] have demonstrated N-CQDs as fluorescent probe for detection of Hg^{2+} ions. Similarly, Gao et al. [146] have demonstrated that carbon nanodots can be used for Ag^+ detection. Fe^{+3} detection was reported by Ju et al. [147] using N-GQDs. Ju et al. [138, 148] have demonstrated that N-GQDs with high QY can be used as efficient fluorescent probes for the detection of glutathione (GSH) and cellular imaging.

5.5.4 Miscellaneous applications of GQDs

Although, for molecular devices/molecular scale electronics, most QDs reported till date operate at cryogenic temperatures, limiting their use in such applications while GQDs have shown room temperature response [149]. The size of the GQDs islands is estimated to be in the 1 nm range and the large addition energies (1.6 eV) of GQDs allow for Coulomb blockade at room temperature, with possible application to single-electron devices [149]. In practice, the two most important conditions for room-temperature operation are addition energies much larger than the thermal energy at 300 K, (i. e. 26 meV), and stable device performance. Individual molecules in between two electrical contacts can act as QDs, and addition energies are usually in excess of 100 meV [150, 151].

5.6 Critical safety and environmental considerations

Safety is of prime importance before carrying out any chemical experiment. Most of the chemicals come with their own hazards. So it is always recommended for researchers, especially learners to check out the hazards from the MSDS of the chemical before using it. There are few chemicals such as harsh acids like sulphuric/nitric acid, hydrazine hydrate, organic solvents like DMF, benzene, toluene, etc., used for oxidation/reduction/functionalization or doping with hetero atoms like N of the chemicals for the synthesis of GQDs are highly toxic and should be handled with care. One carrying out the laboratory experiment is expected to follow the primary lab safety rule like wearing lab coat, safety glass, mask if required, gloves, trousers, fully covered shoes. More importantly, all the experiments should be carried out inside fume hood. In case if one is operating microwave for synthesis, in case of fire, the main power supply should be shut down immediately, in the condition of one can strictly compliance with experimental safety regulations.

5.7 Conclusion and future perspectives

Some of the most recent progresses in the intriguing aspects of GQDs have been discussed in this review with particular emphasis on their preparation, characterization, properties and various applications. The opening of bandgap for graphene, together with the tweaking of solubility due to surface functionalization opens up wider application domains for GQDs from biological applications to energy. Despite the huge and ever-growing applications and demand of GQDs, further studies are still needed to understand its inherent properties. Though generally its surface area, active edge sites and modifiable structure are key players to its versatility, concern remains for these materials to be viable alternatives in future due to the lack of the production of high-quality GQDs with bulk amount still. A method that allows for the large-scale synthesis of GQDs of well-defined size and chemical functionality is still lacking and would certainly boost research in this area and will open doors for the future of its industrial applicability. Although research has been undertaken on graphene and GQDs for several decades mainly theoretically, the luminescence mechanisms and electronic properties are not fully understood yet. There are ample scope for GQDs to be engaged in future applications like

- Bioimaging
- Temperature sensing
- Drug delivery
- Cancer therapeutics
- Energy fields like LEDs, OPV solar cells
- Surfactants
- Photoluminescent material
- Biosensors
- Catalysis

It is expected that as synthesis methods improve, both the spectroscopic and theoretical understanding of the underlying mechanisms will evolve and at the same time commercialization can also be expected. Indeed, GQDs constitute an exciting next generation nanocarbon materials, especially for the future of environmentally benign devices across a wide range of areas.

References

[1]   Demming A. King of the elements? Nanotechnology. 2010;21:300201.

[2]   Mcdonough W. Carbon is not the enemy. Nature. 2016;539:349–51.

[3]   Ajayan PM, Ebbesen TW. Nanometre-size tubes of carbon. Rep Prog Phys. 1997;1025:1025–62.

[4]   Wallace PR. The band theory of graphite. Phys Rev. 1947;71:622–634.

[5]   Bacon M, Bradley SJ, Nann T. Graphene quantum dots. Part Part Syst Charact. 2014;31:415–28.

[6]   Geim AK, Novoselov KS. The Rise of Graphene. Nat Mater. 2007;6:183–91.

[7]   Cao Y, Fatemi V, Fang S, Watanabe K, Taniguchi T, Kaxiras E, et al. Unconventional superconductivity in magic-angle graphene superlattices. Nature. 2018;556:43–50.

[8]   Sk MA, Ananthanarayanan A, Huang L, Lim KH, Chen P. Revealing the tunable photoluminescence properties of graphene quantum dots. J Mater Chem C. 2014;2:6954–60.

[9]   Yan Y, Chen J, Li N, Tian J, Li K, Jiang J, et al. Systematic bandgap engineering of graphene quantum dots and applications for photocatalytic water splitting and $CO2$ reduction. ACS Nano. 2018;12:3523–32.

[10]   Shinde DB, Pillai VK. Electrochemical resolution of multiple redox events for graphene quantum dots. Angew Chem Int Ed Engl. 2013;52:2482–5.

[11]   Ponomarenko LA, Schedin F, Katsnelson MI, Yang R, Hill EW, Novoselov KS, et al. Chaotic dirac billiard in graphene quantum dots. Science (80-.). 2008;320:356–8.

[12]   Velasco J, Lee J, Wong D, Kahn S, Tsai HZ, Costello J, et al. Visualization and control of single-electron charging in bilayer graphene quantum dots. Nano Lett. 2018;18:5104–10.

[13]   Kundu S, Yadav RM, Shelke MV, Narayanan TN, Vajtai R, Ajayan PM, et al. Synthesis of N, F and S co-doped graphene quantum dots. Nanoscale. 2015;7:11515–9.

[14]   Kundu S, Sarojinijeeva P, Karthick R, Anantharaj G, Saritha G, Bera R, et al. Enhancing the efficiency of DSSCs by the modification of $TiO2$ photoanodes using N, F and S, co-doped graphene quantum dots. Electrochim Acta. 2017;242:337–43.

[15]   Ye R, Xiang C, Lin J, Peng Z, Huang K, Yan Z, et al. Coal as an abundant source of graphene quantum dots. Nat Commun. 2013;4:1–6.

[16]   Ju J, Chen W. Graphene quantum dots as fluorescence probes for sensing metal ions: synthesis and applications. Curr Org Chem. 2015;19:1150–62.

[17]   Du Y, Gao S. Chemically doped fluorescent carbon and graphene quantum dots for bioimaging, sensor, catalytic and photoelectronic applications. Nanoscale. 2016;8:2532–43.

[18]   Zhu S, Song Y, Zhao X, Shao J, Zhang J, Yang B. The photoluminescence mechanism in carbon dots (Graphene Quantum Dots, Carbon Nanodots, and Polymer Dots): current state and future perspective. Nano Res. 2015;8:355–81.

[19]   Gao X, Du C, Zhuang Z, Chen W. Carbon quantum dots-based nanoprobes for metal ions detection. J Mater Chem C. 2016;4:6927–45.

[20]   Musselman KP, Ibrahim KH, Yavuz M, Musselman KP, Ibrahim KH, Yavuz M. Research update: beyond graphene — synthesis of functionalized quantum dots of 2D materials and their applications research update: beyond graphene — synthesis of functionalized quantum dots of 2D materials and their applications. APL Mater. 2018;6:120701.

[21]   Xu Y, Wang X, Zhang L, Lv F. Chem soc rev recent progress in two-dimensional inorganic. Chem Soc Rev. 2018;47:586–625.

[22]   Slonczewski JC, Weiss PR. Band structure of graphite. Phys Rev. 1958;330:272–9.

[23]   Li Y, Shu H, Niu X, Wang J. Electronic and optical properties of edge-functionalized graphene quantum dots and the underlying mechanism. J Phys Chem. 2015;119:24950–7.

[24] Li Y, Shu H, Wang S, Wang J. Electronic and optical properties of graphene quantum dots: the role of many-body effects. J Phys Chem C. 2015;119:4983–9.

[25] Morgenstern M, Freitag N, Vaid A, Pratzer M, Liebmann M. Graphene quantum dots probed by scanning tunneling spectroscopy and transport spectroscopy after local anodic oxidation. arXiv preprint arXiv:1505.04092. 2015;1–14.

[26] Wurm J, Rycerz A, Adagideli I, Wimmer M, Richter K, Baranger HU. Symmetry classes in graphene quantum dots: universal spectral statistics, weak localization, and conductance fluctuations. Phys Rev Lett. 2009;102:6–9.

[27] Okamoto T, Sasagawa A, Harada Y, Nakano S, Norimatsu W, Kusunoki M, et al. Visualization of zero-dimensional plasmons in graphene quantum dots with near-field infrared microscopy. Int. Conf. Infrared, Millimeter, Terahertz Waves, IRMMW-THz. 2018–Septe. 2018:1–2.

[28] Tian W, Li W, Yu W, Liu X. A review on lattice defects in graphene: types, generation, effects and regulation. Micromachines. 2017;8:1–15.

[29] Terrones M, Botello-Méndez AR, Campos-Delgado J, López-Urías F, Vega-Cantú YI, Rodríguez-Macías FJ, et al. Graphene and graphite nanoribbons: morphology, properties, synthesis, defects and applications. Nano Today. 2010;5:351–72.

[30] Libisch F, Stampfer C, Burgdörfer J. Graphene quantum dots: beyond a dirac billiard. Phy Rev B. 2009;79:1–6.

[31] Ziatdinov M, Lim H, Fujii S, Kusakabe K, Kiguchi M, Enoki T, et al. Chemically induced topological zero mode at graphene armchair Edges †. Phys Chem Chem Phys. 2017;19:5145–54.

[32] Ziatdinov M, Fujii S, Kusakabe K, Kiguchi M, Mori T, Enoki T. Visualization of electronic states on atomically smooth graphitic edges with different types of hydrogen termination. Phys Rev B Phys. 2013;87:115427.

[33] Yeh TF, Huang WL, Chung CJ, Chiang IT, Chen LC, Chang HY, et al. Elucidating quantum confinement in graphene oxide dots based on excitation-wavelength-independent photoluminescence. J Phys Chem Lett. 2016;7:2087–92.

[34] Li L, Peng J, Zhao J, Zhu J. Focusing on luminescent graphene quantum dots: current status and future perspectives. Nanoscale 2013;5:4015–39.

[35] Shen J, Zhu Y, Chen C, Yang X, Li C. Facile preparation and upconversion luminescence of graphene quantum dots. Chem Commun. 2011;47:2580–2.

[36] Zhuo S, Shao M, Lee S-T. Upconversion and downconversion fluorescent graphene quantum dots: ultrasonic preparation and photocatalysis. ACS Nano. 2012;6:1059–64.

[37] Wang Z, Zeng H, Sun L. Graphene quantum dots: versatile photoluminescence for energy, biomedical, and environmental applications. J Mater Chem C Mater Opt Elec Dev. 2015;3:1157–65.

[38] Kundu S, Malik B, Pattanayak DK, Pitchai R, Pillai VK. Role of specific N-containing active sites in interconnected graphene quantum dots for the enhanced electrocatalytic activity towards oxygen evolution reaction. Chem Sel. 2017;2:9943–6.

[39] Kundu S, Malik B, Pattanayak DK, Pitchai R, Pillai VK. Unraveling the hydrogen evolution reaction active sites in N-functionalized interconnected graphene quantum dots. Chem Sel. 2017;2:4511–5.

[40] kundu S, Malik B, Pattanayak DK, Pillai VK. Effect of dimensionality and doping in quasi "one dimensional (1-D)" N – doped graphene nanoribbons on the oxygen reduction reaction. ACS Appl Mater Interfaces. 2017;9:38409–18.

[41] Shinde DB, Dhavale VM, Kurungot S, Pillai VK. Electrochemical preparation of nitrogen-doped graphene quantum dots and their size-dependent electrocatalytic activity for oxygen reduction. Bull Mater Sci. 2015;38:435–42.

[42] Zhang R, Adsetts JR, Nie Y, Sun X, Ding Z. Electrochemiluminescence of nitrogen- and sulfur-doped graphene quantum dots. Carbon N Y. 2018;129:45–53.

[43] Li LL, Ji J, Fei R, Wang CZ, Lu Q, Zhang JR, et al. A facile microwave avenue to electrochemiluminescent two-color graphene quantum dots. Adv Funct Mater. 2012;22:2971–9.

[44] Zhang D, Zhang Z, Wu Y, Fu K, Chen Y, Li W, et al. Biomaterials systematic evaluation of graphene quantum dot toxicity to male mouse sexual behaviors, reproductive and o Ff spring health. Biomaterials. 2019;194:215–32.

[45] Wang M, Sun Y, Cao X, Peng G, Javed I, Kakinen A, et al. Graphene quantum dots against human IAPP aggregation and toxicity: in vivo. Nanoscale. 2018;10:19995–20006.

[46] Jin K, Gao H, Lai L, Pang Y, Zheng S, Niu Y, et al. Preparation of highly fluorescent sulfur doped graphene quantum dots for live cell imaging. J Lumin. 2018;197:147–52.

[47] Xie Y, Wan B, Yang Y, Cui X, Xin Y, Guo LH. Cytotoxicity and autophagy induction by graphene quantum dots with different functional groups. J Environ Sci (China). 2018;77:198–209.

[48] Chandra A, Deshpande S, Shinde DB, Pillai VK, Singh N. Mitigating the cytotoxicity of graphene quantum dots and enhancing their applications in bioimaging and drug delivery. ACS Macro Lett. 2014;3:1064–8.

[49] Shen J, Zhu Y, Yang X, Li C. Graphene quantum dots: emergent nanolights for bioimaging, sensors, catalysis and photovoltaic devices. Chem Commun (Camb). 2012;48:3686–99.

[50] Shen J, Zhu Y, Yang X, Zong J, Zhang J, Li C. One-pot hydrothermal synthesis of graphene quantum dots surface-passivated by polyethylene glycol and their photoelectric conversion under near-infrared light. New J Chem. 2012;36:97–101.

[51] Eda G, Lin YY, Mattevi C, Yamaguchi H, Chen HA, Chen IS, et al. Blue photoluminescence from chemically derived graphene oxide. Adv Mater. 2010;22:505–9.

[52] Fan L, Hu Y, Wang X, Zhang L, Li F, Han D, et al. Fluorescence resonance energy transfer quenching at the surface of graphene quantum dots for ultrasensitive detection of TNT. Talanta. 2012;101:192–7.

[53] Li H, He X, Liu Y, Huang H, Lian S, Lee ST, et al. One-step ultrasonic synthesis of water-soluble carbon nanoparticles with excellent photoluminescent properties. Carbon N Y. 2011;49:605–9.

[54] Zhu Y, Murali S, Stoller MD, Velamakanni A, Piner RD, Ruoff RS. Microwave assisted exfoliation and reduction of graphite oxide for ultracapacitors. Carbon N Y. 2010;48:2118–22.

[55] Massabeau S, Riccardi E, Rosticher M, Valmorra F, Huang P, Tignon J, et al. THz band gap in encapsulated graphene quantum dots. IEEE. 2018. DOI:10.1109/IRMMW-THz.2018.8509964.

[56] Zhou J, Booker C, Li R, Zhou X, Sham T. An electrochemical avenue to blue luminescent nanocrystals from multiwalled carbon nanotubes (MWCNTs). J Am Chem Soc. 2007;8:744–5.

[57] Baker SN, Baker GA. Luminescent carbon nanodots: emergent nanolights. Angew Chemie Int Ed. 2010;49:6726–44.

[58] Zhao Q-L, Zhang Z-L, Huang B-H, Peng J, Zhang M, Pang D-W. Facile preparation of low cytotoxicity fluorescent carbon nanocrystals by electrooxidation of graphite. Chem Commun. 2008;281:5116–8.

[59] Li Y, Hu Y, Zhao Y, Shi G, Deng L, Hou Y, et al. An electrochemical avenue to green-luminescent graphene quantum dots as potential electron-acceptors for photovoltaics. Adv Mater. 2011;23:776–80.

[60] Shinde DB, Pillai VK. Electrochemical preparation of luminescent graphene quantum dots from multiwalled carbon nanotubes. Chem A Eur J. 2012;18:12522–8.

[61] Gokus T, Nair RR, Bonetti A, Böhmler M, Lombardo A, Novoselov KS, et al. Making graphene luminescent by oxygen plasma treatment. ACS Nano. 2009;3:3963–8.

[62] Casiraghi C, Hartschuh A, Lidorikis E, Qian H. Rayleigh imaging of graphene and graphene layers. Nano Nano Lett. 2007;7:2711–7.
[63] Ferrari AC, Meyer JC, Scardaci V, Casiraghi C, Lazzeri M, Mauri F, et al. Raman spectrum of graphene and graphene layers. Phys Rev Lett. 2006;97:1–4.
[64] Chen J-L, Yan X-P. Ionic strength and ph reversible response of visible and near-infrared fluorescence of graphene oxide nanosheets for monitoring the extracellular PH. Chem Commun (Camb). 2011;47:3135–7.
[65] Tang D, Liu J, Yan X, Kang L. The graphene oxide derived graphene quantum dots with different photoluminescence properties and peroxidase-like catalytic activity duosi. RSC Adv. 2016;6:50609–17.
[66] Zhou X, Zhang Y, Wang C, Wu X, Yang Y, Zheng B, et al. Photo-fenton reaction of graphene oxide: a new strategy to prepare graphene quantum dots for DNA cleavage. ACS Nano. 2012;6:6592–9.
[67] Lin L, Zhang S. Creating high yield water soluble luminescent graphene quantum dots via exfoliating and disintegrating carbon nanotubes and graphite flakes. Chem Commun (Camb). 2012;48:10177–9.
[68] Tang L, Ji R, Cao X, Lin J, Jiang H, Li X, et al. Deep ultraviolet photoluminescence graphene quantum dots. ACS Nano. 2012;6:5102–10.
[69] Kim J, Suh JS. Size-controllable and low-cost fabrication of graphene quantum dots using thermal plasma jet. ACS Nano. 2014;8:4190–6.
[70] Li Q, Zhang S, Dai L, Li L. Nitrogen-doped colloidal graphene quantum dots and their size-dependent electrocatalytic activity for the oxygen reduction reaction. J Am Chem Soc. 2012;134:18932–5.
[71] Lu J, Yeo PSE, Gan CK, Wu P, Loh KP. Transforming C60 molecules into graphene quantum dots. Nat Nanotechnol. 2011;6:247–52.
[72] Yan X, Cui X, Li B, Li L. Large, solution-processable graphene quantum dots as light absorbers for photovoltaics. Nano Lett. 2010;10:1869–73.
[73] Yan X, Cui X, Li L. Synthesis of large, stable colloidal graphene quantum dots with tunable size. J Am Chem Soc. 2010;132:5944–5.
[74] Liu R, Wu D, Feng X. Dots with uniform morphology. J Am Chem Soc. 2011;133:15221–3.
[75] Dong Y, Guo CX, Chi Y, Li CM. Reply to comment on "one-step and high yield simultaneous preparation of single- and multi-layer graphene quantum dots from CX-72 carbon black". J Mater Chem. 2012;22:21777–8.
[76] Peng J, Gao W, Gupta BK, Liu Z, Romero-Aburto R, Ge L, et al. Graphene quantum dots derived from carbon fibers. Nano Lett. 2012;12:844–9.
[77] Dong Y, Chen C, Zheng X, Gao L, Cui Z, Yang H, et al. One-step and high yield simultaneous preparation of single- and multi-layer. J Mater Chem. 2012;22:8764–6.
[78] Ding Z, Li F, Wen J, Wang X, Sun R. Gram-scale synthesis of single-crystalline graphene quantum dots derived from lignin biomass. Green Chem. 2018;20:1383–90.
[79] Xu Y, Wang S, Hou X, Sun Z, Jiang Y, Dong Z, et al. Coal-derived nitrogen, phosphorus and sulfur co-doped graphene quantum dots: a promising ion fluorescent probe. Appl Surf Sci. 2018;445:519–26.
[80] Kundu S, Ragupathy P, Pillai VK. Effect of reversible lithium ion intercalation on the size-dependent optical properties of graphene quantum dots. J Electrochem Soc. 2016;163: A1112–A1119.
[81] Pan D, Guo L, Zhang J, Xi C, Xue Q, Huang H, et al. Cutting Sp2 clusters in graphene sheets into colloidal graphene quantum dots with strong green fluorescence. J Mater Chem. 2012;22:3314–8.

[82] Tetsuka H, Asahi R, Nagoya A, Okamoto K, Tajima I, Ohta R, et al. Optically tunable amino-functionalized graphene quantum dots. Adv Mater. 2012;24:5333–8.

[83] Zhu S, Zhang J, Qiao C, Tang S, Li Y, Yuan W, et al. Strongly green-photoluminescent graphene quantum dots for bioimaging applications. Chem Commun (Camb). 2011;47:6858–60.

[84] Shin Y, Lee J, Yang J, Park J, Lee K, Kim S, et al. Mass production of graphene quantum dots by one-pot synthesis directly from graphite in high yield. Small. 2014;10:866–70.

[85] Li W, Li M, Liu Y, Pan D, Li Z, Wang L, et al. Three minute ultrarapid microwave-assisted synthesis of bright fluorescent graphene quantum dots for live cell staining and white LEDs. ACS Appl Nano Mater. 2018;1:1623–30.

[86] Zhu Y, Wang G, Jiang H, Chen L, Zhang X. One-step ultrasonic synthesis of graphene quantum dots with high quantum yield and their application in sensing alkaline phosphatase. Chem Commun (Camb). 2014;51:948–51.

[87] Zhang M, Bai L, Shang W, Xie W, Ma H, Fu Y, et al. Facile synthesis of water-soluble, highly fluorescent graphene quantum dots as a robust biological label for stem cells. J Mater Chem. 2012;22:7461–7.

[88] Dong Y, Shao J, Chen C, Li H, Wang R, Chi Y, et al. Blue luminescent graphene quantum dots and graphene oxide prepared by tuning the carbonization degree of citric acid. Carbon N Y. 2012;50:4738–43.

[89] Zhang L, Zhang Z-Y, Liang R-P, Li Y-H, Qiu J-D. Boron-doped graphene quantum dots for selective glucose sensing based on the "abnormal" aggregation-induced photoluminescence enhancement. Anal Chem. 2014;86:4423–30.

[90] Li S, Li Y, Cao J, Zhu J, Fan L, Li X. Sulfur-doped graphene quantum dots as a novel fluorescent probe for highly selective and sensitive detection of Fe 3+. Anal Chem. 2014;86:10201–7.

[91] Li Y, Zhao Y, Cheng H, Hu Y, Shi G, Dai L, et al. Nitrogen-doped graphene quantum dots with oxygen-rich functional groups. J Am Chem Soc. 2012;134:15–18.

[92] Qu D, Zheng M, Zhang L, Zhao H, Xie Z, Jing X, et al. Formation mechanism and optimization of highly luminescent N-doped graphene quantum dots. Sci Rep. 2014;4. DOI:10.1038/srep05294.

[93] Qu D, Zheng M, Du P, Zhou Y, Zhang L, Li D, et al. Highly luminescent S, N Co-doped graphene quantum dots with broad visible absorption bands for visible light photocatalyst. Nanoscale. 2013;5:12272–7.

[94] Zhang B-X, Gao H, Li X-L. Synthesis and optical properties of nitrogen and sulfur co-doped graphene quantum dots. New J Chem. 2014;38:4615–21.

[95] Wang L, Wang Y, Xu T, Liao H, Yao C, Liu Y, et al. Quantum dots with superior optical properties. Nat Commun. 2014;5:1–9.

[96] Kumar GS, Roy R, Sen D, Ghorai UK, Thapa R, Mazumder N, et al. Amino-functionalized graphene quantum dots: origin of tunable heterogeneous Photoluminescence. Nanoscale. 2014;6:3384–91.

[97] Kosynkin DV, Higginbotham AL, Sinitskii A, Lomeda JR, Dimiev A, Price BK, et al. Longitudinal unzipping of carbon nanotubes to form graphene nanoribbons. Nature. 2009;458:872–6.

[98] Li Z, Zhang W, Luo Y, Yang J, Hou JG. How graphene is cut upon oxidation. J Am Chem Soc. 2009;131:6320–1.

[99] Li JL, Kudin KN, McAllister MJ, Prud'homme RK, Aksay IA, Car R. Oxygen-driven unzipping of graphitic materials. Phys Rev Lett. 2006;96:5–8.

[100] Achadu OJ, Nyokong T. Application of graphene quantum dots functionalized with thymine and thymine-appended zinc phthalocyanine as novel photoluminescent nanoprobes. New J Chem. 2016;40:8727–36.

[101] Cui P. Effect of boron and nitrogen doping on carrier relaxation dynamics of graphene quantum dots. Mater Res Express. 2018;5:1–13.

[102] Basak T, Basak T. Effect of carrier doping and external electric field on the optical properties of graphene quantum dots. IOP Conf Ser Mater Sci Eng. 2018;310:1–8.

[103] Kundu S, Ghosh S, Fralaide M, Narayanan TN, Pillai VK, Talapatra S. Fractional photo-current dependence of graphene quantum dots prepared from carbon nanotubes. Phys Chem Chem Phys. 2015;17:24566–9.

[104] Kundu S, Ragupathy P, Pillai VK. Effect of reversible lithium ion intercalation on the size-dependent optical properties of graphene quantum dots. J Electrochem Soc. 2016;163:1112–9.

[105] Kim S, Shin DH, Kim CO, Kang SS, Sin S. Size-dependence of raman scattering from graphene quantum dots: interplay between shape and thickness size-dependence of raman scattering from graphene quantum dots: interplay between shape and thickness. Appl Phys Lett. 2013;102:053108–1–053108–3.

[106] Liu Q, Zhang J, He H, Huang G, Xing B, Jia J. Green preparation of high yield fluorescent graphene quantum dots from coal-tar-pitch by mild oxidation. Nanomaterials. 2018;8:844–54.

[107] Feng J, Dong H, Pang B, Chen Y, Yu L, Dong L. Tuning electronic and optical properties of graphene quantum dots by selective boronization. J Mater Chem C. 2019;7:237–48.

[108] Kundu S, Yadav RM, Narayanan TN, Shelke MV, Vajtai R, Ajayan PM, et al. Synthesis of N, F and S co-doped graphene quantum dots. Nanoscale. 2015;7:11515–9.

[109] Zhang Z, Zhang J, Chen N, Qu L. Graphene quantum dots: an emerging material for energy-related applications and beyond. Energy Environ Sci. 2012;5:8869–90.

[110] Zhou J, Li C, Yin L, Wang L, Zhang J. Facile way to fabricate high quality white led with yellow graphene quantum dots. Proc. - 2018 19th Int Conf Electron Packag Technol ICEPT. 2018;2018:1598–1601.

[111] Li X, Rui M, Song J, Shen Z, Zeng H. Carbon and graphene quantum dots for optoelectronic and energy devices: a review. Adv Funct Mater. 2015;25:4929–47.

[112] Dinari M, Mohsen M, Meysam M. Dye-sensitized solar cells based on nanocomposite of polyaniline/graphene quantum dots. J Mater Sci. 2016;51:2964–71.

[113] Kim J, Lee B, Kim YJ, Hwang SW. Enhancement of dye-sensitized solar cells Ef Fi ciency using graphene quantum dots as photoanode. Bull Korean Chem Soc. 2018;40:56–61.

[114] Song SH, Jang M-H, Chung J, Jin SH, Kim BH, Hur S-H, et al. Highly efficient light-emitting diode of graphene quantum dots fabricated from graphite intercalation compounds. Adv Opt Mater. 2014;2:1016–23.

[115] Wang S, Li Z, Xu X, Zhang G, Li Y, Peng Q. Amino-functionalized graphene quantum dots as cathode interlayer for efficient organic solar cells: quantum dot size on interfacial modification ability and photovoltaic performance. Adv Mater Interf. 2018;1801480:1–9.

[116] Chao D, Zhu C, Xia X, Liu J, Zhang X, Wang J, et al. Graphene quantum dots coated VO 2 arrays for highly durable electrodes for Li and Na Ion batteries. Nano Lett. 2015;15:565–73.

[117] Zhang Y, Zhang K, Jia K, Liu G, Ren S, Li K, et al. Preparation of coal-based graphene quantum dots/α -Fe 2 O 3 nanocomposites and their lithium-ion storage properties. Fuel. 2019;241:646–52.

[118] Guo J, Zhu H, Sunb Y, Tanga L, Zhanga X. Boosting the lithium storage performance of MoS2 with graphene quantum dots. J Mater Chem A. 2016;4:4783–9.

[119] Liu W, Feng Y, Yan X, Chen J, Xue Q. Superior micro-supercapacitors based on graphene quantum dots. Mater Views. 2013;23:4111–22.

[120] Li Z, Cao L, Qin P, Liu X, Chen Z, Wang L, et al. Nitrogen and oxygen co-doped graphene quantum dots with high capacitance performance for micro-supercapacitors. Carbon N Y. 2018;139:67–75.

[121] Jin H, Huang H, He Y, Feng X, Wang S, Dai L, et al. Graphene quantum dots supported by graphene nanoribbons with ultrahigh electrocatalytic performance for oxygen reduction. J Am Chem Soc. 2015;137:7588–91.

[122] Fei H, Ye R, Ye G, Gong Y, Peng Z, Fan X, et al. Boron- and nitrogen-doped graphene quantum dots/graphene hybrid nanoplatelets as E Ffi cient electrocatalysts for oxygen reduction. ACS Nano. 2014;8:10837–43.

[123] Guo J, Zhu H, Sun Y, Tang L, Zhang X. Doping MoS2 with graphene quantum dots: structural and electrical engineering towards enhanced electrochemical hydrogen evolution. Electrochim Acta. 2016;211:603–10.

[124] Li Y, Shu H, Niu X, Wang J. Electronic and optical properties of edge-functionalized graphene quantum dots and the underlying mechanism. J Phys Chem C. 2015;119:24950–7.

[125] Yan Y, Liang Q-F, Zhao H, Wu C-Q. Thermoelectric properties of hexagonal graphene quantum dots. Phys Lett A. 2012;376:1154–8.

[126] Guo X, Lu GL, Chen J. Graphene-based materials for photoanodes in dye-sensitized solar cells. Front Energy Res. 2015;3:1–15.

[127] Gupta V, Chaudhary N, Srivastava R, Sharma GD, Bhardwaj R, Chand S. Luminscent graphene quantum dots for organic photovoltaic devices. J Am Chem Soc. 2011;133:9960–3.

[128] Lim SP, Pandikumar A, Lim HN, Ramaraj R, Huang NM. Boosting photovoltaic performance of dye-sensitized solar cells using silver nanoparticle-decorated N,S-Co-Doped-TiO2 photoanode. Sci Rep. 2015;5:1–14.

[129] Khan F, Kim JH. N-functionalized graphene quantum dots with ultrahigh quantum yield and large stokes shift: efficient downconverters for CIGS solar cells. ACS Photonics. 2018;5:4637–743.

[130] Kuila T, Bose S, Mishra AK, Khanra P, Kim NH, Lee JH. Chemical functionalization of graphene and its applications. Prog Mater Sci. 2012;57:1061–105.

[131] Hai X, Feng J, Chen X, Wang J. Tuning the optical properties of graphene quantum dots for biosensing and bioimaging. J Mater Chem B. 2018;6:3219–34.

[132] Tang W. Fluorescent graphene quantum dots as traceable, PH-sensitive drug delivery systems. Int J Nanomed. 2015;10:6709–24.

[133] Cai Z, Li F, Wu P, Ji L, Zhang H, Cai C, et al. Synthesis of nitrogen-doped graphene quantum dots at low temperature for electrochemical sensing trinitrotoluene. Anal Chem. 2015;87:11803–11.

[134] Liu Q, Wang K, Huan J, Zhu G, Qian J, Mao H, et al. Graphene quantum dots enhanced electrochemiluminescence of cadmium sulfide nanocrystals for ultrasensitive determination of pentachlorophenol. Analyst. 2014;139:2912–8.

[135] Markovic ZM, Ristic BZ, Arsikin KM, Klisic DG, Harhaji-Trajkovic LM, Todorovic-Markovic BM, et al. Graphene quantum dots as autophagy-inducing photodynamic agents. Biomaterials. 2012;33:7084–92.

[136] Shiddiky MJA, Rauf S, Kithva PH, Trau M. Graphene/quantum dot bionanoconjugates as signal amplifiers in stripping voltammetric detection of EpCAM biomarkers. Biosens Bioelectron. 2012;35:251–7.

[137] Tang J, Ma X, Liu J, Zheng S, Wang J. Simultaneous determination of hydroquinone and catechol using carbon glass electrode modified with graphene quantum dots. Int J Electrochem Sci Int. 2018;13:11250–62.

[138] Ju J, Zhang R, He S, Chen W. Nitrogen-doped graphene quantum dots-based fluorescent probe for the sensitive turn-on detection of glutathione and its cellular imaging. RSC Adv. 2014;4:52583–9.

[139] Akbarnia A, Zare HR. A voltammetric assay for microRNA-25 based on the use of amino-functionalized graphene quantum dots and Ss- and Ds-DNAs as gene probes. Microchim Acta. 2018;185. DOI:10.1007/s00604-018-3037-6.

[140] Bharathi S, John SA. Highly selective naked eye detection of Vitamin B1 in the presence of other vitamins using graphene quantum dots capped gold nanoparticles. New J Chem. 2019;43:2111–7.

[141] Ben Aoun S. Nanostructured carbon electrode modified with n-doped graphene quantum dots – Chitosan nanocomposite: a sensitive electrochemical dopamine sensor. R Soc Open Sci. 2017;4:1–12.

[142] Wang W, Wang Z, Liu J, Peng Y, Yu X, Wang W, et al. Materials and interfaces one-pot facile synthesis of graphene quantum dots from rice husks for Fe3 + sensing one-pot facile synthesis of graphene quantum dots from rice husks for Fe 3 + sensing. Ind Eng Chem Res. 2018;57:9144–50.

[143] Tang Y, Li J, Guo Q, Nie G. Sensors and actuators B: chemical an ultrasensitive electrochemiluminescence assay for Hg 2 + through graphene quantum dots and poly (5-Formylindole) nanocomposite. Sens Actuat B Chem. 2019;282:824–30.

[144] Ju J, Chen W. In situ growth of surfactant-free gold nanoparticles on nitrogen- doped graphene quantum dots for electrochemical detection of hydrogen peroxide in biological environments. Anal Chem. 2015;87:1903–10.

[145] Zhang R, Chen W. Nitrogen-doped carbon quantum dots: facile synthesis and application as a "turn-off" fluorescent probe for detection of Hg2+ Ions. Biosens Bioelectron. 2014;55:83–90.

[146] Gao X, Lu Y, Zhang R, He S, Ju J, Liu M. One-pot synthesis of carbon nanodots for Fl uorescence turn-on detection of Ag + based on the Ag + -induced enhancement of Fl Uorescence †. J Mater Chem C. 2015;3:2302–9.

[147] Ju J, Chen W. Synthesis of highly fluorescent nitrogen-doped graphene quantum dots for sensitive, label-free detection of Fe (III) in aqueous media. Biosens Bioelectron. 2014;58:219–25.

[148] Ju J, Zhang R, Chen W. Photochemical deposition of surface-clean silver nanoparticles on nitrogen-doped graphene quantum dots for sensitive colorimetric detection of glutathione. Sens Actuat B. 2016;228:66–73.

[149] Barreiro A, Van Der Zant HSJ, Vandersypen LMK. Quantum dots at room temperature carved out from few-layer graphene. Nano Lett. 2012;12:6096–100.

[150] Kubatkin S, Danilov A, Hjort M, Cornil J, Brédas J-L, Stuhr-Hansen N, et al. Single-electron transistor of a single organic molecule with access to several redox states. Nature. 2003;425:698–701.

[151] Osorio EA, O'Neill K, Stuhr-Hansen N, Nielsen OF, Bjørnholm T, Van Der Zant HSJ. Addition energies and vibrational fine structure measured in electromigrated single-molecule junctions based on an oligophenylenevinylene derivative. Adv Mater. 2007;19:281–5.

Bionotes

Sumana Kundu is currently a post doctoral research fellow since December 2018 in the department of Material Science and Engineering, Technion, Israel Institute of Technology, Israel. Recently, she has received PBC post doctoral fellowship in Technion. She has received her PhD degree in Chemical Science in April, 2018 under the supervision of Prof Vijayamohanan K. Pillai from CSIR-Central Electrochemical Research Institute (*CSIR-CECRI*), Tamil Nadu, India. Her PhD work focused on "synthesis and characterizations of graphene based nanomaterials for energy applications". Her current research interest includes synthesis of metal chalcogenides for magnesium ion batteries, graphene-metal hybrid for electrocatalysis, etc.

Vijayamohanan K Pillai is currently Dean (R&D) Professor & Chair, Chemistry, Indian Institute of Science Education and Research (IISER)-Tirupati. His research work of more than 20 years is primarily focused on batteries, fuel cells, bio-electrochemistry, electrochemical sensors, chemically modified electrodes, application of graphene-based nanomaterials for energy applications, anodization, electro-deposition, electro-organic synthesis, etc. Pillai has more than 220 research papers and 20 patents to his credit. Under his guidance, 20 students have received PhD degrees. He is a recipient of many prestigious awards including Medals of the MRSI and CRSI. He is a Fellow of the Indian Academy of Sciences and Affiliate Member of the IUPAC. He is also in the editorial boards of several prestigious journals. Pillai assumed charge as Director, CSIR-CECRI on 24 April 2012 and continued as Director, CSIR-CECRI till 23 October 2018. In addition to CSIR-CECRI, he held the additional charge as Director, CSIR-NCL from 01 June 2015 to 29 February 2016.

Index

https://doi.org/10.1515/9783110345001-006